An Introduction to Polymer Physics

No previous knowledge of polymers is assumed in this book which provides a general introduction to the physics of solid polymers.

The book covers a wide range of topics within the field of polymer physics, beginning with a brief history of the development of synthetic polymers and an overview of the methods of polymerisation and processing. In the following chapter, David Bower describes important experimental techniques used in the study of polymers. The main part of the book, however, is devoted to the structure and properties of solid polymers, including blends, copolymers and liquid-crystal polymers.

With an approach appropriate for advanced undergraduate and graduate students of physics, materials science and chemistry, the book includes many worked examples and problems with solutions. It will provide a firm foundation for the study of the physics of solid polymers.

DAVID BOWER received his D.Phil. from the University of Oxford in 1964. In 1990 he became a reader in the Department of Physics at the University of Leeds, retiring from this position in 1995. He was a founder member of the management committee of the IRC in Polymer Science and Technology (Universities of Leeds, Durham and Bradford), and co-authored *The Vibrational Spectroscopy of Polymers* with W. F. Maddams (CUP, 1989). His contribution to the primary literature has included work on polymers, solid-state physics and magnetism.

An Introduction to Polymer Physics

David I. Bower
Formerly at the University of Leeds

CAMBRIDGE UNIVERSITY PRESS
Cambridge, New York, Melbourne, Madrid, Cape Town, Singapore, São Paulo

Cambridge University Press
The Edinburgh Building, Cambridge CB2 8RU, UK

Published in the United States of America by Cambridge University Press, New York

www.cambridge.org
Information on this title: www.cambridge.org/9780521631372

First published 2002
Reprinted 2004

The right of David I. Bower to be identified as the author of this
work has been asserted by him in accordance with the Copyright,
Designs and Patents Act 1988.

A catalogue record for this publication is available from the British Library

Library of Congress Cataloguing in Publication data

Bower, David I., 1937–
   An introduction to polymer physics / David I. Bower.
      p. cm.
   Includes bibliographical references and index.
   ISBN 0 521 63137 8–ISBN 0 521 63721 X (pb.)
   1. Polymers. 2. Polymerization.   I. Title.
QD381.8.B67 2002
547′.7–dc21   2001037966

ISBN 978-0-521-63137-2 hardback
ISBN 978-0-521-63721-3 paperback

Transferred to digital printing 2008

Contents

Preface

There are already a fairly large number of textbooks on various aspects of polymers and, more specifically, on polymer physics, so why another? While presenting a short series of undergraduate lectures on polymer physics at the University of Leeds over a number of years I found it difficult to recommend a suitable textbook. There were books that had chapters appropriate to some of the topics being covered, but it was difficult to find suitable material at the right level for others. In fact most of the textbooks available both then and now seem to me more suitable for postgraduate students than for undergraduates. This book is definitely for undergraduates, though some students will still find parts of it quite demanding.

In writing any book it is, of course, necessary to be selective. The criteria for inclusion of material in an undergraduate text are, I believe, its importance within the overall field covered, its generally non-controversial nature and, as already indicated, its difficulty. All of these are somewhat subjective, because assessing the importance of material tends to be tainted by the author's own interests and opinions. I have simply tried to cover the field of solid polymers widely in a book of reasonable length, but some topics that others would have included are inevitably omitted. As for material being non-controversial, I have given only rather brief mentions of ideas and theoretical models that have not gained general acceptance or regarding which there is still much debate. Students must, of course, understand that all of science involves uncertainties and judgements, but such matters are better left mainly for discussion in seminars or to be set as short research tasks or essays; inclusion of too much doubt in a textbook only confuses.

Difficulty is particularly subjective, so one must judge partly from one's own experiences with students and partly from comments of colleagues who read the text. There is, however, no place in the modern undergraduate text for long, very complicated, particularly mathematically complicated, discussions of difficult topics. Nevertheless, these topics cannot be avoided altogether if they are important either practically or for the general development of the subject, so an appropriate simplified treatment must be given. Comments from readers have ranged from 'too difficult' to 'too easy' for various parts of the text as it now stands, with a large part 'about right'. This seems to me a good mix, offering both comfort and challenge, and I have not, therefore, aimed at greater homogeneity.

It is my experience that students are put off by unfamiliar symbols or symbols with a large number of superscripts or subscripts, so I have attempted where possible to use standard symbols for all quantities. This means that, because the book covers a wide range of areas of physics, the same symbols sometimes have different meanings in different places. I have therefore, for instance, used θ to stand for a wide range of different angles in different parts of the book and only used subscripts on it where absolutely necessary for clarity. Within a given chapter I have, however, tried to avoid using the same symbol to mean different things, but where this was unavoidable without excess complication I have drawn attention to the fact.

It is sometimes said that an author has simply compiled his book by taking the best bits out of a number of other books. I have certainly *used* what I consider to be some of the best or most relevant bits from many more specialised books, in the sense that these books have often provided me with general guidance as to what is important in a particular area in which my experience is limited and have also provided many specific examples of properties or behaviour; it is clearly not sensible to use poor examples because somebody else has used the best ones! I hope, however, that my choice of material, the way that I have reworked it and added explanatory material, and the way that I have cross-referenced different areas of the text has allowed me to construct a coherent whole, spanning a wider range of topics at a simpler level than that of many of the books that I have consulted and made use of. I therefore hope that this book will provide a useful introduction to them.

Chapters 7 and 8 and parts of chapter 11, in particular, have been influenced strongly by the two more-advanced textbooks on the mechanical properties of solid polymers by Professor I. M. Ward, and the section of chapter 12 on liquid-crystal polymers has drawn heavily on the more-advanced textbook by Professors A. Donald and A. H. Windle. These books are referred to in the sections on further reading in those chapters and I wish to acknowledge my debt to them, as to all the books referred to there and in the corresponding sections of other chapters.

In addition, I should like to thank the following for reading various sections of the book and providing critical comments in writing and sometimes also in discussion: Professors D. Bloor, G. R. Davies, W. J. Feast, T. C. B. McLeish and I. M. Ward and Drs P. Barham, R. A. Duckett, P. G. Klein and D. J. Read. In addition, Drs P. Hine and A.

P. Unwin read the whole book between them and checked the solutions to all the examples and problems. Without the efforts of all these people many obscurities and errors would not have been removed. For any that remain and for sometimes not taking the advice offered, I am, of course, responsible.

Dr W. F. Maddams, my co-author for an earlier book, *The Vibrational Spectroscopy of Polymers* (CUP 1989), kindly permitted me to use or adapt materials from that book, for which I thank him. I have spent considerable time trying to track down the copyright holders and originators of the other figures and tables not drawn or compiled by me and I am grateful to those who have given permission to use or adapt material. If I have inadvertently not given due credit for any material used I apologise. I have generally requested permission to use material from only one of a set of co-authors and I hope that I shall be excused for using material without their explicit permission by those authors that I have not contacted and authors that I have not been able to trace. Brief acknowledgements are given in the figure captions and fuller versions are listed on p. xv. This list may provide useful additional references to supplement the books cited in the further reading sections of each chapter. I am grateful to The University of Leeds for permission to use or adapt some past examination questions as problems.

Finally, I should like to thank my wife for her support during the writing of this book.

D. I. B., Leeds, November 2001

Acknowledgements

Full acknowledgements for use of figures are given here. Brief acknowledgements are given in figure captions.

1.1 Reproduced by permission of Academic Press from Atalia R. H. in *Wood and Agricultural Residues*, ed J. Soltes, Academic Press, 1983, p.59.

1.5, **1.6**, **2.11**, **3.3**, **4.2**, **4.7(a)**, **4.16** and **5.8** Reproduced from *The Vibrational Spectroscopy of Polymers* by D. I. Bower and W. F. Maddams. © Cambridge University Press 1989.

New topologies, 1(a) Adapted by permission from Tomalia, D. A. *et al.*, Macromolecules **19**, 2466, Copyright 1986 American Chemical Society; 1(b) reproduced by permission of the Polymer Division of the American Chemical Society from Gibson, H. W. *et al.*, *Polymer Preprints*, **33(1)**, 235 (1992).

1.7 Data from *World Rubber Statistics Handbook* for 1900-1960, 1960-1990, 1975-1995 and *Rubber Statistical Bulletin* **52**, No.11, 1998, The International Rubber Study Group, London.

1.8 (a) Reprinted by permission of Butterworth Heinemann from *Handbook for Plastics Processors* by J. A. Brydson; Heineman Newnes, Oxford, 1990; (b) courtesy of ICI.

1.9 Reproduced by permission of Oxford University Press from *Principles of Polymer Engineering* by N. G. McCrum, C. P. Buckley and C. B. Bucknall, 2nd Edn, Oxford Science Publications, 1997. © N G. McCrum, C. P. Buckley and C. B. Bucknall 1997.

2.2 Reproduced by permission of PerkinElmer Incorporated from *Thermal Analysis Newsletter* **9** (1970).

2.12 Adapted by permission from *Identification and Analysis of Plastics* by J. Haslam and H. A. Willis, Iliffe, London, 1965. © John Wiley & Sons Limited.

2.14 and **5.26(a)** reproduced, **2.16** adapted from *Nuclear Magnetic Resonance in Solid Polymers* by V. J. McBrierty and K. J. Packer , © Cambridge University Press, 1993.

2.17 (a) and **5.22** reproduced from *Multidimensional Solid State NMR and Polymers* by K. Schmidt-Rohr and H. W Spiess, Academic Press, 1994.

2.17 (b) adapted from Box, A., Szeverenyi, N. M. and Maciel, G. E., *J. Mag. Res.* **55**, 494 (1983). By permission of Academic Press.

2.19 Reproduced by permission of Oxford University Press from *Physical Properies of Crystals* by J. F. Nye, Oxford, 1957.

2.20(a) and **2.21** Adapted by permission of Masaki Tsuji from 'Electron Microscopy' in *Comprehensive Polymer Science*, Eds G. Allen *et al.*, Pergamon Press, Oxford, 1989, Vol 1, chap 34, pp 785-840.

2.20(b) and **5.17(b)** adapted from *Principles of Polymer Morphology* by D. C. Bassett. © Cambridge University Press 1981.

3.4 Reproduced from *The Science of Polymer Molecules* by R. H. Boyd and P. J. Phillips, © Cambridge University Press 1993.

3.5 (a) Reproduced by permission of Springer-Verlag from Tschesche, H. 'The chemical structure of biologically important macromolecules' in *Biophysics*, eds W. Hoppe, W. Lohmann, H. Markl and H. Ziegler, Springer-Verlag, Berlin 1983, Chap. 2, pp.20-41, fig. 2.28, p.35.

4.3 Reproduced by permission from Tadokoro, H. *et al.*, *Reps. Prog. Polymer Phys Jap.* **9**, 181 (1966).

4.5 Reproduced from King, J., Bower, D. I., Maddams, W. F. and Pyszora, H., *Makromol. Chemie* **184**, 879 (1983).

4.7 (b) reprinted by permission from Bunn, C. W. and Howells, E. R., *Nature* **174**, 549 (1954). Copyright (1954) Macmillan Magazines Ltd.

4.12 Adapted and **4.17(a)** reproduced by permission of Oxford University Press from *Chemical Crystallography*, by C. W. Bunn, 2nd Edn, Clarendon Press, Oxford, 1961/3. © Oxford University Press 1961.

4.14 Reprinted with permission from Ferro, D. R., Bruckner, S., Meille, S. V. and Ragazzi, M., *Macromolecules* **24**, 1156, (1991). Copyright 1991 American Chemical Society.

4.19 Reproduced by permission of the Royal Society from Daubeny, R. de P., Bunn, C. W. and Brown, C. J., 'The crystal structure of polyethylene terephthalate', *Proc. Roy. Soc. A* **226**, 531 (1954), figs.6 & 7, p. 540.

4.20 Reprinted by permission of John Wiley & Sons, Inc. from Holmes, D. R., Bunn, C. W. and Smith, D. J. 'The crystal structure of polycaproamide: nylon 6', *J. Polymer Sci.* **17**, 159 (1955). Copyright © 1955 John Wiley & Sons, Inc.

4.21 Reproduced by permission of the Royal Society from Bunn, C. W. and Garner, E. V., 'The crystal structure of two polyamides ('nylons')', *Proc. Roy. Soc. A* **189**, 39 (1947), 16, p.54.

5.1 Adapted by permission of IUCr from Vonk, C. G., *J. Appl. Cryst.* **6**, 148 (1973).

5.3 Reprinted by permission of John Wiley & Sons, Inc. from **(a)** Holland, V. F. and Lindenmeyer, P. H., 'Morphology and crystal growth rate of polyethylene crystalline complexes' *J. Polymer Sci.* **57**, 589 (1962); **(b)** Blundell, D. J., Keller, A. and Kovacs, A. I., 'A new self-nucleation phenomenon and its application to the growing of polymer crystals from solution', *J. Polymer Sci. B* **4**, 481 (1966). Copyright 1962, 1966 John Wiley & Sons, Inc.

5.5 Reprinted from Keller, A., 'Polymer single crystals', *Polymer* **3**, 393-421. Copyright 1962, with permission from Elsevier Science.

5.6 and **12.9** Reproduced by permission from *Polymers: Structure and Bulk Properties* by P. Meares, Van Nostrand, London. 1965 © John Wiley & Sons Limited.

5.7 Reproduced by permission of IUPAC from Fischer, E. W., *Pure and Applied Chemistry*, **50**, 1319 (1978).

5.9 Reproduced by permission of the Society of Polymer Science, Japan, from Miyamoto, Y., Nakafuku, C. and Takemura, T., *Polymer Journal* **3**, 122 (1972).

5.10 (a) and **(c)** reprinted by permission of Kluwer Academic Publishers from Ward, I. M. 'Introduction' in *Structure and Properties of Oriented Polymers*, Ed. I. M. Ward, Chapman and Hall, London, 1997, Chap.1, pp.1-43; **(b)** reprinted by permission of John Wiley & Sons, Inc. from Pechhold, W. 'Rotational isomerism, microstructure and molecular motion in polymers', *J. Polym. Sci. C, Polymer Symp.* **32**, 123-148 (1971). Copyright © 1971 John Wiley & Sons, Inc.

5.11 Reproduced by permission of Academic Press, from *Macromolecular Physics* by B. Wunderlich, Vol.1, Academic Press, New York, 1973.

5.13 Reproduced with the permission of Nelson Thornes Ltd from *Polymers: Chemistry & Physics of Modern Materials*, 2nd Edition, by J. M. G. Cowie, first published in 1991.

5.16 (b) reproduced by permission of the American Institute of Physics from Stein, R. S. and Rhodes, M. B, *J. Appl. Phys* **31**, 1873 (1960).

5.17 (a) Reprinted by permission of Kluwer Academic Publishers from Barham P. J. 'Structure and morphology of oriented polymers' in *Structure and Properties of Oriented Polymers*, Ed. I. M. Ward, Chapman and Hall, London, 1997, Chap 3, pp.142-180.

5.18 Adapted by permission of John Wiley & Sons, Inc. from Pennings, A. J., 'Bundle-like nucleation and longitudinal growth of fibrillar polymer crystals from flowing solutions', *J. Polymer Sci. (Symp)* **59**, 55 (1977). Copyright © 1977 John Wiley & Sons, Inc.

5.19 (a) Adapted by permission of Kluwer Academic Publishers from Hoffman, J.D., Davis, G.T. and Lauritzen, J.I. Jr, 'The rate of crystallization of linear polymers with chain folding' in *Treatise on Solid State Chemistry*, Vol.3, Ed. N. B. Hannay, Ch.7, pp.497-614, Plenum, NY, 1976.

5.21 Reproduced by permission of Academic Press from Hagemeyer, A., Schmidt-Rohr, K. and Spiess, H. W., *Adv. Mag. Reson.* **13**, 85 (1989).

5.24 and **5.25** Adapted with permission from Jelinski, L. W., Dumais, J. J. and Engel, A. K., *Macromolecules* **16,** 492 (1983). Copyright 1983 American Chemical Society.

5.26 (b) reproduced by permission of IBM Technical Journals from Lyerla, J. R. and Yannoni, C. S., *IBM J. Res Dev.*, **27**, 302 (1983).

5.27 Adapted by permission of the American Institute of Physics from Schaefer, D. and Spiess, H. W., *J. Chem. Phys* **97**, 7944 (1992).

6.6 and **6.8** Reproduced by permission of the Royal Society of Chemistry from Treloar, L. R. G., *Trans Faraday Soc.* **40**, 59 (1944).

6.11 Reproduced by permission of Oxford University Press from *The Physics of Rubber Elasticity* by L. R. G. Treloar, Oxford, 1949.

7.12 Reproduced by permission from *Physical Aging in Amorphous Polymers and Other Materials* by L. C. E. Struik, Elsevier Scientific Publishing Co., Amsterdam, 1978.

7.14 Adapted by permission of Academic Press from Dannhauser, W., Child, W. C. Jr and Ferry, J. D., *J. Colloid Science* **13**, 103 (1958).

7.17 Reproduced by permission of Oxford University Press from *Mechanics of Polymers* by R. G. C. Arridge, Clarendon Press, Oxford, 1975. © Oxford University Press 1975.

7.18 Adapted by permission of John Wiley & Sons, Inc. from McCrum, N. G., 'An internal friction study of polytetrafluoroethylene', *J. Polym Sci.* **34**, 355 (1959). Copyright © 1959 John Wiley & Sons, Inc.

8.2 and **11.6** Adapted by permission from *Mechanical Properties of Solid Polymers* by I. M. Ward, Wiley, London, 1971.

8.5 Reprinted by permission of Kluwer Academic Publishers from Bowden, P.B. and Jukes, J.A., *J. Mater Sci.* **3**, 183 (1968).

8.6 Reprinted by permission of John Wiley & Sons, Inc. from Bauwens-Crowet, C., Bauens, J. C. and Holmès, G., 'Tensile yield-stress behaviour of glassy polymers', *J. Polymer Sci. A-2*, **7**, 735 (1969). Copyright 1969, John Wiley & Sons, Inc.

8.9 Adapted with permission of IOP Publishing Limited from Schinker, M.G. and Doell, W. 'Interference optical measurements of large deformations at the tip of a running crack in a glassy thermoplastic' in *IOP Conf. Series No.47*, Chap 2, p.224, IOP, 1979.

8.10 Adapted by permission of Marcel Dekker Inc., N. Y.. from Brown, H. R. and Kramer, E. J., *J. Macromol. Sci.:Phys,* **B19**, 487 (1981).

8.11 Adapted by permission of Taylor & Francis Ltd. from Argon, A. S. and Salama, M. M., *Phil. Mag.* **36**, 1217, (1977). http://www.tandf.co.uk/journals.

8.13 (a) reprinted **(b)** adapted, with permission, from the *Annual Book of ASTM Standards*, copyright American Society for Testing and Materials, 100 Barr Harbour Drive, West Conshohocken, PA 19428.

9.3 Adapted by permission of Carl Hanser Verlag from *Dielectric and Mechanical Relaxation in Materials*, by S. Havriliak Jr and S. J. Havriliak, Carl Hanser Verlag, Munich, 1997.

9.4 Adapted by permission from *Dielectric Spectroscopy of Polymers* by P. Hedvig, Adam Hilger Ltd, Bristol. © Akadémiai Kiadó, Budapest 1977.

9.5 Reproduced by permission of the American Institute of Physics from Kosaki, M., Sugiyama, K. and Ieda, M., *J. Appl. Phys* **42**, 3388 (1971).

9.6 Adapted by permission from Conwell, E.M. and Mizes, H.A. 'Conjugated polymer semiconductors: An introduction', p.583-65 in *Handbook of Semiconductors*, ed. T.S. Moss, Vol. 1, ed. P.T. Landsberg, North-Holland, 1992, Table 1, p.587.

9.7 (a), (b) Adapted from *Organic Chemistry: Structure and Function* by K. Peter C. Vollhardt and Neil E. Schore © 1987, 1994, 1999 by W. H. Freeman and Company. Used with permission. **(c)** Adapted by permission of John Wiley

& Sons, Inc. from *Fundamentals of Organic Chemistry* by T. W. G. Solomons, 3rd edn, John Wiley & Sons, NY, 1990, fig. 10.5, p.432. Copyright © 1990 John Wiley & Sons Inc.

9.8 Adapted and **9.11** reproduced by permission from *One-dimensional Metals: Physics and Materials Science* by Siegmar Roth, VCH, Weinheim, 1995. © John Wiley & Sons Limited.

9.10 Adapted by permission of the American Physical Society from Epstein, A. *et al.*, *Phys Rev. Letters* **50**, 1866 (1983).

10.3 Reprinted by permission of Kluwer Academic Publishers from Schuur, G. and Van Der Vegt, A. K., 'Orientation of films and fibrillation' in *Structure and Properties of Oriented Polymers*, Ed. I. M. Ward, Chap. 12, pp.413-453, Applied Science Publishers, London, 1975).

10.8 Reprinted with permission of Elsevier Science from Karacan, I., Bower, D. I. and Ward, I. M., 'Molecular orientation in uniaxial, one-way and two-way drawn poly(vinyl chloride) films', *Polymer* **35**, 3411-22. Copyright 1994.

10.9 Reprinted with permission of Elsevier Science from Karacan, I., Taraiya, A. K., Bower, D. I. and Ward, I. M. 'Characterization of orientation of one-way and two-way drawn isotactic polypropylene films', *Polymer* **34**, 2691-701. Copyright 1993.

10.11 Reprinted with permission of Elsevier Science from Cunningham, A., Ward, I. M., Willis, H. A. and Zichy, V., 'An infra-red spectroscopic study of molecular orientation and conformational changes in poly(ethylene terephthalate)', *Polymer* **15**, 749-756. Copyright 1974.

10.12 Reprinted with permission of Elsevier Science from Nobbs, J. H., Bower, D.I., Ward, I. M. and Patterson, D. 'A study of the orientation of fluorescent molecules incorporated in uniaxially oriented poly(ethylene terephthalate) tapes', *Polymer* **15**, 287-300. Copyright 1974.

10.13 Reprinted from Robinson, M. E. R., Bower, D. I. and Maddams, W. F., *J. Polymer Sci.: Polymer Phys* **16**, 2115 (1978). Copyright © 1978 John Wiley & Sons, Inc.

11.4 Reproduced by permission of the Royal Society of Chemistry from Pinnock, P. R. and Ward, I. M., *Trans Faraday Soc.*, **62**, 1308, (1966).

11.5 Reproduced by permission of the Royal Society of Chemistry from Treloar, L.R.G., *Trans Faraday Soc.*, **43**, 284 (1947).

11.7 Reprinted by permission of Kluwer Academic Publishers from Hadley, D. W., Pinnock, P. R. and Ward, I. M., *J. Mater. Sci.* **4**, 152 (1969).

11.8 Reproduced by permission of IOP Publishing Limited from McBrierty, V. J. and Ward, I. M., *Brit. J. Appl. Phys: J. Phys D, Ser. 2*, **1**, 1529 (1968).

12.4 Adapted by permission of Marcel Dekker Inc., N. Y. from Coleman, M. M. and Painter, P. C., *Applied Spectroscopy Reviews*, **20**, 255 (1984).

12.5 Adapted with permission of Elsevier Science from Koning, C., van Duin, M., Pagnoulle, C. and Jerome, R., 'Strategies for compatibilization of polymer blends', *Prog. Polym. Sci.* **23**, 707-57. Copyright 1998.

12.10 Adapted with permission from Matsen, M.W. and Bates, F.S., *Macromolecules* **29**, 1091 (1996). Copyright 1996 American Chemical Society.

12.11 Reprinted with permission of Elsevier Science from Seddon J. M., 'Structure of the inverted hexagonal (HII) phase, and non-lamellar phase-transitions of lipids' *Biochim. Biophys. Acta* **1031**, 1-69. Copyright 1990).

12.12 Reprinted with permission from (**a**) Khandpur, A. K. *et al.*, *Macromolecules* **28**, 8796 (1995); (**b**) Schulz, M. F *et al.*, *Macromolecules* **29**, 2857 (1996); (**c**) and (**d**) Förster, S. *et al.*, *Macromolecules* **27**, 6922 (1994). Copyright 1995, 1996, 1994, respectively, American Chemical Society.

12.13 Reprinted by permission of Kluwer Academic Publishers from *Polymer Blends and Composites* by J. A. Manson and L. H. Sperling, Plenum Press, NY, 1976.

12.14 Reprinted by permission of Kluwer Academic Publishers from Abouzahr, S. and Wilkes, G. L. 'Segmented copolymers with emphasis on segmented polyurethanes' in *Processing, Structure and Properties of Block Copolymers*, Ed. M. J. Folkes, Elsevier Applied Science, 1985, Ch.5, pp.165-207.

12.15 Adapted by permission of John Wiley & Sons, Inc. from Cooper, S. L. and Tobolsky, A.V., 'Properties of linear elastomeric polyurethanes', *J. Appl. Polym. Sci.* **10**, 1837 (1966). Copyright © 1966 John Wiley & Sons, Inc.

12.17(b), **12.25**, and **12.26(a)** and (**b**) reproduced and **12.22** adapted from *Liquid Crystal Polymers* by A. M. Donald and A. H. Windle. © Cambridge University Press 1992.

12.21 Re-drawn from Flory, P.J. and Ronca, G., *Mol. Cryst. Liq. Cryst.*, **54**, 289 (1979), copyright OPA (Overseas Publishers Association) N.V., with permission from Taylor and Francis Ltd.

12.26 (c), (**d**) and (**e**) reproduced by permission from Shibaev, V. P. and Platé, N. A. 'Thermotropic liquid-crystalline polymers with mesogenic side groups' in *Advs Polym. Sci.* **60/61** (1984), Ed. M. Gordon, Springer-Verlag, Berlin, p.173, figs 10a,b,c, p.193. © Springer Verlag 1984.

12.27 Adapted by permission of Kluwer Academic Publishers from McIntyre, J. E. 'Liquid crystalline polymers' in *Structure and Properties of Oriented Polymers*, Ed. I. M. Ward, Chapman and Hall, London, 1997, Chap.10, pp. 447-514).

Chapter 1
Introduction

1.1 Polymers and the scope of the book

Although many people probably do not realise it, everyone is familiar
with polymers. They are all around us in everyday use, in rubber, in
plastics, in resins and in adhesives and adhesive tapes, and their common
structural feature is the presence of long covalently bonded chains of
atoms. They are an extraordinarily versatile class of materials, with
properties of a given type often having enormously different values for
different polymers and even sometimes for the same polymer in different
physical states, as later chapters will show. For example, the value of
Young's modulus for a typical rubber when it is extended by only a few
per cent may be as low as 10 MPa, whereas that for a fibre of a
liquid-crystal polymer may be as high as 350 GPa, or 35 000 times higher.
An even greater range of values is available for the electrical conductivity
of polymers: the best insulating polymer may have a conductivity as low
as 10^{-18} Ω^{-1} m^{-1}, whereas a sample of polyacetylene doped with a few
per cent of a suitable donor may have a conductivity of 10^4 Ω^{-1} m^{-1}, a
factor of 10^{22} higher! It is the purpose of this book to describe and, when
possible, to explain this wide diversity of properties.

The book is concerned primarily with *synthetic polymers*, i.e. materials
produced by the chemical industry, rather than with *biopolymers*, which are
polymers produced by living systems and are often used with little or no
modification. Many textile fibres in common use, such as silk, wool and
linen, are examples of materials that consist largely of biopolymers. Wood
is a rather more complicated example, whereas natural rubber is a bio-
polymer of a simpler type. The synthetic polymers were at one time thought
to be substitutes for the natural polymers, but they have long outgrown this
phase and are now seen as important materials in their own right. They are
frequently the best, or indeed only, choice for a wide variety of applications.
The following sections give a brief history of their development, and indicate some of the important properties that make polymers so versatile.

A further restriction on the coverage of this book is that it deals predominantly with polymers in the *solid state*, so it is helpful to give a definition of a solid in the sense used here. A very simple definition that might be considered is that a solid is a material that has the following property: under any change of a set of stresses applied to the material it eventually takes up a new equilibrium shape that does not change further unless the stresses are changed again.

It is, however, necessary to qualify this statement in two ways. The first qualification is that the word *any* must be interpreted as *any within a certain range*. If stresses outside this range are used the material may *yield* and undergo a continuous change of shape or it may *fracture*. This restriction clearly applies to solids of almost any type. The yield and fracture of polymers are considered in chapter 8. The second qualification is that the words *does not change further* need to be interpreted as meaning that *in a time long compared with that for the new so-called equilibrium shape to be reached, the shape changes only by an amount very much smaller than that resulting from the change in the applied stresses*. This restriction is particularly important for polymers, for which the time taken to reach the equilibrium shape may be much longer than for some other types of solids, for example metals, which often appear to respond instantaneously to changes in stress.

Whether a material is regarded as solid may thus be a matter of the time-scale of the experiment or practical use to which the material is put. This book will consider primarily only those polymer systems that are solids on the time-scales of their normal use or observation. In this sense a block of pitch is a solid, since at low stresses it behaves elastically or viscoelastically provided that the stress is not maintained for extremely long times after its first application. If, however, a block of pitch is left under even low stresses, such as its own weight, for a very long time, it will flow like a liquid. According to the definition, a piece of rubber and a piece of jelly are also solid; the properties of rubbers, or *elastomers* as they are often called, forms an important topic of chapter 6. Edible jellies are structures formed from biopolymers and contain large amounts of entrapped water. Similar *gels* can be formed from synthetic polymers and suitable solvents, but they are not considered in any detail in this book, which in general considers only macroscopic systems containing predominantly polymer molecules.

1.2 A brief history of the development of synthetic polymers

Some of the synthetic polymers were actually discovered during the nineteenth century, but it was not until the late 1930s that the manufacture and use of such materials really began in earnest. There were several reasons for this. One was the need in the inter-war years to find replacements for natural materials such as rubber, which were in short supply. A second reason was that there was by then an understanding of the nature of these materials. In 1910, Pickles had suggested that rubber was made up of long chain molecules, contrary to the more generally accepted theory that it consisted of aggregates of small ring molecules. During the early 1920s, on the basis of his experimental research into the structure of rubber, Staudinger reformulated the theory of chain molecules and introduced the word *Makromolekül* into the scientific literature in 1922. This idea was at first ridiculed, but at an important scientific meeting in Düsseldorf in 1926, Staudinger presented results, including his determinations of molar masses, which led to the gradual acceptance of the idea over the next few years. This made possible a more rational approach to the development of polymeric materials. Other reasons for the accelerated development were the fact that a new source of raw material, oil, was becoming readily available and the fact that great advances had been made in processing machinery, in particular extruders and injection moulders (see section 1.5.3). In the next few pages a brief summary of the development of some of the more important commercial polymers and types of polymer is given.

The first synthetic polymer, cellulose nitrate, or *celluloid* as it is usually called, was derived from natural cellulosic materials, such as cotton. The chemical formula of cellulose is shown in fig. 1.1. The formula for cellulose nitrate is obtained by replacing some of the —OH groups by —ONO_2 groups. Cellulose nitrate was discovered in 1846 by Christian Frederick Schönbein and first produced in a usable form by Alexander Parkes in 1862. It was not until 1869, however, that John Wesley Hyatt took out his patent on celluloid and shortly afterwards, in 1872, the Celluloid Manufacturing Company was set up. It is interesting to note, in view of the current debates on the use of ivory, that in the 1860s destruction of the elephant herds in Africa was forcing up the price of ivory and it was Hyatt's interest in finding a substitute that could be used for billiard balls that led to his patenting of celluloid. In the end the material unfortunately turned out to be too brittle for this application.

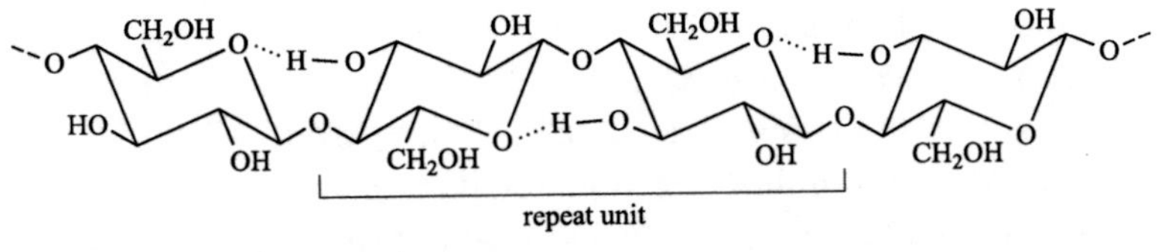

repeat unit

Fig. 1.1 The structure of cellulose. (Reproduced by permission of Academic Press.)

The second important plastic to be developed was *Bakelite*, for which the first patents were taken out by Leo Baekeland in 1907. This material is obtained from the reaction of phenol, a product of the distillation of tar, and formaldehyde, which is used in embalming fluid. Resins formed in this way under various chemical conditions had been known for at least 30 years. Baekeland's important contribution was to produce homogeneous, mouldable materials by careful control of the reaction, in particular by adding small amounts of alkali and spreading the reaction over a fairly long time. It is interesting how frequently important discoveries are made by two people at the same time; a striking example is the fact that the day after Baekeland had filed his patents, Sir James Swinburne, an electrical engineer and distant relative of the poet Swinburne, attempted to file a patent on a resin that he had developed for the insulation of electrical cables, which was essentially Bakelite. The properties of such lacquers were indicated in the punning, pseudo-French name that Swinburne gave to one of his companies – The Damard Lacquer Company. Baekeland is commemorated by the Baekeland Award of the American Chemical Society and Swinburne by the Swinburne Award of the Institute of Materials (London).

A second polymer based on modified cellulose, *cellulose acetate*, was also one of the earliest commercial polymers. This material is obtained by

replacing some of the —OH groups shown in fig. 1.1 by $-\text{O}-\text{C} \overset{\displaystyle \parallel \text{O}}{\underset{\displaystyle \text{CH}_3}{\diagdown}}$

groups. Although the discovery of cellulose acetate was first reported in 1865 and the first patents on it were taken out in 1894, it was only 30 years later that its use as a plastics material was established. Its development was stimulated by the 1914–18 war, during which it was used as a fire-proof dope for treating aircraft wings, and after the war an artificial silk was perfected using it. By 1927 good-quality sheet could be made and until the end of World War II it was still by far the most important injection-moulding material, so the need to process cellulose acetate was a great contributor to the development of injection moulders. Cellulose acetate is still used, for example, in the manufacture of filter tips for cigarettes and in packaging materials.

Before leaving the early development of the cellulosic polymers it is worth mentioning that the first artificial silk, called *rayon*, was made from reconstituted cellulose. The first patents were taken out in 1877/8 and the *viscose process* was patented by Cross, Bevan and Beadle in 1892. It involves the conversion of the cellulose from wood-pulp into a soluble derivative of cellulose and its subsequent reconstitution. The material is thus not a synthetic polymer but a processed natural polymer.

The first of what may be called the modern synthetic polymers were developed during the inter-war years. The first commercial manufacture of polystyrene took place in Germany in 1930, the first commercial sheet of poly(methyl methacrylate), 'Perspex', was produced by ICI in 1936 and the first commercial polyethylene plant began production shortly before the beginning of World War II. Poly(vinyl chloride), or PVC, was discovered by Regnault in 1835, but it was not until 1939 that the plasticised material was being produced in large quantities in Germany and the USA. The production of rigid, unplasticised PVC also took place in Germany from that time. The chemical structures of these materials are described in section 1.3.3.

Apart from the rather expensive and inferior methyl rubber produced in Germany during World War I, the first industrial production of synthetic rubbers took place in 1932, with polybutadiene being produced in the USSR, from alcohol derived from the fermentation of potatoes, and neoprene (polychloroprene) being produced in the USA from acetylene derived from coal. In 1934 the first American car tyre produced from a synthetic rubber was made from neoprene. In 1937 butyl rubber, based on polyisobutylene, was discovered in the USA. This material has a lower resilience than that of natural rubber but far surpasses it in chemical resistance and in having a low permeability to gases. The chemical structures of these materials are shown in fig. 6.10.

In 1928 Carothers began to study condensation polymerisation (see section 1.3.3), which leads to two important groups of polymers, the polyesters and the polyamides, or nylons. By 1932 he had succeeded in producing aliphatic polyesters with high enough molar masses to be drawn into fibres and by 1925 he had produced a number of polyamides. By 1938 nylon-6,6 was in production by Du Pont and the first nylon stockings were sold in 1939. Nylon moulding powders were also available by 1939; this was an important material for the production of engineering components because of the high resistance of nylon to chemicals and abrasion and the low friction shown by such components, in combination with high strength and lightness.

The years 1939–41 brought important studies of polyesters by Whinfield and Dickson and led to the development of poly(ethylene terephthalate) as an example of the deliberate design of a polymer for a specific purpose, the production of fibres, with real understanding of what was required. Large-scale production of this extremely important polymer began in 1955. Its use is now widespread, both as a textile fibre and for packaging in the form of films and bottles. Polymers of another class, the polyurethanes, are produced by a type of polymerisation related to condensation polymerisation and by 1941 they were being produced commercially in Germany, leading to the production of polyurethane foams.

A quite different class of polymer was developed during the early 1940s, relying on a branch of chemistry originated by Friedel and Crafts in 1863, when they prepared the first organosilicon compounds. All the polymers described so far (and in fact the overwhelming majority of polymeric materials in use) are based on chain molecules in which the atoms of the main chain are predominantly carbon atoms. The new polymers were the silicone polymers, which are based on chain molecules containing silicon instead of carbon atoms in the main chain. Silicone rubbers were developed in 1945, but they and other silicones are restricted to special uses because they are expensive to produce. They can withstand much higher temperatures than the organic, or carbon-based, rubbers.

The 1950s were important years for developments in the production of polyolefins, polymers derived from olefins (more properly called alkenes), which are molecules containing one double bond and having the chemical formula C_nH_{2n}. In 1953 Ziegler developed the low-pressure process for the production of polyethylene using catalysts. This material has a higher density than the type produced earlier and also a greater stiffness and heat resistance. The chemical differences among the various types of polyethylene are described in section 1.3.3. The year 1954 saw the first successful polymerisation of propylene to yield a useful solid polymer with a high molar mass. This was achieved by Natta, using Ziegler-type catalysts and was followed shortly afterwards by the achievement of stereospecific polymerisation (see section 4.1) and by 1962 polypropylene was being manufactured in large volume.

Another important class of polymers developed in these years was the polycarbonates. The first polycarbonate, a cross-linked material, was discovered in 1898, but the first linear thermoplastic polycarbonate was not made until 1953 and brought into commercial production in 1960. The polycarbonates are tough, engineering materials that will withstand a wide range of temperatures.

The first verification of the theoretical predictions of Onsager and of Flory that rod-like molecular chains might exhibit liquid-crystalline properties (see section 1.3.2 and chapter 12) was obtained in the 1960s and fibres from para-aramid polymers were commercialised under the name of Kevlar in 1970. These materials are very stiff and have excellent thermal stability; many other materials of this class of rigid main-chain liquid-crystal polymers have been developed. They cannot, however, be processed by the more conventional processing techniques and this led to the development in the 1980s of another group of liquid-crystal polymers, the thermoplastic co-polyesters.

The development and bringing into production of a new polymer is an extremely expensive process, so any method of reducing these costs or the cost of the product itself is important. For these reasons a great interest developed during the 1970s and 1980s in the blending of polymers of different types to give either cheaper products or products with properties that were a combination of those of the constituent polymers. It was also realised that new properties could arise in the blends that were not present in any of the constituents. The number of polymer blends available commercially is now enormous and developments continue. Even as early as 1987 it was estimated that 60%–70% of polyolefins and 23% of other polymers were sold as blends. Blends are considered in chapter 12.

Another important way in which existing types of polymer can be used to form new types of material and the expense of development of new polymers can be avoided is by influencing their properties by various physical treatments, such as annealing and stretching. As described in later sections, some polymers are non-crystalline and some can partially crystallise under suitable conditions. Heat treatment of both kinds of polymer can affect their mechanical properties quite considerably. An important example of the usefulness of the combination of stretching and heat treatment is to be found in the production of textile fibres from polyester. Stretching improves the tensile strength of the fibre, but unless the fibre is partially crystallised by suitable heat treatment, called 'heat setting', it will shrink under moderate heating as the molecules randomise their orientations. From the 1970s to the present time continuous improvements have been made in the properties of thermoplastic polymers such as polyethylene by suitably orienting and crystallising the molecules, so that even these materials can rival the more expensive liquid-crystal polymers in their stiffnesses.

It must not, however, be thought that the development of new polymers has come to an end. This is by no means the case. Polymer chemists continue to develop both new polymers and new polymerisation processes for older polymers. This leads not only to the introduction of polymers for special uses, which are often expensive, but also to the production of polymers specially constructed to test theoretical understanding of how specific features of structure affect physical properties. Totally novel types of polymer are also synthesised with a view to investigating whether they might have useful properties. These developments are considered further in section 1.3.4, and the following section describes the chemical nature of polymers in more detail than has so far been considered.

1.3   The chemical nature of polymers

1.3.1   Introduction

In this book the term *polymer* is used to mean a particular class of macro-molecules consisting, at least to a first approximation, of a set of regularly repeated chemical units of the same type, or possibly of a very limited number of different types (usually only two), joined end to end, or sometimes in more complicated ways, to form a *chain molecule*. If there is only one type of chemical unit the corresponding polymer is a *homopolymer*; if there is more than one type it is a *copolymer*. This section deals briefly with some of the main types of chemical structural repeat units present in the more widely used synthetic polymers and with the polymerisation methods used to produce them. Further details of the structures of individual polymers will be given in later sections of the book.

It should be noted that the term *monomer* or *monomer unit* is often used to mean either the *chemical repeat unit* or the small molecule which polymerises to give the polymer. These are not always the same in atomic composition, as will be clear from what follows, and the chemical bonding must of course be different even when they are.

The simplest polymers are chain-like molecules of the type

$$-A-A-A-A-A-A-A-A-A-A-A-A-A-$$

where A is a small group of covalently bonded atoms and the groups are covalently linked. The simplest useful polymer is polyethylene

$$-CH_2-CH_2-CH_2-CH_2-CH_2-CH_2-CH_2-CH_2- \text{ or } (CH_2)_n$$

wherein a typical length of chain, corresponding to $n \sim 20\,000$ (where $\sim$ means 'of the order of'), would be about 3 μm. A piece of string typically has a diameter of about 2 mm, whereas the diameter of the polyethylene chain is about 1 nm, so that a piece of string with the same ratio of length to diameter as the polymer chain would be about 1.5 m long. It is the combination of length and flexibility of the chains that gives polyethylene its important properties.

The phrase 'typical length of chain' was used above because, unlike those of other chemical compounds, the molecules of polymers are not all identical. There is a *distribution of relative molecular masses* (M_r) (*often called molecular weights*) and the corresponding molar masses, M. This topic is considered further in section 3.2. The value of M_r for the chain considered in the previous paragraph would be 280 000, corresponding to $M = 280\,000$ g mol^{-1}. Commercial polymers often have average values of M between about 100 000 and 1 000 000 g mol^{-1}, although lower values are not infrequent.

The flexibility of polyethylene chains is due to the fact that the covalent bonds linking the units together, the so-called *backbone bonds*, are non-collinear single bonds, each of which makes an angle of about 112° with the next, and that very little energy is required to rotate one part of the molecule with respect to another around one or more of these bonds. The chains of other polymers may be much less flexible, because the backbone bonds need not be single and may be collinear. A simple example is poly-paraphenylene[†],

for which all the backbone bonds are collinear and also have a partial double-bond character, which makes rotation more difficult. Such chains are therefore rather stiff. It is these differences in stiffness, among other factors, that give different types of polymer their different physical properties.

The chemical structures of the repeat units of some common polymers are shown in fig. 1.2, where for simplicity of drawing the backbone bonds are shown as if they were collinear. The real shapes of polymer molecules are considered in section 3.3. Many polymers do not consist of simple linear chains of the type so far considered; more complicated structures are introduced in the following section.

1.3.2   The classification of polymers

There are many possible classifications of polymers. One is according to the general types of polymerisation processes used to produce them, as considered in the following section. Two other useful classifications are the following.

(i)   Classifications based on structure: *linear, branched* or *network* polymers. Figure 1.3 shows these types of polymer schematically. It should be noted that the real structures are three-dimensional, which is particularly important for networks. In recent years interest in more complicated structures than those shown in fig. 1.3 has increased (see section 1.3.4).

(ii)   Classifications based on properties: *(thermo)plastics, rubbers (elastomers)* or *thermosets*.

[†] It is conventional in chemical formulae such as the one shown here not to indicate explicitly the six carbon atoms of the conjugated benzene ring and any hydrogen atoms attached to ring carbon atoms that are not bonded to other atoms in the molecule. In the molecule under consideration there are four such hydrogen atoms for each ring. Carbon and hydrogen atoms are also often omitted from other formulae where their presence is understood.

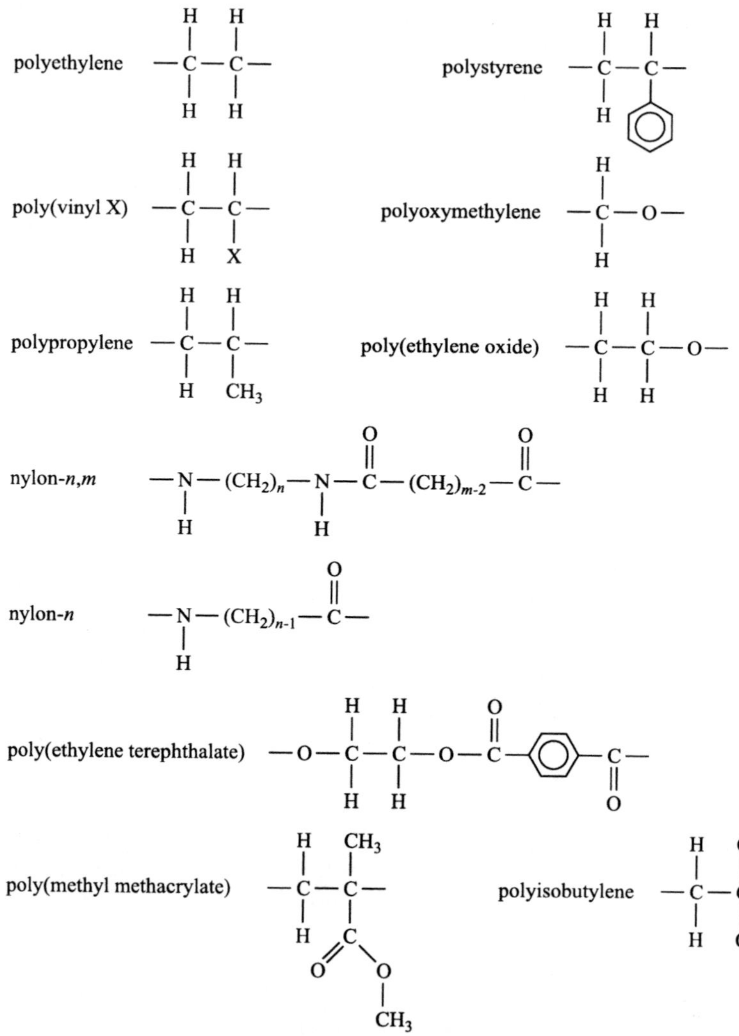

Fig. 1.2 Structures of the repeating units of some common polymers.

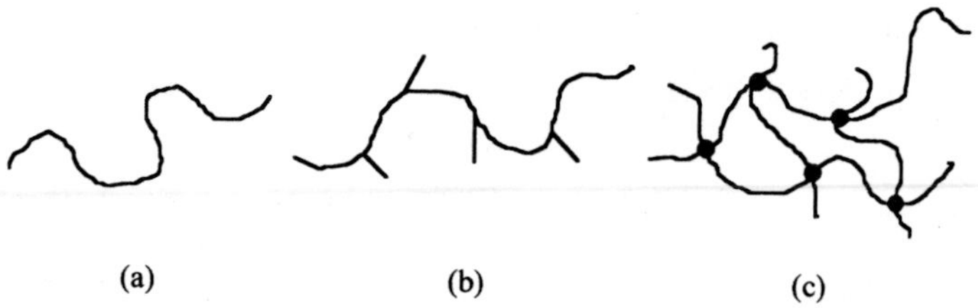

polyethylene

$$-\overset{\overset{\displaystyle H}{|}}{\underset{\underset{\displaystyle H}{|}}{C}}-\overset{\overset{\displaystyle H}{|}}{\underset{\underset{\displaystyle H}{|}}{C}}-$$

polystyrene

poly(vinyl X)

$$-\overset{\overset{\displaystyle H}{|}}{\underset{\underset{\displaystyle H}{|}}{C}}-\overset{\overset{\displaystyle H}{|}}{\underset{\underset{\displaystyle X}{|}}{C}}-$$

polyoxymethylene

polypropylene

$$-\overset{\overset{\displaystyle H}{|}}{\underset{\underset{\displaystyle H}{|}}{C}}-\overset{\overset{\displaystyle H}{|}}{\underset{\underset{\displaystyle CH_3}{|}}{C}}-$$

poly(ethylene oxide)

nylon-n,m

$$-\overset{}{\underset{\underset{\displaystyle H}{|}}{N}}-(CH_2)_n-\overset{}{\underset{\underset{\displaystyle H}{|}}{N}}-\overset{\overset{\displaystyle O}{\|}}{C}-(CH_2)_{m-2}-\overset{\overset{\displaystyle O}{\|}}{C}-$$

nylon-n

$$-\overset{}{\underset{\underset{\displaystyle H}{|}}{N}}-(CH_2)_{n-1}-\overset{\overset{\displaystyle O}{\|}}{C}-$$

poly(ethylene terephthalate)

$$-O-\overset{\overset{\displaystyle H}{|}}{\underset{\underset{\displaystyle H}{|}}{C}}-\overset{\overset{\displaystyle H}{|}}{\underset{\underset{\displaystyle H}{|}}{C}}-O-\overset{\overset{\displaystyle O}{\|}}{C}-\langle\bigcirc\rangle-\overset{}{\underset{\underset{\displaystyle O}{\|}}{C}}-$$

poly(methyl methacrylate)

polyisobutylene

Fig. 1.3 Schematic representations of (a) a linear polymer, (b) a branched polymer and (c) a network polymer. The symbol • represents a cross-link point, i.e. a place where two chains are chemically bonded together.

(a)   (b)   (c)

These two sets of classifications are, of course, closely related, since structure and properties are intimately linked. A brief description of the types of polymer according to classification (ii) will now be given.

Thermoplastics form the bulk of polymers in use. They consist of linear or branched molecules and they soften or melt when heated, so that they can be moulded and remoulded by heating. In the molten state they consist of a tangled mass of molecules, about which more is said in later chapters. On cooling they may form a *glass* (a sort of 'frozen liquid') below a temperature called the *glass transition temperature*, T_g, or they may crystallise. The glass transition is considered in detail in chapter 7. If they crystallise they do so only partially, the rest remaining in a liquid-like state which is usually called *amorphous*, but should preferably be called *non-crystalline*. In some instances, they form a *liquid-crystal phase* in some temperature region (see below and chapter 12).

Rubbers, or *elastomers*, are network polymers that are lightly cross-linked and they are reversibly stretchable to high extensions. When unstretched they have fairly tightly *randomly coiled molecules* that are stretched out when the polymer is stretched. This causes the chains to be less random, so that the material has a lower entropy, and the retractive force observed is due to this lowering of the entropy. The cross-links prevent the molecules from flowing past each other when the material is stretched. On cooling, rubbers become glassy or crystallise (partially). On heating, they cannot melt in the conventional sense, i.e. they cannot flow, because of the cross-links.

Thermosets are network polymers that are heavily cross-linked to give a dense three-dimensional network. They are normally rigid. They cannot melt on heating and they decompose if the temperature is high enough. The name arises because it was necessary to heat the first polymers of this type in order for the cross-linking, or *curing*, to take place. The term is now used to describe this type of material even when heat is not required for the cross-linking to take place. Examples of thermosets are the epoxy resins, such as Araldites, and the phenol- or urea-formaldehyde resins.

Liquid-crystal polymers (LCPs) are a subset of thermoplastics. Consider first non-polymeric liquid crystals. The simplest types are rod-like molecules with aspect ratios greater than about 6, typically something like

In some temperature range the molecules tend to line up parallel to each other, but not in crystal register. This leads to the formation of anisotropic regions, which gives them optical properties that are useful for displays etc. Polymeric liquid-crystal materials have groups similar to these incorporated in the chains. There are two principal types.

(a) *Main-chain LCPs* such as e.g. Kevlar. These are stiff materials that will withstand high temperatures and are usually used in a form in which they have high molecular orientation, i.e. the chains are aligned closely parallel to each other. A schematic diagram of a main-chain LCP is shown in fig 1.4(a).

(b) *Side-chain LCPs* may be used as non-linear optical materials. Their advantage is that it is possible to incorporate into the polymer, as chemically linked side-chains, some groups that have useful optical properties but which would not dissolve in the polymer. A schematic diagram of a side-chain LCP is shown in fig. 1.4(b).

Liquid-crystal polymers are considered in detail in chapter 12.

1.3.3 'Classical' polymerisation processes

In *polymerisation*, monomer units react to give polymer molecules. In the simplest examples the chemical repeat unit contains the same group of atoms as the monomer (but differently bonded), e.g. ethylene → polyethylene

$$n(CH_2=CH_2) \rightarrow -(CH_2-CH_2)_{\overline{n}}$$

Fig. 1.4 Schematic representations of the principal types of liquid-crystal polymers (LCPs): (a) main-chain LCP and (b) side-chain LCP. The rectangles represent long stiff groups. The other lines represent sections of chain that vary in length and rigidity for different LCPs.

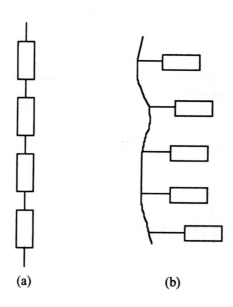

(a)   (b)

More generally the repeat unit is not the same as the monomer or monomers but, as already indicated, it is nevertheless sometimes called the 'monomer'. Some of the simpler, 'classical' processes by which many of the bulk commercial polymers are made are described below. These fall into two main types, *addition polymerisation* and *step-growth polymerisation*.

The sequential addition of monomer units to a growing chain is a process that is easy to visualise and is the mechanism for the production of an important class of polymers. For the most common forms of this process to occur, the monomer must contain a double (or triple) bond. The process of addition polymerisation occurs in three stages. In the *initiation* step an *activated species*, such as a free radical from an initiator added to the system, attacks and opens the double bond of a molecule of the monomer, producing a new activated species. (A free radical is a chemical group containing an unpaired electron, usually denoted in its chemical formula by a dot.) In the *propagation* step this activated species adds on a monomer unit which becomes the new site of activation and adds on another monomer unit in turn. Although this process may continue until thousands of monomer units have been added sequentially, it always terminates when the chain is still of finite length. This *termination* normally occurs by one of a variety of specific chain-terminating reactions, which lead to a corresponding variety of end groups. Propagation is normally very much more probable than termination, so that macromolecules containing thousands or tens of thousands of repeat units are formed.

The simplest type of addition reaction is the formation of polyethylene from ethylene monomer:

$$-(CH_2)_n-CH_2-CH_2^{\cdot} + CH_2{=}CH_2 \rightarrow -(CH_2)_{n+2}-CH_2-CH_2^{\cdot}$$

There are basically three kinds of polyethylene produced commercially. The first to be produced, low-density polyethylene, is made by a high-pressure, high-temperature uncatalysed reaction involving free radicals and has about 20–30 branches per thousand carbon atoms. A variety of branches can occur, including ethyl, $-CH_2CH_3$, butyl, $-(CH_2)_3CH_3$, pentyl, $-(CH_2)_4CH_3$, hexyl, $-(CH_2)_5CH_3$ and longer units. High-density polymers are made by the homopolymerisation of ethylene or the copolymerisation of ethylene with a small amount of higher α-olefin. Two processes, the Phillips process and the Ziegler–Natta process, which differ according to the catalyst used, are of particular importance. The emergence of a new generation of catalysts led to the appearance of linear low-density polyethylenes. These have a higher level of co-monomer incorporation and have a higher level of branching, up to that of low-density material, but the branches in any given polymer are of one type only, which may be ethyl, butyl, isobutyl or hexyl.

Polyethylene is a special example of a generic class that includes many of the industrially important macromolecules, the *vinyl* and *vinylidene* polymers. The chemical repeat unit of a vinylidene polymer is $-(CH_2-CXY)-$, where X and Y represent single atoms or chemical groups. For a vinyl polymer Y is H and for polyethylene both X and Y are H. If X is $-CH_3$, Cl, $-CN$, $-\text{\textcircled{}}$ or $-O(C=O)CH_3$, where $-\text{\textcircled{}}$ represents the mono-substituted benzene ring, or phenyl group, and Y is H, the well-known materials polypropylene, poly(vinyl chloride) (PVC), polyacrylonitrile, polystyrene and poly(vinyl acetate), respectively, are obtained.

When Y is not H, X and Y may be the same type of atom or group, as with poly(vinylidene chloride) (X and Y are Cl), or they may differ, as in poly-(methyl methacrylate) (X is $-CH_3$, Y is $-COOCH_3$) and poly(α-methyl styrene) (X is $-CH_3$, Y is $-\text{\textcircled{}}$). When the substituents are small, polymerisation of a tetra-substituted monomer is possible, to produce a polymer such as polytetrafluoroethylene (PTFE), with the repeat unit $-(CF_2-CF_2)-$, but if large substituents are present on both carbon atoms of the double bond there is usually steric hindrance to polymerisation, i.e. the substituents would overlap each other if polymerisation took place.

Polydienes are a second important group within the class of addition polymers. The monomers have two double bonds and one of these is retained in the polymeric structure, to give one double bond per chemical repeat unit of the chain. This bond may be in the backbone of the chain or in a side group. If it is always in a side group the polymer is of the vinyl or vinylidene type. The two most important examples of polydienes are poly-butadiene, containing 1,4-linked units of type $-(CH_2-CH=CH-CH_2)-$ or 1,2-linked vinyl units of type $-(CH_2-CH(CH=CH_2))-$, and poly-isoprene, containing corresponding units of type $-(CH_2-C(CH_3)=CH-CH_2)-$ or $-(CH_2-C(CH_3)(CH=CH_2))-$. Polymers containing both 1,2 and 1,4 types of unit are not uncommon, but special conditions may lead to polymers consisting largely of one type. Acetylene, $CH\equiv CH$, poly-merises by an analogous reaction in which the triple bond is converted into a double bond to give the chemical repeat unit $-(CH=CH)-$.

Ring-opening polymerisations, such as those in which cyclic ethers poly-merise to give polyethers, may also be considered to be addition polymer-isations:

$$n\overline{CH_2-(CH_2)_{m-1}-O} \rightarrow -((CH_2)_m-O)_n$$

The simplest type of polyether, polyoxymethylene, is obtained by the similar polymerisation of formaldehyde in the presence of water:

$$nCH_2=O \rightarrow -(CH_2-O)_n$$

Step-growth polymers are obtained by the repeated process of joining together smaller molecules, which are usually of two different kinds at the beginning of the polymerisation process. For the production of linear (unbranched) chains it is necessary and sufficient that there should be two reactive groups on each of the initial 'building brick' molecules and that the molecule formed by the joining together of two of these molecules should also retain two appropriate reactive groups. There is usually no specific initiation step, so that any appropriate pair of molecules present anywhere in the reaction volume can join together. Many short chains are thus produced initially and the length of the chains increases both by the addition of monomer to either end of any chain and by the joining together of chains.

Condensation polymers are an important class of step-growth polymers formed by the common condensation reactions of organic chemistry. These involve the elimination of a small molecule, often water, when two molecules join, as in *amidation*:

$$RNH_2 + HOOCR' \rightarrow RNHCOR' + H_2O$$

which produces the *amide linkage*

$$-\underset{\underset{\displaystyle H}{|}}{N}-\underset{\underset{\displaystyle O}{\|}}{C}-$$

and *esterification*

$$RCOOH + HOR' \rightarrow RCOOR' + H_2O$$

which produces the *ester linkage*

$$-\underset{\underset{\displaystyle O}{\|}}{C}-O-$$

In these reactions R and R' may be any of a wide variety of chemical groups.

The amidation reaction is the basis for the production of the *polyamides* or *nylons*. For example, nylon-6,6, which has the structural repeat unit $-(HN(CH_2)_6NHCO(CH_2)_4CO)-$, is made by the condensation of hexamethylene diamine, $H_2N(CH_2)_6NH_2$, and adipic acid, $HOOC(CH_2)_4COOH$, whereas nylon-6,10 results from the comparable reaction between hexamethylene diamine and sebacic acid, $HOOC(CH_2)_8COOH$. In the labelling of these nylons the first number is the number of carbon atoms in the amine residue and the second the number of carbon atoms in the acid residue. Two nylons of somewhat simpler structure, nylon-6 and nylon-11, are obtained, respectively, from the ring-opening polymerisation of the cyclic compound ε-caprolactam:

$$n\overline{OC(CH_2)_5NH} \rightarrow -(OC(CH_2)_5NH)_{\overline{n}}$$

and from the self-condensation of ω-amino-undecanoic acid:

$$nHOOC(CH_2)_{10}NH_2 \rightarrow -(OC(CH_2)_{10}NH)_{\overline{n}} + nH_2O$$

The most important *polyester* is poly(ethylene terephthalate), $-((CH_2)_2OOC-\!\!\!\bigcirc\!\!\!-COO)_{\overline{n}}$, which is made by the condensation of ethylene glycol, $HO(CH_2)_2OH$, and terephthalic acid, $HOOC-\!\!\!\bigcirc\!\!\!-COOH$, or dimethyl terephthalate, $CH_3OOC-\!\!\!\bigcirc\!\!\!-COOCH_3$, where $-\!\!\!\bigcirc\!\!\!-$ represents the *para*-disubstituted benzene ring, or *p*–phenylene group. There is also a large group of unsaturated polyesters that are structurally very complex because they are made by multicomponent condensation reactions, e.g. a mixture of ethylene glycol and propylene glycol, $CH_3CH(OH)CH_2OH$, with maleic and phthalic anhydrides (see fig. 1.5).

An important example of a reaction employed in step-growth polymerisation that does not involve the elimination of a small molecule is the reaction of an isocyanate and an alcohol

$$RNCO + HOR' \rightarrow RNHCOOR'$$

which produces the *urethane linkage*

$$\begin{array}{ccc} -N & -C & -O- \\ | & \| & \\ H & O & \end{array}$$

One of the most complex types of step-growth reaction is that between a di-glycol, $HOROH$, and a di-isocyanate, $O\!=\!C\!=\!NR'N\!=\!C\!=\!O$, to produce a *polyurethane*, which contains the structural unit $-O-R-O-(C\!=\!O)-(NH)-R'-(NH)-(C\!=\!O)-$. Several subsidiary reactions can also take place and, although all of the possible reaction products are unlikely to be present simultaneously, polyurethanes usually have complex structures. Thermoplastic polyurethanes are copolymers that usually incorporate sequences of polyester or polyether segments.

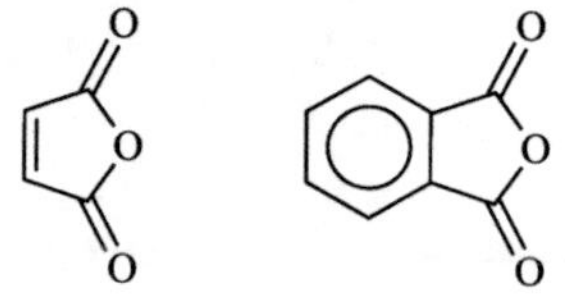

Fig. 1.5 The chemical formulae of (a) maleic anhydride and (b) phthalic anhydride. (Reproduced from *The Vibrational Spectroscopy of Polymers* by D. I. Bower and W. F. Maddams. © Cambridge University Press 1989.)

(a)   (b)

Formaldehyde, $H_2C{=}O$, provides a very reactive building block for step-growth reactions. For example in polycondensation reactions with phenol, ⊚—OH, or its homologues with more than one —OH group, it yields the *phenolic resins*, whereas with urea, $O{=}C(NH_2)_2$, or melamine (see fig. 1.6(a)) it yields the *amino resins*. The products of such condensation reactions depend on the conditions employed but they are usually highly cross-linked. Acid conditions lead to the formation of methylene-bridged polymers of the type shown in figs. 1.6(b) and (c), whereas alkaline conditions give structures containing the methylol group, —CH_2OH, which may condense further to give structures containing ether bridges, of the form R—O—R′ (fig. 1.6(d)).

1.3.4 Newer polymers and polymerisation processes

The polymerisation processes described in the previous section are the classical processes used for producing the bulk commercial polymers. Newer processes have been and are being developed with a variety of aims in mind. These involve the production of novel polymer topologies (see box); precise control over chain length and over monomer sequences in copolymers; control of isomerism (see section 4.1); production of polymers with special reactive end groups, the so-called *telechelic polymers*; production of specially designed thermally stable polymers and liquid-crystal polymers with a variety of different structures and properties. Other developments include the production of polymers with very precisely defined molar masses, and of networks with precisely defined chain lengths

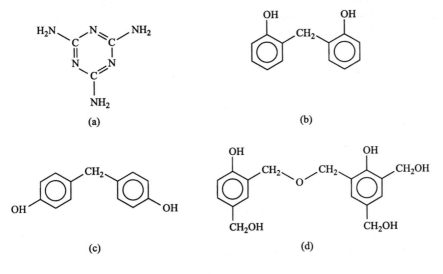

Fig. 1.6 The chemical formulae of (a) melamine; and (b), (c) and (d) various bridging structures in phenolic resins. (Reproduced from *The Vibrational Spectroscopy of Polymers* by D. I. Bower and W. F. Maddams. © Cambridge University Press 1989.)

Examples of new topologies, Figs 1(a) and (b) and Fig. 2. (1(a) Adapted by permission of the American Chemical Society; 1(b) reproduced by permission of the Polymer Division of the American Chemical Society.)

Examples of new topologies

Figure 1(a) shows the topology of a *dendritic polymer*.

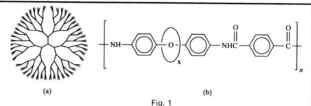

(a)   (b)

Fig. 1

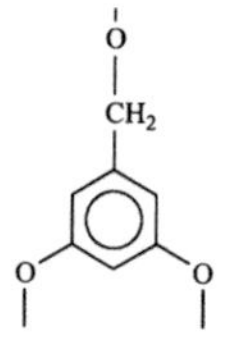

Fig. 2

Each junction point is formed by a chemical group that can join to several other groups, in the case illustrated to three. An example of such a group is shown in fig. 2. Although fig. 1(a) is two-dimensional, whereas the true polymer structure is three-dimensional, it does convey the correct impression that the density of units increases on going outwards from the centre, which puts a limit on the degree of polymerisation for such structures. It also shows that the structure can contain cavities, which could be useful for incorporating small unbonded molecules into the structure. Figure 1(b) shows a *polymeric rotaxane*, in which x represents a ring molecule that is not attached to the chain backbone by any chemical bonds, but which is prevented from leaving the chain by means of large end groups or by chain folds.

between entanglements or made from stiff rather than flexible chains. Some of the developments are already in commercial use, whereas others are still in the experimental stages.

A development of particular importance for the controlled production of block copolymers is the perfection of various so-called *living polymerisation* techniques. In the classical addition polymerisations there was always a termination stage, leading to the production of chains with non-reactive groups at both ends of the polymer chain. Polymerisation could therefore stop before all monomer had been exhausted, although ideally the termination step was of much lower probability than the propagation step. In living polymerisations there is no termination step and the reaction proceeds in the ideal case until all monomer has been exhausted. The chains still have reactive ends and a second type of monomer can then be added to the reaction to produce a block of a different type of polymer.

1.4 Properties and applications

Some of the properties of polymers have already been mentioned in preceding sections and, of course, form the subject matter of the rest of the

Table 1.1. *World production of various materials in 1984 and 1993 (millions of metric tons)*[a]

Material	1984	1993
Polyethylene	19.5	24.7
Poly(vinyl chloride)	12.1	14.8
Polypropylene	7.4	12.5
Polystyrene	6.8	7.5
Synthetic rubbers	9.0	7.4
Phenolic and cresylic plastics	1.4	2.6
Amino plastics	2.3	2.6
Alkyd resins	1.7	1.5
Regenerated cellulose	0.5	0.2
Non-cellulosic fibres	4.6	5.7
Cellulosic fibres	0.9	0.7
Crude steel	717.8	717.9
Unwrought aluminium	18.2	18.3
Unrefined copper	8.7	8.6

[a]Data from *Industrial Commodities Statistics Yearbook*, UN, New York, 1995.

book. Some of the uses have also been mentioned. This section contains some rather general comments on the reasons why the properties of polymers make them so important for a wide variety of applications. Tables 1.1 and 1.2 give first some statistical information about the production and use of polymers.

Table 1.1 illustrates the growth in production of polymers compared with the static state of production of some important metals. When it is remembered that aluminium has a density roughly twice, and copper and steel have densities of order six times, those of even the denser polymeric materials, table 1.1 also illustrates that the volume production of some of the commoner polymers roughly equals or exceeds that of aluminium and copper, and that the total volume production of all the polymers listed is about 60% of the volume production of steel.

The versatility of polymers, already commented on, must be taken to apply not only to these materials as a class, but also to many of its individual members. Poly(ethylene terephthalate) (PET), for instance, is used not only as a textile fibre but also as a packaging material in the form of both film and bottles. Poly(vinyl chloride) (PVC) is used not only as a rigid material for making mouldings but also, in plasticised form, for making flexible tubing and artificial leather.

Table 1.2. *Estimated end uses of plastics by weight*[a]

Packaging	37%[b]
Building and construction	23%
Electrical and electronic	10%
Transport	9%
Furniture	5%
Toys	3%
Housewares	3%
Agriculture	2%
Medical	2%
Sport	2%
Clothing	1%
Others	3%

[a]Reproduced by permission of the Institute of Materials from *Plastics: The Layman's Guide* by James Maxwell, IOM Communications Ltd, 1999.
[b]Since this table was compiled, use for packaging has increased and now probably represents about 50%.

As well as the classification of uses in table 1.2, polymers can also be classified broadly as being used as plastics, rubbers, fibres, coatings, adhesives, foams or speciality polymers. In many of their uses, as in plastics and fibres, it is often the combination of properties such as high strength-to-weight or stiffness-to-weight ratio and high resistance to chemical attack that gives them their importance. In other uses it is flexibility combined with toughness. In yet others it is resistance to chemical attack combined with high electrical resistance. One of the most important properties for many applications is the ability to be cast or moulded into complex shapes, thus reducing machining and assembly costs. Non-medical speciality uses include conducting polymers for rechargeable batteries; polymer sensors for many applications; high-density information storage, including CD and holographic devices; smart windows that can react to levels of light; and liquid-crystal displays, among many others. Medical uses include tooth fillings, components for hip-joint replacement and contact and implant lenses. It is not quite a matter of 'you name it and polymers will solve it', but their uses continue to expand into an ever-increasing variety of fields.

Although rubbers, or elastomers, form only a few per cent of polymers in use, they are vital for many applications, in particular for tyres and tyre products, which now consume about 50% of all rubber produced. The production of both synthetic and natural rubbers has expanded steadily

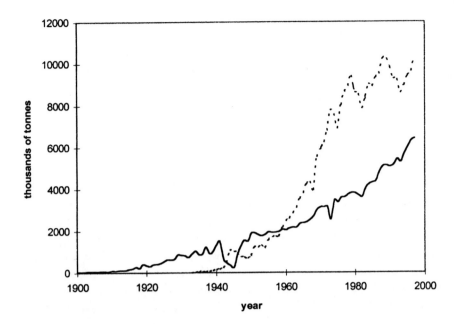

Fig. 1.7 World production of natural (————) and synthetic (– – –) rubber, 1990–1997. (Data by courtesy of The International Rubber Study Group, London.)

over the years, as shown in fig. 1.7. In 1997 synthetic rubbers formed approximately 60% of all rubber used, a decrease from the maximum of about 70% reached in the early 1970s. The fraction of synthetic rubber used in tyres and tyre products has also fallen since the early 1970s. In 1971 the ratio of synthetic to natural rubber used for this purpose in the USA rose to about 70 : 30, whereas by 1995 it had fallen to 61 : 39.

1.5   Polymer processing

1.5.1   Introduction

The major part of this book is concerned with the physics of solid polymers. Physics almost invariably attempts to generalise and, at least in the early stages of understanding, to simplify what happens in the real world. This book certainly follows that recipe. The purpose of this section is therefore to give a very brief introduction to the processing of polymers so that the student is aware that polymeric materials in actual use are often not as simple as is implied in the rest of the book. They frequently contain additives, introduced either to ease processing or to improve performance, and the precise conditions under which a small sample taken from a product was actually processed may be difficult to determine. Sometimes, however, large-scale commercial processing can yield materials that are far more homogeneous than those that can be produced on a small scale in the laboratory.

Further information about methods for producing oriented polymers is given in chapter 10. Further information on all the topics dealt with here can be obtained from the books cited as (7) in sections 1.6.1 and 1.6.2.

1.5.2   Additives and composites

Additives are used for a wide variety of purposes, and may be classified as *fillers, anti-oxidants, stabilisers, plasticisers, fire retardants, pigments* and *lubricants*. *Anti-oxidants* and *stabilisers* are usually used in rather small quantities in order to prevent degradation of the polymer when it is exposed to air, light and heat; the intention here is to maintain the properties of the polymer rather than to modify them. *Fillers* may be used either simply to produce a cheaper product or to improve the properties, in particular the mechanical properties. *Lubricants* may be used externally, to prevent adhesion of the polymer to the processing equipment, or internally, either to aid flow during processing or to reduce friction between the product and other materials.

Composites are materials in which a second component with very different properties is added to the polymer so that both components contribute to the properties of the product. The second component often increases the strength or stiffness of the product and is said to *reinforce* it. Particulate materials such as carbon black are often used to reinforce elastomers, for instance in car tyres, but fibres are usually used for reinforcing other types of polymer and are also used in tyres. Glass or carbon fibres are often used, but polymeric fibres are appropriate for some applications, as are metallic filaments, e.g. again in tyres. Such fibres are often aligned in one direction within a *matrix* of polymer, which gives the material anisotropic properties. Materials that are isotropic in one plane can be produced by using layers with the fibres aligned in different directions within the plane or by using mats of chopped fibres as the reinforcement. In these mats the fibres point randomly in all directions in a plane.

Although composites are a very important class of polymeric materials they form a separate subject in their own right, in which it is necessary to assume an understanding of the properties both of the polymer matrix and of the reinforcing material. They are not discussed further in this book, but it is interesting to note that in some ways semi-crystalline polymers can be considered as self-reinforcing polymers, because the mechanical properties of the crystalline parts are different from those of the non-crystalline parts, which often effectively form a matrix in which the crystals are embedded.

1.5.3 Processing methods

Additives may be introduced into the polymer either before it is processed to give the final product or during the processing. Although it is important to ensure that the components are adequately mixed, there is not space here to deal with the details of the mixing processes. The processing methods are therefore described as if the feed material were homogeneous.

The principal processing methods are (i) *injection moulding*, (ii) *extrusion*, (iii) *blow moulding*, (iv) *calendering*, (v) *thermoforming* and (vi) *reaction moulding*. These are now considered in turn.

Injection moulding allows the automated mass production of articles that may have complex shapes and whose properties can be varied by choosing different materials and the appropriate processing conditions. The basic principle of the method is shown in fig. 1.8(a). Polymer contained in a hopper is fed into a cylinder where it is pushed by a *ram* or plunger through a heated region, in which it melts, into a mould. The filled mould is allowed to cool and the solidified product is ejected. The ram is then drawn back and the process is repeated. This simple process suffers from a number of disadvantages and modern injection moulders are always of the *screw-injection* type illustrated in fig. 1.8(b).

During the injection part of the cycle, the screw is used as a plunger to drive the polymer forwards into the mould. When the mould is full and the

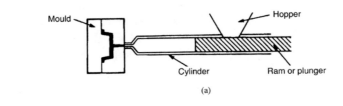

(a)

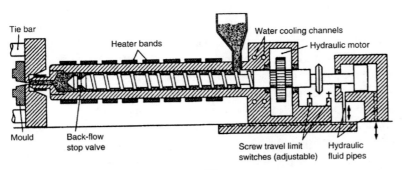

(b)

Fig. 1.8 (a) A schematic outline of a basic ram injection-moulding machine. (b) A single-screw injection-moulding machine, showing the injection unit. ((a) Reprinted by permission of Butterworth Heinemann; (b) courtesy of ICI.)

product is cooling, the screw is caused to turn and is therefore forced back along the barrel as polymer accumulates in front of it ready for the next injection stroke. Among the advantages of this type of machine are that the polymer is fed continuously to the barrel, that the screw action homogenises the polymer and that the heating and melting stage is effectively separated from the injection stage. Injection moulding can be done with either thermoplastic or thermosetting materials. For the latter the appropriate mixture is fed to the screw and injected into the mould. The chemical reaction takes place in the mould, which usually has to be heated.

Extrusion consists in principle of forcing molten polymer through a *die* to produce an *extrudate* of constant cross-sectional shape. Unlike injection moulding, extrusion is a continuous process. The shape of the product is controlled by the shape of the die. The most important types of extruders are *single-screw extruders* and in many ways they resemble the screw injector. The difference is that the screw rotates continuously and does not move along its length, so that polymer is continuously fed from the hopper along the screw and out through the die. The extrudate must be removed from the vicinity of the die, which requires a *haul-off* mechanism. Most of the heat necessary for melting and homogenising the polymer is generated by frictional forces due to shearing between the screw and the barrel. Extrusion is used to produce plastic products such as film, sheet and many kinds of *profiles* such as rods and pipes, but it is also used to produce plastic pellets in preparation for other manufacturing processes.

Sheet material, i.e. material thicker than about 0.25 mm, is usually produced by using a slit-shaped die, whereas thinner material is often produced by a *blown film extrusion process* in which an annular die is used and air is blown into the centre of the tubular extrudate to blow it into a sort of bubble. At a certain distance from the die the polymer is sufficiently cool to solidify into a film, which is then flattened and collected on rollers. Figure 1.9 illustrates a blown-film system.

Another important class of extruded materials consist of *filaments* and *yarns*. These are produced by *spinning*, which is the extrusion of molten material through fine holes in a *spinneret*. After passing through the spinneret the filaments cool, either in air or in water. If single-filament material is required, as for instance with nylon fishing line, the different filaments are wound up separately. If a *textile yarn* is required finer filaments produced simultaneously are twisted together before wind-up.

Blow moulding is a process used for making hollow products, such as bottles, by inflating a precursor tube, called a *parison*. In the simplest method the tubular parison is produced by extrusion and passes between the two halves of a mould, which are then clamped together, pinching off

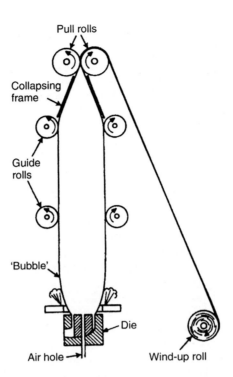

Pull rolls

Collapsing frame

Guide rolls

'Bubble'

Die

Air hole

Wind-up roll

Fig. 1.9 A schematic representation of blown tubular film extrusion. (Reproduced by permission of Oxford University Press.)

the parison at one end, the bottom of a bottle for instance. The parison is then inflated by compressed air so that it is forced into the shape of the mould.

1.6 Further reading

1.6.1 Some general polymer texts

The list below gives a selection of the many polymer textbooks available. Some of these contain more about polymerisation processes and about polymers in solution and the melt than the present book, but their coverage of polymers in the solid state is often rather narrower. All are at a similar or slightly more advanced level than the present book, with the exception of (1). Although it is now quite old, this elementary book can profitably be read cover to cover by the complete beginner.

(1) *Introduction to Polymer Science* by L. R. G. Treloar, Wykeham Publications, London, 1970.
(2) *Introduction to Synthetic Polymers* by I. M. Campbell, 2nd Edn, OUP, 2000.
(3) *Introduction to Polymers* by R. J. Young and P. A. Lovell, 2nd Edn, Chapman & Hall, London, 1991.
(4) *Introduction to Physical Polymer Science* by L. H. Sperling, 3rd Edn, John Wiley & Sons Inc., New York, 2001.

(5) *Polymer Physics* by U. F. Gedde, Chapman & Hall, London, 1995.
(6) *The Physics of Polymers: Concepts for Understanding their Structure and Behaviour* by G. R. Strobl, 2nd Edn, Springer Verlag, Berlin, 1997.
(7) *The Principles of Polymer Engineering* by N. G. McCrum, C. P. Buckley and C. B. Bucknall, 2nd Edn, Oxford Science Publications, 1997.

1.6.2 Further reading specifically for chapter 1

(1) *Giant Molecules,* by H. F. Mark *et al.,* Time-Life International (Nederland) NV, revised 1973. An excellent, profusely illustrated layman's introduction to the early development of synthetic polymers.
(2) *Landmarks of the Plastics Industry,* Imperial Chemical Industries Ltd, Plastics Division, Welwyn, 1962. An anonymously authored volume published to mark the centenary of Alexander Parke's invention of the world's first man-made plastic. A very interesting account of the subject to the year of publication. Well illustrated and containing references to many sources for the history of the subject.
(3) *History of Polymer Science and Technology* by R. B. Seymour, Marcel Dekker, New York, 1982. A series of chapters on the history of various categories of polymers.
(4) *Polymer Science and Engineering: the Shifting Research Frontiers,* National Academy Press, Washington, 1994. A volume produced for the National Research Council (USA). It contains a very readable account of the state of polymer science and engineering immediately prior to the date of publication, including the description of many applications of polymers, and suggests ways in which the subject can be developed further.
(5) *Polymers: Chemistry and Physics of Modern Materials* by J. M. G. Cowie, 2nd Edn, Blackie, Glasgow, 1991. Six chapters of this book are devoted to methods of polymerisation. The remaining eleven cover various topics in solution and solid-state properties.
(6) *Polymers* by D. Walton and P. Lorimer, Oxford Chemistry Primers, OUP, 2000. This book, as the series to which it belongs suggests, is largely devoted to the polymerisation and physical chemistry of polymers, but it also includes some history and a section on conducting polymers.
(7) *Handbook for Plastics Processors,* by J. A. Brydson, Heineman Newnes, Oxford, 1990 (in association with the Plastics and Rubber Institute). An excellent introduction with many detailed diagrams of equipment, a survey of major plastics materials and a brief section on the principles of product design.

Chapter 2
Some physical techniques for studying polymers

2.1 Introduction

Many physical techniques are used in the study of solid polymers and some, such as NMR spectroscopy, can give information about a wide variety of features of the structure or properties. On the other hand, some of these features, for instance the degree of crystallinity, can be studied by a wide variety of techniques that give either similar or complementary information. The techniques described in this chapter are those that have a wide applicability within polymer physics or where the description of the technique is most easily introduced by reference to non-polymeric materials. The reader who is already familiar with a particular technique, e.g. X-ray diffraction, may wish to omit the corresponding section of this chapter. It is suggested that other readers should read through the chapter fairly quickly to gain an overall view of the techniques and then look back when applications are referred to in later chapters if they wish to gain a deeper understanding. Techniques that are more specific to polymers are described in the appropriate chapters and only a few, which are generally available only in specialised laboratories, are mentioned without further discussion.

2.2 Differential scanning calorimetry (DSC) and differential thermal analysis (DTA)

If unit mass of any substance is heated slowly at constant pressure in such a way that the rate of supply of energy Q is dQ/dt and the rate of rise of temperature T is dT/dt, the specific heat of the substance at constant pressure C_p is given by

$$C_p = dQ/dT = dQ/dt \div dT/dt. \tag{2.1}$$

It follows that if dT/dt is maintained constant and equal to a and dQ/dt is plotted against T, the graph will show the value of aC_p as a function of T.

This will, however, be true only provided that there are no first-order phase transitions, such as melting, in the temperature range scanned. If there is such a transition, energy will be required to change the phase without any change of temperature taking place, so that, for constant dT/dt, the value of dQ/dt would have to be infinite. A schematic diagram of the corresponding plot is shown in fig. 2.1(a). It has been assumed for simplicity that the value of C_p is constant on each side of the transition but has slightly different values on the two sides.

In reality, the best that can be done experimentally is to place the sample in a holder and to measure the temperature of the holder, rather than that of the sample. This has two effects on what is observed. The first is that the holder will have its own heat capacity, which must be allowed for in deducing the specific heat of the sample, and the second is that heat must be transferred from the holder to the sample, so that there will be a lag between the average temperature of the sample and the temperature of the holder. It is now possible to raise the temperature of the holder at a constant rate even when the sample is undergoing a first-order transition. Once the transition is complete, dQ/dt becomes proportional to the new specific heat of the sample, assuming that the heat capacity of the holder has been allowed for. The plot of dQ/dt against T during a first-order transition will therefore appear as shown schematically in fig. 2.1(b). Because T changes uniformly with time, the area under the curve of dQ/dt against T for any temperature range is proportional to the total energy supplied in the corresponding time.

It follows that this simple type of experiment can give the values of three important quantities: (i) the temperature dependence of the specific heat C_p, (ii) the temperature of any first-order transition and (iii) the change in enthalpy during the transition. Commercial instruments are available for DSC measurements and work on the principle of comparing the energy supplied to the sample holder with that supplied to a reference, empty holder. Figure 2.2 shows a schematic diagram of such a system.

Fig. 2.1 Rate of supply of energy dQ/dt plotted against temperature T or time t for a substance undergoing a first-order transition: (a) the theoretical plot for a substance uniformly and directly heated with a constant value of the rate of change of temperature dT/dt; (b) an idealised plot for a substance heated indirectly with a thermal resistance between it and a holder whose temperature is changed at the constant rate dT/dt.

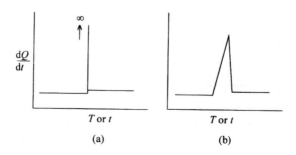

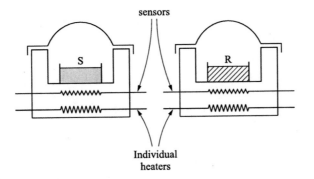

sensors

Individual
heaters

Fig. 2.2 A schematic diagram of a DSC apparatus. S is the sample pan, R is the reference pan and the sensors are platinum resistance thermometers. For technical reasons a small constant difference in temperature is maintained between the sample and reference holder blocks. (Reproduced by permission of PerkinElmer Incorporated.)

In practice, the sample is placed within a *sample pan* which is then placed within the sample holder block. A similar empty pan is placed in the reference holder block. The instrument then allows T, the common temperature of the two holders, to be changed at a constant rate $dT/dt = \dot{T}$, while the two holders are ideally (see caption to fig. 2.2) maintained at the same temperature by a feedback loop. Extra energy Q must be supplied to the sample holder at a rate dQ/dt to maintain its temperature the same as that of the reference holder; this rate is registered by the instrument and plotted either against T or against time t. It is generally possible to assume that, away from any transitions in the sample, the sample and the sample pan are at the same temperature and that the sample and reference pans are identical. It then follows that $dQ/dt = mC_p$, where m is the mass of the sample and C_p is the specific heat per unit mass.

When the sample passes through a first-order transition, such as melting, heat must be supplied to the sample while its temperature does not change until the transition is complete. Since experiments are usually performed at constant (atmospheric) pressure, this heat is the enthalpy of transition, ΔH. Because the temperature of the sample *holder* is changing at a constant rate, there must be a difference in temperature ΔT between the sample and the holder, and the rate of transfer of energy dQ/dt to the sample must therefore be equal to $\kappa \Delta T$, where κ is the thermal conductance between the sample holder and the sample. During the transition, ΔT increases uniformly with time at the rate $\dot{T}$, because the temperature of the sample remains constant, so that

$$\frac{dQ}{dt} = \kappa \Delta T = \kappa \dot{T}(t - t_0) + \frac{dQ}{dt}\bigg|_{t_0} \qquad (2.2)$$

where the transition starts at time t_0. The rate of transfer of energy to the sample, dQ/dt, thus varies linearly with time, as shown schematically in fig. 2.1(b). Such a plot is obtained only from a pure substance with a very well-defined melting point. Such a substance is 99.999% pure indium, which is often used as a standard in DSC work. Using indium instead of the sample allows $\kappa \dot{T}$ and hence κ to be determined. When a less pure substance is used, or a polymer with a range of different crystal melting temperatures, a much more rounded peak is obtained, and it is then usually assumed that the mean melting temperature is given by drawing a line of this slope back from the maximum value in the melting peak to the baseline, as shown in fig. 2.3.

Figure 2.3 also illustrates two other features of the DSC trace often seen for polymers. One is a relatively sudden change in the value of C_p. This corresponds in the simplest case to the glass transition temperature, T_g, of the polymer, which is discussed in chapter 7. The second feature is a negative-going peak due to the crystallisation of a polymer that had previously been cooled from the melt sufficiently quickly to suppress crystallisation. Heat is being given out by the sample in this transition; such transitions are said to be *exothermic*, whereas transitions like the melting transition, in which heat is being absorbed, are *endothermic*.

Another technique related to DSC is DTA, *differential thermal analysis*. In this method sample and reference are heated by a single source and temperatures are measured by thermocouples embedded in the sample and reference or attached to their pans. Because heat is now supplied to the two holders at the same rate, a difference in temperature between the sample and the reference develops, which is recorded by the instrument. The difference in temperature depends, among other things, on the value of κ, which needs to be low to obtain large enough differences in temperature to measure accurately. The area under a transition peak now depends on κ and it is difficult to determine this accurately or to maintain it at a constant

Fig. 2.3 The DSC trace for a sample of a thermoplastic polyester that had previously been cooled very rapidly, so that it was initially non-crystalline. On heating it undergoes the glass transition, followed by crystallisation and finally by melting. (Courtesy of PerkinElmer Incorporated.)

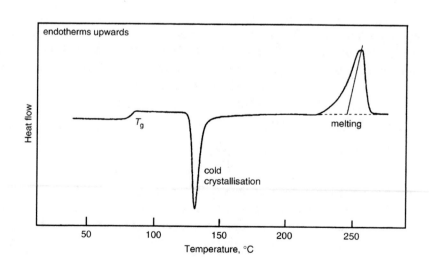

value. For this reason, DSC is usually preferred to DTA for quantitative work.

2.3 Density measurement

For most pure substances other than polymers the density does not vary significantly from one sample to another. As discussed in chapter 5, however, many polymers can be partially crystalline and the degree of crystallinity can have a noticeable effect on their density. Conversely, a knowledge of the density can be useful in assessing the degree of crystallinity. One of the most important ways that densities of polymers are determined is by the use of a *density-gradient column, or density column.*

A density column is made by carefully injecting a varying mixture of two miscible liquids of different density into the bottom of a vertical tube. When this is done in the correct way the density of the resulting column of liquid increases almost linearly with distance from the top of the column. The liquids must be chosen both to cover the required range of density and to be compatible with the samples to be studied, i.e. the liquids must not attack or penetrate the polymer. Small pieces of the material for which knowledge of the density is required are wetted with the upper liquid and very carefully introduced to the top of the column. Each piece is then observed until it ceases to fall down the column, when its density must be equal to that of the liquid at that level, and its position is then determined. A scale is placed in front and a mirror behind the column, so ensuring that the piece of polymer is viewed normally to the scale when its position is noted. The column must be calibrated to give the density-versus-distance relationship, which is done by carefully inserting specially calibrated floats into it, each of which has a known effective density.

It is essential to use several small pieces from any sample of material in order to check for consistency, particularly since it may be difficult to see small bubbles attached to such a piece. It is also essential that the column be kept within a transparent constant-temperature bath, because the densities both of the liquid and of the polymer samples are highly temperature-dependent. Once the measurements have been made, the pieces of polymer or the floats are removed from the column very slowly by means of a small wire-mesh basket. This can be done without causing significant disturbance of the density gradient.

Other methods of determining density can be used, such as the specific gravity bottle or pyknometer. However, the density-column method is usually far simpler and of greater accuracy for the small samples usually used in polymer research, for which an inaccuracy of 0.1 kg m^{-3} can be achieved.

2.4   Light scattering

The scattering of waves from any structure can give information about the details of the structure provided that the wavelength of the waves is neither very much smaller nor very much larger than the important features of the structure. Various kinds of waves are used in studying the structures of polymers, including light waves, X-rays and the waves associated with electrons or neutrons according to wave mechanics. Of these waves the longest are light waves, which typically have wavelengths of about 0.5 μm and can therefore give information about structures ranging in size from about 0.05–5 μm. Aggregates of crystallites in many polymers are within this range of sizes and can thus be studied by light scattering (see section 5.5.2).

In the simplest possible low-angle light-scattering experiment with a solid polymer, a thin sheet of polymer is placed at right angles to a laser beam and a piece of photographic film is placed several centimetres behind the polymer, also at right angles to the laser beam, as shown in fig. 2.4. A small hole in the film allows the undeviated laser beam to pass through the film without blackening it. After a suitable exposure time the film is developed and the degree of blackening at any point depends on the amount of light scattered in the corresponding direction. If the laser is powerful enough, the pattern can be observed in a darkened room directly on a white screen replacing the film.

It is not easy to convert blackening on a photographic film to intensity of light, so the film is usually replaced either by a light sensitive detector that can be rotated to intercept light scattered at different angles or by a multi-channel detector, which simultaneously detects light scattered at an array of angles determined by the spacing of the detector elements and their distance from the sample. The use of polarised incident light and an analyser, behind the sample, that can be rotated to have its transmission axis parallel or perpendicular to the polarisation of the incident light can give further information about the structures. The fact that laser beams are usually polarised is thus an advantage.

Light scattering is also used extensively for the study of polymers in solution, one application of which is the determination of molar masses (see section 3.2.2).

Fig. 2.4 A schematic diagram showing the arrangement for the observation of low-angle light scattering from a polymer film.

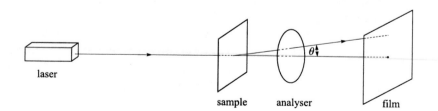

laser

sample   analyser   film

2.5   X-ray scattering

X-rays are electromagnetic waves of very short wavelength; the X-rays used in polymer studies have wavelengths of about 0.1–0.2 nm. When waves of any kind are scattered from structures with which they interact the angles of scatter are large when the lengths within the structure are comparable to the wavelength and small when the lengths are large compared with the wavelength of the waves. Two types of X-ray scattering are therefore used in the study of polymers, *wide-angle X-ray scattering*, or WAXS, and *small-angle X-ray scattering*, or SAXS, depending on the scale of the features studied.

Scattering from structures of any size is regular, i.e. takes place at well defined angles, only when the structures are periodic. The scattering is then usually called *diffraction*. The most important periodic structures suitable for WAXS investigation are *crystals*, which are periodic in three dimensions.

2.5.1   Wide-angle scattering (WAXS)

Imagine two sets of parallel planes with equal spacing a measured normal to the planes and let each set be vertical and inclined to the other at right angles. These planes cut any horizontal plane at right angles and the intersections with such a plane form a series of identical squares of side a. If, instead of a single horizontal plane they cut a series of horizontal planes with spacing a, the intersections of all three sets of planes divide space into a series of cubes of edge a, each identical to the others except for its position and each having an identical cube touching it on each of its six faces. Any corner of each cube is also the corner of seven other cubes. These corners form a *lattice*, a *simple cubic lattice*.

Now imagine that the three sets of planes are inclined to each other at angles different from 90° and that the spacings are different for the three sets, while each set remains equally spaced. Each cube now becomes a parallelepiped with edges a, b and c and angles α, β and γ (see fig. 2.5). The direction along which the edge length or *lattice translational period* is a is called the a-axis, and similarly for the b- and c-axes. The angle γ is defined to be the angle between the a- and b-axes, the angle β the angle between the c- and a-axes and the angle α the angle between the b- and c-axes. By varying the six quantities a, b, c, α, β and γ every possible lattice can be obtained, and if $\boldsymbol{a}$, $\boldsymbol{b}$ and $\boldsymbol{c}$ are chosen as vectors parallel to the a-, b- and c-axes, respectively, with lengths a, b and c, the *lattice vector* joining any lattice point to an origin O chosen to coincide with a lattice point can be written $l\boldsymbol{a} + m\boldsymbol{b} + n\boldsymbol{c}$, where l, m and n are integers.

Fig. 2.5 A general three-
dimensional lattice,
showing the definitions of
a, *b* and *c* and α, β and γ.
The lines represent the
intersections of two lattice
planes and the filled circles
represent the intersections
of three lattice planes, the
lattice points.

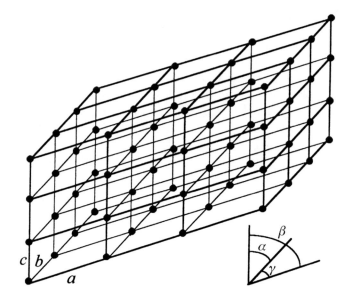

A crystal can now be imagined to be formed by placing identical *single atoms*, *groups of atoms* or *molecules* at each lattice point. In order that the structure should remain periodic, the groups of atoms or molecules must be placed in the same orientation at each lattice point.

If the angles α, β and γ are all different from 90° and from each other the crystal is said to be *triclinic*. The only simplification that may be possible in describing the structure is to find another set of parallelepipeds with angles closer to 90° that still describe the structure. For certain special values of α, β and γ and special relationships among *a*, *b* and *c* it is possible to find other descriptions of the lattice that conform to the symmetry of the structure. A particular example is the *face-centred cubic* lattice, illustrated in fig. 2.6, in which the cube of edge *a* clearly exhibits the full symmetry of the structure, but the rhombohedron of edges $a/\sqrt{2}$ does not. This parallelepiped is one of the infinite number of possible types of *primitive unit cell* that can be chosen so as to contain only one lattice point per cell. The cubic unit cell is not primitive; there are four lattice points per cubic cell. Care must be taken when describing diffraction from non-primitive cells, which can occur both for polymeric and for non-polymeric crystals.

When a narrow parallel beam of monochromatic X-rays, i.e. X-rays of a single wavelength, is incident on a crystal the atom, group of atoms or molecule at each lattice point will scatter the X-rays in exactly the same way, i.e. with the same intensity in any given direction, but with different intensities in different directions. Consider scattering in a particular direction. Although the structure at each lattice point in the

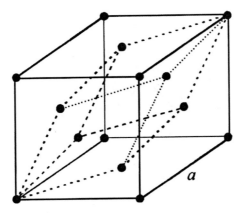

Fig. 2.6 The face-centred cubic unit cube, showing also the rhombohedral primitive cell.

crystal scatters in this direction, the crystal as a whole will only scatter strongly in this direction if the scattering from all structures is in phase. This means that there must be a path difference of a whole number of wavelengths for scattering from any two lattice points in the crystal. Suppose that in a particular direction of strong scattering there are integral numbers h, k and l of wavelengths path difference for scattering from any two adjacent points along the a-, b- and c-axes, respectively. This condition is not only sufficient, but clearly necessary for there to be a path difference of a whole number of wavelengths for scattering from any two lattice points. Scattering in such a direction will produce a *diffraction spot* on an X-ray film; the numbers h, k and l are then called the *diffraction indices* for the spot. Assigning the correct values of h, k and l to a spot is called *indexing* it.

A different way of looking at diffraction from crystals that is often used is to consider the various possible sets of parallel planes that can be drawn through the lattice points so that every lattice point lies on one of the planes. This leads to the idea of 'Bragg reflection', by analogy with the reflection of light from a mirror.

It is well known that, when a parallel beam of light falls on a mirror at any angle of incidence, it is reflected on the opposite side of the normal to the mirror at an equal angle. This is because there is then no path difference between different rays within the beam (see fig. 2.7(a)). Similarly, there is no difference in path between X-rays scattered from different lattice points on a crystal plane if the rays make equal angles θ with the plane and lie on opposite sides of the normal to it. For a crystal, however, there are a very large number of lattice planes from which the X-rays are scattered, and there is thus a further condition for strong 'reflection': the beams 'reflected' from different parallel planes must be in phase. It is easy to see from fig. 2.7(b) that this phase difference is $2d \sin\theta$ for adjacent lattice planes. It

Fig. 2.7 (a) Reflection from
a mirror. The path from A
to A' is the same as that
from B to B' wherever the
points P_1 and P_2 are
situated on the mirror. (b)
'Bragg reflection'. The path
difference for the two rays
shown is easily seen to be
$2d \sin \theta$ and this will be the
same for rays 'reflected' at
the angle θ from any points
on the two different planes,
as is shown by considering
the result for (a).

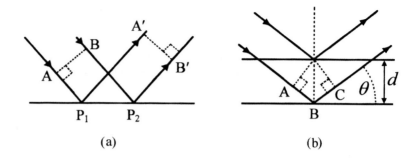

(a)   (b)

should be noted that, in X-ray diffraction it is conventional to measure the
angle θ from the surface of the plane rather than from its normal, as is
usual with optical reflection. The condition for strong 'reflection' from the
set of planes with spacing d is thus

$$2d \sin \theta = n\lambda \qquad (2.3)$$

where n is an integer and λ is the wavelength of the X-rays. Equation (2.3)
is called *Bragg's law* and θ is then the *Bragg angle*. The integer n, which is
the number of wavelengths path difference for scattering from adjacent
planes, is called the *order* of diffraction or 'reflection'.

It is convenient, however, to regard all scattering as if it were of first
order, i.e. to regard nth-order scattering from a set of planes with spacing d
that actually all pass through lattice points as if it were first order ($n = 1$)
scattering from a set of planes separated by $d^* = d/n$, some of which do
not pass through lattice points. It has already been seen above that every
diffraction spot can be associated with a set of diffraction indices (hkl), so
that the separation d^* can be replaced by d_{hkl}. It then follows that

$$2d_{hkl} \sin \theta = \lambda \qquad (2.4)$$

The spacing d_{hkl} is easily calculated for a given measured value of θ and
planes with spacing d_{hkl} give 'reflection' with one wavelength path difference ($n = 1$) between any two points on adjacent planes. However, by
definition, the (hkl) reflection has h, k and l wavelengths path difference
for scattering from adjacent lattice points along the three axes. Hence, the
(hkl) planes are separated by a/h, b/k and c/l along the a-, b- and c-axes.
Figure 2.8 shows a two dimensional section through a crystal and indicates
the labelling of planes with the corresponding diffraction indices.

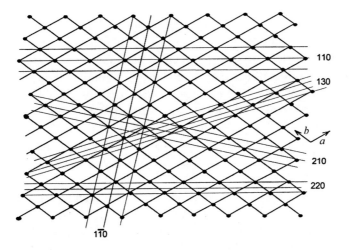

110

130

210

220

1$\overline{1}$0

Fig. 2.8 Crystal plane indices. All the planes shown are assumed to be parallel to the *c*-axis, which is normal to the plane of the page. A bar over an integer indicates that, if any particular lattice point on a starting plane of a given set is taken as the origin, it is necessary to move in the negative direction of the corresponding axis to find the adjacent plane for which the intersections with the other two axes lie on their positive directions. Note that the (1$\overline{1}$0) planes could equally well be labelled the ($\overline{1}$10) planes.

In general, the value of d_{hkl} depends on the values of a, b, c, α, β and γ. For an orthorhombic unit cell, for which all the angles are right angles, $d_{hkl} = 1\sqrt{h^2/a^2 + k^2/b^2 + l^2/c^2}$. Figure 2.9 shows some important planes for the special case of a lattice with a rectangular projection on a plane perpendicular to the *c*-axis.

The sets of integers (*hkl*) can be imagined to label the points of intersection of three sets of equally spaced parallel planes. It is straightforward to show that these planes can be chosen so that this new lattice, which is then called the *reciprocal lattice*, has the following property: for all values of *h*, *k* and *l* the line joining the origin of the reciprocal lattice to the point (*hkl*) is of length $1/d_{hkl}$ and is normal to the (*hkl*) planes of the real lattice.

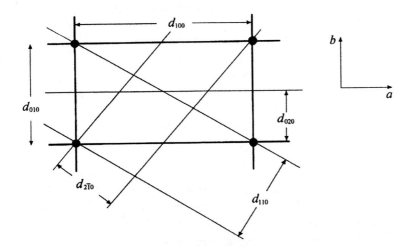

d_{100}

d_{010}

d_{020}

$d_{2\overline{1}0}$

d_{110}

b

a

Fig. 2.9 Some important planes for a lattice with a rectangular projection on a plane perpendicular to the *c*-axis.

It is also possible to show that the reciprocal-lattice plane for a given value of l and all values of h and k is perpendicular to the c-axis and is distant l/c from the origin of the reciprocal lattice, i.e. reciprocal-lattice planes for different values of l are separated by $1/c$. Similarly, reciprocal-lattice planes for different values of h are perpendicular to the a-axis and are separated by $1/a$, whereas reciprocal-lattice planes for different values of k are perpendicular to the b-axis and are separated by $1/b$. The two lattices are in fact mutually reciprocal and knowledge of the shape and dimensions of the reciprocal lattice thus provides immediately the dimensions and shape of the real lattice.

The experimental arrangement used for WAXS studies of polymers depends on the purpose of the studies and will be discussed in the appropriate later chapters. These studies include the determination of crystal structure, the determination of crystallinity and studies of molecular orientation, among others.

2.5.2 Small-angle scattering (SAXS)

SAXS studies are performed on polymers for the investigation of structures on a much larger scale than the separations of crystal planes (see section 3.4.2), which implies scattering angles much smaller than those observed with WAXS. Typical angles involved are of order $1°$ and, in order to observe scattering through such small angles, a fairly large distance between the sample and the detector is required.

SAXS experiments are often performed with apparatus that consists simply of a long box, at one end of which is a set of pinholes to collimate the X-ray beam from a generator and at the other end of which is a piece of film or a two-dimensional X-ray detector. The sample is placed in the X-ray beam near the collimator end and a beam-stop to block the direct, undiffracted X-ray beam is placed near the film or detector. The box can usually be evacuated to avoid scattering of the X-rays by air. More elaborate cameras are also sometimes used.

2.6 Infrared and Raman spectroscopy
2.6.1 The principles of infrared and Raman spectroscopy

All molecules, whether polymeric or not, consist of atoms bound together by chemical bonds. The bonds and the angles between them are not rigid. To a first approximation the force required to make a small change in the length of a bond, or a small change in the angle between two bonds, is proportional to the change produced; similarly, the torque required to twist one part of a molecule around a bond through a small angle with respect to the rest of the molecule is approximately proportional to the angle of twist. The molecule thus consists of a set of *coupled harmonic oscillators* and, if it is disturbed from its equilibrium state, it will vibrate in such a way that the motion can be considered to be a superposition of a number of simple-harmonic vibrations. In each of these so-called *normal modes* every atom in the molecule vibrates with the same frequency, and in the simplest molecules, all atoms pass through their respective positions of zero displacement simultaneously.

There are three principal methods by which the vibrations can be studied, *infrared* (IR) *spectroscopy, Raman spectroscopy* and *inelastic neutron scattering*. The first two methods are available in very many laboratories, because the equipment required is relatively small and cheap. Neutron scattering is less readily available, because the technique requires a neutron source, which is usually a nuclear reactor, and relatively specialised and expensive equipment to analyse the energies of the neutrons scattered from the sample. Neutron scattering is not considered in any detail in this book, although it will be mentioned occasionally.

A vibrating molecule can interact in two distinctly different ways with electromagnetic radiation of appropriate frequency. If the radiation has the same frequency as one of the normal modes of vibration, which usually means that it will be in the infrared region of the electromagnetic spectrum, it may be possible for the molecule to *absorb* the radiation. The energy absorbed will later be lost by the molecule either by re-radiation or, more usually, by transfer to other molecules of the material in the form of heat energy. An *infrared absorption spectrum* of a material is obtained simply by allowing infrared radiation to pass through the sample and determining what fraction is absorbed at each frequency within some particular range. The frequency at which any peak in the absorption spectrum appears is equal to the frequency of one of the normal modes of vibration of the molecules of the sample.

The second way in which electromagnetic radiation can interact with the molecule is by being *scattered*, with or without a change of frequency. If light is allowed to fall on a sample both types of scattering will in general take place. The scattering without change of frequency includes scattering from inhomogeneities and the scattering that would still occur in a homogeneous medium even if the molecules were not vibrating, which is called *Rayleigh scattering*. The scattering with change of frequency is called *Raman scattering*, for which the change in frequency is equal to the frequency of one of the normal modes of vibration of the molecules. The strongest scattering is at frequencies lower than that of the incident light and this is called the *Stokes Raman scattering*.

The frequencies, ν, of the normal vibrational modes of small molecules generally lie in the region $(6\text{–}120) \times 10^{12}$ Hz or 6–120 THz and this is generally also the most important region for polymers. It is usual to work in reciprocal-wavelength units, or wavenumber units, $\tilde{\nu} = 1/\lambda$, which specify the number of waves per metre or, more usually, per centimetre. In terms of wavenumbers the range of the infrared spectrum, including overtones, is roughly 10–8000 cm^{-1} or 10^3 to 8×10^5 m^{-1}. In the Raman spectrum the difference in frequency between the incident light and the Raman-scattered light is equal to the frequency of the vibration. These differences span exactly the same ranges as the IR frequencies or wavenumbers and only the differences in frequency, or more usually the corresponding differences in wavenumber $\Delta\tilde{\nu}$, are used in describing the spectrum.

In general, many but not all of the modes of vibration of a particular type of molecule can be observed by means of infrared spectroscopy and these are said to be *infrared-active* modes. Similarly, some but not all modes are *Raman-active*. Which modes are active for which process depends on the symmetry of the molecule. This means that, if the maximum possible amount of information is required, both types of spectroscopy should be used, but it also means that one of the two methods may be more useful than the other for a particular type of study.

For a mode to be IR-active the molecule must have an oscillating electric dipole moment, called the *transition dipole* or *transition moment*, when it vibrates in this mode. For a vibration to be Raman-active the molecule must have an oscillating electrical polarisability (see section 9.2) when it vibrates in the mode. This means essentially that the shape of the molecule must be different in opposite phases of the vibration. These requirements constitute the *selection rules* for the two kinds of spectroscopy. Figure 2.10 shows the four modes of vibration of the CO_2 molecule and their IR and Raman activities.

Fig. 2.10 Vibrations of the CO_2 molecule. The plus and minus signs indicate the partial charges on the atoms and the arrows indicate the directions of motion, with the amplitude of motion greatly exaggerated: (a) the symmetric stretching mode, (b) the anti-symmetric stretching mode and (c) the bending mode. This mode can take place in two independent directions at right angles and is said to be doubly-degenerate.

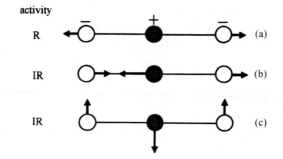

2.6.2 Spectrometers for infrared and Raman spectroscopy

Infrared spectrometers used for polymer work consist of (i) a source of radiation with a continuous spectrum over a wide range of infrared wavelengths, (ii) a means of dispersing the radiation into its constituent wavelengths, (iii) an arrangement for allowing the radiation to pass through the sample or be reflected from its surface, (iv) a means of measuring intensities using a detector that responds over the whole range of wavelengths of interest and (v) a method of displaying, and possibly performing calculations on, the spectrum. There are two distinct types of infrared spectrometer that may be used for studying polymers, *dispersive* instruments and *Fourier-transform* (FT) instruments. In dispersive instruments the radiation is physically split up into its constituent wavelengths by a monochromator, either before, or more usually after, it passes through the sample, whereupon the various wavelengths are processed in sequence. In the Fourier-transform instrument radiation of all wavelengths passes through the sample to the detector simultaneously. The detector measures the total transmitted intensity as a function of the displacement of one of the mirrors in a double-beam interferometer, usually of the Michelson type, and the separation of the various wavelengths is subsequently done mathematically by performing a Fourier transform on the intensity versus displacement data, using a dedicated computer. FT instruments are now the dominant type.

In principle, all that is needed for obtaining a Raman spectrum is a source of high-intensity monochromatic light, which is used to irradiate the sample, a lens system to collect scattered radiation and a spectrometer to disperse it into its component wavelengths and record the intensity as a function of wavelength or wavenumber. A block diagram of such a system is shown in fig. 2.11. The main difficulties arise because the efficiency of Raman scattering is very low, the scattered energy being typically about 10^{-8} to 10^{-6} of the incident energy and about 10^{-3} times that of the elastic

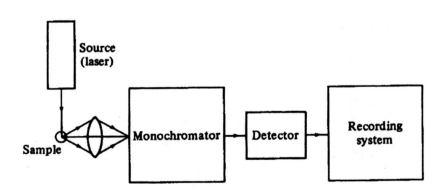

Fig. 2.11 A Block diagram of a minimal system for Raman spectroscopy. (Reproduced from *The Vibrational Spectroscopy of Polymers* by D. I. Bower and W. F. Maddams. © Cambridge University Press 1989.)

Rayleigh scattering. A laser source is therefore used, with a very sensitive detector.

The average time taken to record a spectrum over the range 50–2000 cm^{-1} is about 20 min with older types of Raman spectrometer. Modern instruments can record Raman spectra in times as short as those permitted by FTIR spectroscopy. This has been achieved by the use of multidetector arrays that record the signal within each interval of 1 cm^{-1} simultaneously with the others. By this means it is possible, for instance, to obtain a satisfactory spectrum from a polyethylene pellet in a fraction of a second, which makes it possible to undertake dynamic measurements. Other novel forms of Raman spectrometer have also been introduced, including those with which microscopic images of samples can be obtained in the Raman scattered light at a particular frequency. Fourier-transform instruments, using infrared lasers as sources of excitation are also available for special types of investigation.

2.6.3   The infrared and Raman spectra of polymers

A small molecule consisting of N atoms has a total of $3N$ degrees of freedom. If all the atoms of the molecule do not lie on a single straight line six of these represent the translations and rotations of the whole molecule. There remain $3N - 6$ *internal degrees of freedom* and this leads to $3N - 6$ normal modes of vibration. A polymer molecule may contain tens of thousands of atoms and may thus have tens to hundreds of thousands of normal modes. The infrared or Raman spectrum of a polymer might thus be expected to be impossibly complicated. Figure 2.12 illustrates that this is not so. The basic reason is that the molecule of a homopolymer consists of a large number of chemically identical units, each of which usually contains only tens of atoms or fewer. This leads to a considerable reduction in the complexity of the infrared or

Fig. 2.12 The infrared spectrum of a sample of 1,2-polybutadiene. (Adapted by permission of John Wiley & Sons Limited.)

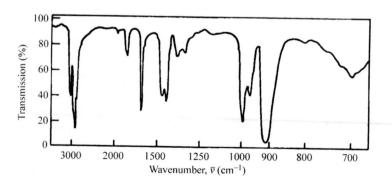

Raman spectrum because, to a first approximation, the spectra obtained are simply those that would be obtained from the chemical units if they were separate molecules.

This is particularly so when the polymer chain is disordered or when the vibration concerned involves mainly the motion of side-group atoms and little motion of the atoms of the chain backbone. The spectrum then consists of broad infrared absorption or Raman scattering peaks due to *group vibrations*, the broadening arising because the groups of atoms of the same kind in different repeat units have slightly different frequencies of vibration as a consequence of the different physical environments of the units to which they belong. For crystalline materials, selection rules show that only modes in which all unit cells vibrate in phase with each other can be IR- or Raman-active.

The IR or Raman spectrum of a polymer thus usually contains a number of peaks which is of order $3n$ or less, where n is the number of atoms in the repeat unit. Because the spectrum of a polymer is, to a first approximation, that of its repeat unit, the spectrum is an aid to qualitative analysis, that is, to finding out what kinds of repeat units are present in a sample. Additives and impurities will also have their own characteristic peaks in the spectrum and can therefore be identified.

In addition to providing information about the chemical structure of a polymer, vibrational spectroscopy can also give very useful information about the physical structure, because any two regions of the polymer that differ in the way the repeat units are arranged may exhibit detectable differences in their spectra. Furthermore, measurements of the strength of IR absorption or of Raman scattering can give quantitative information about the composition of any mixture.

2.6.4 Quantitative infrared spectroscopy – the Lambert–Beer law

When infrared radiation passes through a sample of any substance the intensity is reduced by the same factor for each equal increase in distance travelled. The intensity thus decays exponentially and, if the decay is due only to absorption, rather than to scattering and absorption, then

$$I = I_0 e^{-ax \ln 10} = I_0 \times 10^{-ax} \tag{2.5}$$

where x is the distance travelled from a reference surface within the substance, I_0 is the intensity at the reference surface and a is the *absorption coefficient*. Equation (2.5) is the mathematical expression of *Lambert's law*.

The quantity I/I_o is the *transmittance*, T, of thickness x and a quantity A given by

$$A = \log_{10}(I_o/I) \tag{2.6}$$

is called the *absorbance* of the thickness x of the substance. This definition means that the intensity drops by a factor of 10 if the absorbance is unity, and generally it drops by a factor 10^A when the absorbance is A.

The infrared-absorption spectrum of a substance is ideally a plot of the transmittance or the absorbance of a suitable thickness against the wavelength or wavenumber of the infrared radiation. In practice the losses of intensity by reflection at the surfaces of a sample can often be neglected and then, from equations (2.5) and (2.6)

$$A = ax = \log_{10}(I_o/I) \tag{2.7}$$

with x equal to the thickness of the sample, I_o equal to the intensity of the radiation incident on the surface of the sample and I equal to the intensity of the radiation emerging from the other side of the sample. The absorption coefficient a can be written as kc, where c is the concentration of the absorbing species and k is called the *extinction coefficient* or *absorptivity* of the substance. Substitution of kc for a in equation (2.5) leads to the *Lambert–Beer Law*.

If the sample is very thick the transmittance will approach zero and the absorbance will tend to infinity for all wavelengths. In order to avoid too high an absorbance for a solid polymer it is usually necessary to work with samples of thickness between about 30 and 300 μm.

All modern spectrometers are computer-controlled and are not only capable of producing their output in the form of absorbance or transmittance spectra plotted on a chart but also permit other forms of display and various kinds of mathematical processing of the data. A simple example of such processing is the ability to add together scaled versions of the absorbance spectra of two or more substances to find a match with the spectrum of a mixture of the substances and so determine the composition of the mixture, provided that interactions in the mixture do not change the individual spectra.

2.7 Nuclear magnetic resonance spectroscopy (NMR)

2.7.1 Introduction

Nuclear magnetic resonance spectroscopy is another technique that can give a large amount of information about both the structure of polymers and the nature of the molecular motions taking place within them. As for infrared and Raman spectroscopy, the spectra are due to transitions between quantised energy levels caused by the interaction of the material with electromagnetic radiation. Although the energy levels are of a quite different type from those involved in vibrational spectroscopy, it is the influence of the structure and internal motions of the polymer on the precise form of the spectrum that provides the information about the structure and motions, just as for vibrational spectroscopy.

Most NMR studies, including those on polymers, are carried out on solutions. The spectra are then greatly simplified, as pointed out in section 2.7.3, and are invaluable for the elucidation of chemical structure. Attention here is, however, focused on the use of NMR for studying solid polymers, although the underlying principles are the same.

The method relies on the fact that some atomic nuclei have a spin angular momentum and an associated magnetic moment. For many nuclei the spin is $\frac{1}{2}$, as for the proton and the ^{13}C nucleus, both of which are particularly important for the NMR spectroscopy of polymers. When such a nucleus is placed in a magnetic field B_o the spin can take up two orientations with respect to the field, 'parallel' and 'anti-parallel'. According to quantum mechanics this means that although the magnitude of the spin magnetic moment μ is $(\sqrt{3}/2)\gamma\hbar$, where γ is the *magnetogyric ratio*, the value of its component parallel or anti-parallel to B_o is only $\frac{1}{2}\gamma\hbar$ and only the square of the component in the perpendicular direction is defined. The interaction energy with the field is thus $\mp\frac{1}{2}\gamma\hbar B_o$, depending on whether the spin component along the direction of the field is parallel or anti-parallel to B_o. The semi-classical interpretation of the undetermined individual components of the magnetic moment perpendicular to B_o is that μ precesses around B_o with an angular velocity $\omega_o = \gamma B_o$, as shown in fig. 2.13, which also shows the energy-level diagram. The angular frequency ω_o is equal to the frequency of the radiation necessary to cause the transition from the lower to the upper state and is called the *resonance frequency*.

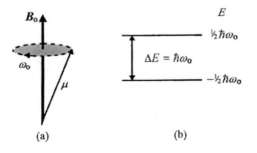

(a) (b)

Fig. 2.13 (a) Precession of μ around B_o. (b) The energy-level diagram for a spin $\frac{1}{2}$ nucleus. The direction of precession depends on the sign of γ; the direction shown is for γ positive.

If all the nuclei of each element present in a sample were isolated the corresponding spectrum would consist of a series of sharp lines, one for each type of nucleus that has a non-zero spin. In a material each nucleus is actually under magnetic influences from the electrons surrounding it, including those binding the atom to other atoms, and from the dipoles due to the nuclei of other atoms. These influences cause splittings, broadenings and shifts of the lines in the spectrum, from which information is obtained about the structure and internal motions of the material. In the simplest kinds of experiments the contribution that each spin of a given type makes to the magnitude of the NMR signal is, however, independent of its environment, which makes quantitative interpretation more straightforward than for IR or Raman spectroscopy.

2.7.2 NMR spectrometers and experiments

The values of B_0 used in NMR are in the range 1–12 T, corresponding to values of $\omega_0/(2\pi)$ for protons in the radio-frequency (rf) range 40–500 MHz. The earliest form of NMR spectroscopy used the *continuous-wave* (CW) method, in which a rf field of small amplitude is applied to the sample continuously and either the frequency of this field or the value of B_0 is varied to scan the spectrum. The transitions are usually detected by a separate coil surrounding the sample. Very little NMR work now uses the CW method, which has largely been replaced by the use of short high-power pulses of rf radiation to disturb the distribution of nuclei in the various energy levels from their values in thermal equilibrium. Fourier-transform (FT) methods are then used to extract spectral information from the signals induced in a detector coil as the distribution *relaxes* after the perturbing pulses. The pulsed FT method has many advantages over the earlier method, including speed of acquisition of spectra, easier determination of *relaxation times* and the possibility of using many specialised sequences of pulses to extract various kinds of information about the sample.

It is important to realise that, in thermal equilibrium, the difference between the number of spins parallel and anti-parallel to B_0 is very small because the difference in energy $\hbar\omega$ is very small compared with kT. This means that the NMR signals are very weak and require sensitive detectors. Figure 2.14 shows a block diagram of the essential features of a spectrometer for pulsed NMR studies. The sample sits within the probe which is surrounded by the magnet that produces the field B_0, which is usually a superconducting electromagnet. The probe contains a coil surrounding the sample that is used to produce the rf field and to detect the induced signal. It also contains mechanisms for rotating the sample and

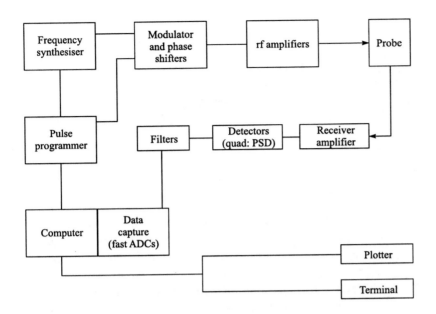

Fig. 2.14 A schematic diagram showing the essential features of a NMR spectrometer for pulsed studies. (Reproduced from *Nuclear Magnetic Resonance in Solid Polymers* by V. J. McBrierty and K. J. Packer. © Cambridge University Press, 1993.)

for varying the temperature. Further details can be found in the books referred to as (3) in section 2.9.

The simplest experiment uses a single pulse of rf radiation lasting for a time of the order of microseconds. Any pulse of radiation of length Δt contains a range of frequencies of about $1/\Delta t$, so that a 1 μs rf pulse 'of frequency 250 MHz' actually contains a range of frequencies spread over about 1 MHz and is thus only defined to about one part in 250, or about 4000 parts per million (ppm). For a constant value of B_0 the values of the resonance frequencies for a given type of nucleus vary with their environment by only a few hundred parts per million, so that such a pulse can excite all the resonances equally, to a good approximation. A so-called 90° pulse is frequently used. To understand what this means it is useful to consider a rotating frame of reference, as shown in fig. 2.15.

According to the semi-classical picture, each spin precesses around B_0 at the frequency ω_0 because the magnetic moment μ is not parallel to the field, which therefore exerts a torque on it perpendicular to both μ and B_0. This torque continuously changes the direction of the spin angular-momentum vector and hence of μ, so causing the precession. For a set of isolated spins of the same kind, ω_0 has the same value for each spin, but the phases of the precessions are random, as illustrated in fig 2.15(a). If a rf field B_1 is applied at right angles to B_0 in such a way that its magnitude remains constant but it rotates with angular velocity ω_0 in the same direction as the precession, an additional torque is applied to each spin, which changes its precessional motion.

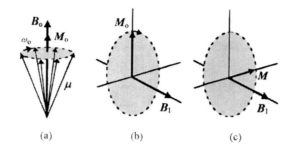

(a)   (b)   (c)

Fig. 2.15 (a) The Laboratory frame with all spins precessing independently, with random relative phases, around B_0 at the frequency ω_0 giving rise to the net magnetisation M_0. (For simplicity only spins with components parallel to B_0 are shown; there are nearly equally many anti-parallel.) In the rotating frame, B_0 effectively vanishes and the spins are static. (b) The rotating frame immediately after application of B_1, with magnetisation M_0 still parallel to B_0. (c) The rotating frame after B_1 has been applied for the time required to produce a 90° pulse. The individual spins have all been rotated around B_1 and, after removal of B_1, are static in the rotating frame, but are precessing together around B_0, with related phases, in the laboratory frame. The magnitude of M is equal to that of M_0.

Imagine now a frame of reference rotating at exactly ω_0. Each spin would, in the absence of B_1 and of any interactions between the spins, remain stationary in this frame of reference, so that in this frame B_0 has effectively been reduced to zero. Switching on B_1, which remains in a fixed direction in this new frame, causes all the spins to precess around B_1. Their phases in this precession are not random, because they all start to precess around this new direction at the same time, so it is easier to visualise what is happening by considering the behaviour of the total magnetisation M_0. Before application of B_1, the magnetisation M_0 is parallel to B_0, because the components of μ in the direction of B_0 for all spins add together, whereas the components perpendicular to B_0 cancel out. On application of B_1, M_0 starts to precess around the fixed direction of B_1 in the rotating frame. If B_1 is applied for just the right length of time, this precession will rotate the magnetisation M through 90° into the plane perpendicular to B_0, maintaining its magnitude M_0. This constitutes the application of a 90° pulse.

Immediately after a 90° pulse has been applied and B_1 has been returned to zero, all the spins (and thus M) are precessing around B_0 in the laboratory frame. This is not, however, an equilibrium state, because at equilibrium the magnetisation is M_0 parallel to B_0, with all the constituent spins precessing with random phases around B_0. Over time the spin distribution relaxes back to this equilibrium state. Two things must happen for this state to be reached. One is that the spins must on average lose energy to regain their original components parallel to the field which add up to give the final value M_0 for the component of magnetisation parallel to B_0. This energy is lost to the thermal motions of the polymer. The process is called *spin–lattice relaxation* or *longitudinal relaxation* and a characteristic relaxation time, T_1, is associated with it. In addition, the net component $M_\perp$ of the magnetisation perpendicular to B_0, which is equal to M_0 immediately after application of the pulse, must fall to zero, corresponding to the randomisation of the phases of the precessions of the spins around B_0. This relaxation process, caused by the mutual interaction of the spins, is called the *spin–spin* or *transverse relaxation* and has a characteristic relaxation time T_2 different from the spin–lattice relaxation time T_1. In solids T_2 is generally very much smaller than T_1. The transverse relaxation of the spin distribution causes the magnitude of the signal induced in the detector by the rotation of $M_\perp$ around B_o to decay with the characteristic time T_2. The observed signal is therefore called the *free-induction decay* (FID), from which the required spectral information is obtained by Fourier transformation.

The single 90° pulse is only one of a vast range of pulse sequences that are used in modern NMR spectroscopy. The following section indicates briefly why it is necessary to modify the experiment so far described in order to extract the required information about solid materials.

2.7.3 Chemical shifts and spin–spin interactions

In a molecule of any kind a particular nuclear spin is surrounded by a large number of electrons and may interact with other nuclear spins in the molecule. The presence of the electrons leads to a *shielding* of the externally applied magnetic field that is different for the nuclei of atoms in different chemical environments, so that, for example, the frequency ω_o in a particular external magnetic field for a ^{13}C atom that is part of a —CH_3 group is different from that of a ^{13}C atom in the backbone of a molecule. The difference between the frequency observed for a spin in a particular molecular environment and that which would be observed for the free spin is called the *chemical shift*. The sensitivity to chemical environment is in fact such that atoms several chemical bonds away can influence the precise frequency of resonance. In the solid state, however, the resonances are broadened by several effects and overlap.

The first broadening effect is that the shielding that gives rise to the chemical shift is anisotropic and is in fact describable by a second-rank tensor (see the appendix). This means that the chemical shifts for corresponding nuclei in polymer segments with different orientations with respect to B_o are different. A second effect is that the interactions with the surrounding nuclear-spin dipoles also broaden the resonances. If information about chemical shifts is sought, this broadening is a nuisance. On the other hand, the precise nature of the broadening can provide structural information about the material. A third broadening effect that is particularly important for polymers is due to the disorder of the structure that is always present. In solution the rapid tumbling and flexing of the molecules causes all these interactions to be averaged out, so that the resonances remain sharp and the spectra can be used to elucidate chemical structure.

The interaction, or coupling, of the spins of different nuclei in the solid state can be understood to a first approximation as follows. Imagine two spins S_1 and S_2 separated by a distance r and let the line joining them make an angle θ with $\boldsymbol{B}_o$. The spin S_1 has a dipole moment with component μ_{B1} parallel to $\boldsymbol{B}_o$ of magnitude $\gamma_1 I_{B1} \hbar$, where I_{B1} represents the spin quantum number for the component of spin parallel to $\boldsymbol{B}_o$ and is equal to $\pm\frac{1}{2}$ for nuclei with spin $\frac{1}{2}$. The component of the dipole perpendicular to $\boldsymbol{B}_o$ is undefined. It follows from the standard expressions for the field due to a dipole that the dipole S_1 produces a magnetic field at S_2 with a component B_{S1} parallel to $\boldsymbol{B}_o$ of magnitude

$$B_{S1} = \frac{\mu_o}{4\pi r^3} \gamma_1 I_{B1} \hbar (3\cos^2\theta - 1) \tag{2.8}$$

The net field experienced by the spin S_2 is thus $\boldsymbol{B}_o + \boldsymbol{B}_{S1}$ and the energy of the spin S_2 is therefore changed from what it would be in the absence of S_1 by the amount

$$\Delta_{12} = \frac{\mu_o}{4\pi r^3} \gamma_1 \gamma_2 I_{B1} I_{B2} \hbar (3\cos^2\theta - 1) \tag{2.9}$$

In a solid polymer sample each nucleus is usually surrounded by a large number of different spins with different values of r and θ, which leads to a broadening of the resonance, as already mentioned.

Because different spins experience different net fields parallel to $\boldsymbol{B}_o$ and have correspondingly different resonant frequencies, they precess around the direction of $\boldsymbol{B}_o$ at different rates and this is one of the reasons for the randomisation of the phases of their precessions after the application of a 90° pulse, i.e. for the transverse relaxation associated with the relaxation time T_2. The greater the spread of resonance frequencies the smaller the value of T_2.

The next section explains how the effects of the anisotropy of the chemical shift and of this interaction between spins can be removed.

2.7.4   Magic-angle spinning, dipolar decoupling and cross polarisation

The effects of the anisotropy of the chemical shift and of the spin–spin interactions in the spectra of solids can be eliminated by the use of a technique called *magic-angle spinning* (MAS). Equation (2.9) shows that the frequency of the transition for a particular spin S_2 interacting with another spin S_1 can be written

$$\omega = \omega_o + A P_2(\cos\theta) \tag{2.10}$$

where A is a constant for given types of spins, $P_2(\cos\theta) = \frac{1}{2}(3\cos^2\theta - 1)$ is the second-order Legendre polynomial in $\cos\theta$ and θ is the angle between the line joining S_1 and S_2 and the direction of $\boldsymbol{B}_o$. The value of $P_2(\cos\theta)$

always lies between -0.5 and 1, so that the width of the broadened line is of order A.

Imagine now that the sample is spun at high angular frequency ω_s about an axis making the angle θ_s with $\textbf{\textit{B}}_o$. It is possible to show from the properties of the Legendre polynomials that the time-average value of $P_2(\cos\theta)$ is given by $P_2(\cos\theta_r) \times P_2(\cos\theta_s)$, where θ_r is the angle between the line joining S_1 and S_2 and the axis of spin. If θ_s is chosen equal to $54.7°$, the so-called *magic angle*, $P_2(\cos\theta_s) = 0$, so that the time-average value of ω is equal to ω_o, the unperturbed value. The value of ω oscillates about this time average, which leads to the existence of *spinning side-bands* separated from ω_o by integral multiples of ω_s. For a sufficiently large value of ω_s, i.e. $\omega_s \gg A$, the amplitudes of the spinning side-bands fall to zero and the signal that remains has only the frequency ω_o.

Although the above discussion shows only that magic-angle spinning can remove the effects of spin–spin interactions, it can be shown that it also removes the effects of the anisotropy of the shielding that gives rise to chemical shifts.

Another method that is used to remove the spin–spin coupling in experiments on ^{13}C NMR is called *dipolar decoupling* (DD). The natural abundance of ^{13}C nuclei is very low, so the spins of such nuclei are well separated from each other and do not generally interact. They can, however, interact with the spins of nearby protons. This interaction can be effectively suppressed by irradiating the sample during the FID with a rf field of the correct frequency to cause transitions between the proton-spin levels and a high enough amplitude to cause upward and downward transitions with a frequency much greater than the precessional frequency, so that their mean components parallel to $\textbf{\textit{B}}_o$ are zero. The corresponding pulse sequence is illustrated in fig. 2.16.

In any pulsed NMR experiment the pulse sequence is repeated many times and the results are added to increase the signal-to-noise ratio. The repetition must not take place until thermal equilibrium has been restored, which is usually regarded as implying a clear time of about $3T_1$ between sequences. The value of T_1 for ^{13}C is often very long, so that the simple DD

Fig. **2.16** The pulse sequence used to determine ^{13}C spectra under dipolar decoupling conditions. Note that only the envelope curve of the FID is shown, not the high-frequency oscillations within it. This is what is usually recorded; Fourier analysis then yields the spectrum as frequency shifts from the irradiating frequency. For all pulse sequences, the recovery time T_R between repeated applications of the sequence must be long compared with T_1. (Adapted from *Nuclear Magnetic Resonance in Solid Polymers* by V. J. McBrierty and K. J. Packer. © Cambridge University Press, 1993.)

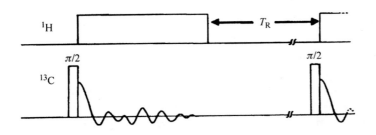

method cannot be used. In these cases a more complicated method called *cross polarisation* (CP) can be used. Essentially the method involves the excitation of the ^{1}H spins and the transfer of the excitation to the ^{13}C spins. The FID of the ^{13}C spins is then observed under DD conditions. Because of the transfer of excitation from ^{1}H to ^{13}C spins, the rate at which the sequence can be repeated is determined by the value of T_1 for ^{1}H, which is much shorter than T_1 for ^{13}C. This means that the pulse sequence can be repeated more frequently. The efficiency of excitation of the ^{13}C spins is also greater in this method, but the simple proportionality of the signals to the numbers of spins in different environments is lost. The CP technique is often combined with MAS to remove the anisotropic chemical shift broadening and the combined technique is called CPMAS.

2.7.5   Spin diffusion

In considering the interaction between spins in section 2.7.3, the component of the dipole of spin S_1 perpendicular to B_o was considered to be undefined. In the semi-classical picture, however, the spin is precessing around B_o and therefore produces a rotating component of magnetic field perpendicular to B_o. If S_2 is a spin of the same kind as S_1 this is at just the right frequency to induce a transition between the two possible energy states of spin S_2. This action is, however, mutual to the two spins, so that a *flip–flop* or *spin exchange* takes place, with no net change of energy. If there is a region within the material where there are initially more spins with a positive component of magnetisation along the direction of the field than there are in an immediately surrounding region, the effect of these spin-flips will be to cause this excess magnetisation to diffuse outwards. This phenomenon of *spin diffusion* is of great use in studying structural inhomogeneities in polymers and is discussed further in section 5.4.3. The spin exchange also contributes to the dephasing of the precessions of the spins and thus reduces T_2 for the interaction of like spins.

2.7.6   Multi-dimensional NMR

The previous section explains how the various broadening mechanisms can be counteracted to obtain high resolution spectra of solids. In earlier sections it is pointed out that the broadening does, however, contain useful information about structure and motion. Because all the broadening effects can be of similar magnitude, very complex line shapes occur and it is therefore desirable to have methods that permit, for instance, the lines due to nuclei in specific chemical environments, i.e. with specific isotropic chemical shifts, to produce completely separated spectra in which the other broadening mechanisms, such as the anisotropic chemical shifts, can be studied. It is possible to do this using so-called *two-dimensional spectroscopy*.

Figure 2.17(a) illustrates schematically the general form of the two-dimensional experiment. The spin system is first disturbed from its equilibrium state and the FID is allowed to take place for an *evolution time, t_1* under a particular set of conditions, for instance with spinning at a particular angle, and subsequently allowed to continue under different conditions during a *detection time, t_2*. The whole experiment is repeated for a range of different values of t_1 and a two-dimensional Fourier transform is performed on the resulting data. In some experiments a *mixing time* is allowed between the evolution and detection times, but this is not required in the simpler experiments. Figure 2.17(b) shows the separation of anisotropic chemical shift patterns for different isotropic chemical shifts obtained by setting the axis of the sample spinner at the magic angle during the detection time and at 90° during the evolution time, with no mixing time.

The result may be understood as follows. Consider the Fourier transform of the FID that takes place during the detection time for a particular value of t_1. Because this decay takes place under MAS conditions the result is a high-resolution spectrum showing the isotropic chemical shifts. This is true for any value of t_1. For different values of t_1 however, the *phases* of the components of different frequencies ω_2 in this spectrum are different, as well as their amplitudes. This is because they 'remember' the phases of the corresponding spins at the end of the evolution time t_1. A second Fourier transform over t_1 performed at each frequency ω_2 of the first transform thus gives the broadened spectrum due to the anisotropic part of the chemical shift for all those spins that have the isotropic chemical shift ω_2.

Many such two-dimensional experiments can be performed to correlate various aspects of the behaviour of the spin system. By introducing a second evolution time, three-dimensional spectra can be obtained and so on.

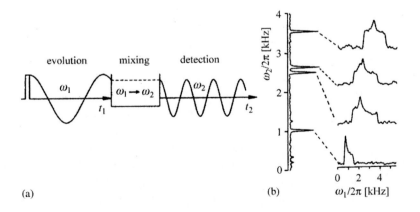

(a)

(b)

Fig. 2.17 (a) A schematic representation of the generic two-dimensional NMR pulse sequence, with evolution time t_1, mixing period t_m and detection time t_2. (b) The separation of anisotropic chemical shift patterns for different chemical shifts (see the text). ((a) reproduced and (b) adapted by permission of Academic Press.)

2.7.7   Quadrupolar coupling and ^{2}H spectra

A very important technique for studying motion in polymers that has not so far been mentioned relies on the fact that the deuterium nucleus ^{2}H has a spin of 1 and possesses a *quadrupole moment*. This means that the charge distribution in the nucleus is not spherically symmetric but has a distribution corresponding to that of two electric dipoles that point in opposite directions along the spin axis and are separated by a small distance. Such a quadrupole has an interaction energy with an electric-field *gradient* that depends on the angle between the axis of the quadrupole and the direction of the field gradient but is unchanged if the spin direction reverses. For the C—D and O—D bonds, where D represents the deuterium atom, the field gradient is cylindrically symmetric about the bond direction and the interaction energy of the quadrupole with the field gradient is small compared with the interaction of the spin of the ^{2}H nucleus with the applied magnetic field $\boldsymbol{B}_0$.

A spin 1 nucleus has three energy levels in the field $\boldsymbol{B}_0$, corresponding to defined components $\gamma\hbar$, 0 and $-\gamma\hbar$ parallel to $\boldsymbol{B}_0$ and to energies $-\gamma\hbar B_0$, 0 and $\gamma\hbar B_0$. The quadrupole interaction affects the first and last of these states equally, but affects the middle state differently, so that the resonance is split into two lines. The splitting $\Delta\nu_q$ depends on the angle θ between the C—D or O—D bond and $\boldsymbol{B}_0$ according to the expression

$$\Delta\nu_q = \tfrac{3}{4}C_q(3\cos^2\theta - 1) \tag{2.11}$$

where C_q is the *quadrupolar coupling constant* and is equal to 85 kHz for C—D. For a random sample this expression must be averaged over all values of θ, which gives rise to a so-called *Pake line-shape*, as illustrated in fig. 2.18.

The quadrupolar line broadening is generally large in comparison with other broadening effects but these need to be taken into account in calculating the exact form of the observed line-shape. The Pake line-shape corresponds to what would be observed if there were no molecular motion. If there is motion, the line-shape is modified in a way that depends on the type and frequency of the motion in a way that can be calculated. The effect is somewhat similar to the narrowing caused by magic-angle spinning (see section 2.7.4). The motions involved are, to a good approximation, jumps between discrete molecular conformations (see section 3.3.1) and, if the jump frequency is very much greater than the line-width, i.e. if the average time τ between jumps is $\ll 1/C_q$, i.e. $\ll 10^{-5}$ s, the line-shape will not change with further reduction in jump time. It will correspond to a line-shape for an averaged electric-field gradient that need not be axially symmetric around the C—D or O—D bond, so that the line-

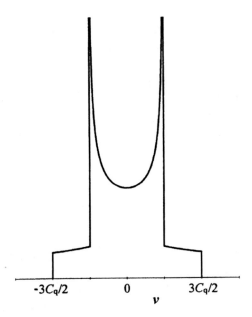

Fig. 2.18 The Pake line-
shape for ^2H quadrupolar
broadening. The infinite
values at $v = \pm 3C_q/4$ and
the discontinuities at
$v = \pm 3C_q/2$ are removed
in practice by other
broadening effects.

shape will not be Pake-like. On the other hand, if the jump time is $\gg 10^{-5}$ s, the change of shape will not be significant, so that the method can be used to study motions in the approximate range 10^{-7} s $> \tau > 10^{-3}$ s. The method has the disadvantage that deuterated samples must be prepared, but has the advantage that these can be selectively deuterated to study the motion of specific parts of the molecule.

A similar Pake line-shape would arise for a pair of like spins interacting by magnetic dipole coupling, as shown by equation (2.9), but in this case observed lines are usually broadened even in the absence of motion because each spin in a solid does not interact with only one neighbouring spin but with a large number of surrounding spins.

2.8   Optical and electron microscopy

2.8.1   Optical microscopy

It is well known that conventional optical microscopy has a limit of resolution of approximately half the wavelength of the light used. For visible light this limit is about 250 nm. Structures to be investigated must thus be at least 1 μm across if any useful information about their detailed arrangements is to be obtained using this method. As described in later chapters, aggregates of crystallites with sizes ranging up to several micrometres can occur in partially crystalline polymers, so such structures can therefore in principle be studied with the optical microscope. In practice, however, if such a transparent sample is viewed in the ordinary microscope, nothing will be seen, because neither the non-crystalline regions nor the crystalline regions absorb a significant amount of light, so that there is no contrast between them. Contrast can be introduced in a variety of ways, but in the microscopy of polymers it is frequently introduced by the use of polarised light. In the *polarising microscope* a polariser is placed below the sample and an analyser above it. The direction of transmission of the analyser is at right angles to that of the polariser.

In order to understand how the polarising microscope makes the crystalline structures visible, it is necessary to understand the propagation of light through a crystal. The refractive index n of a crystal depends in general on the direction of the electric vector D (the *polarisation direction*) of the light wave with respect to the crystal axes. The variation of the refractive index n with the direction of propagation and polarisation can be shown by the *indicatrix* or *refractive-index ellipsoid*. Its radius in any direction is proportional to n for light with D parallel to that direction. If light propagates through the crystal in any direction (wave-normal) there are only two possible polarisation (D vector) directions and these are parallel to the principal axes of the central section of the indicatrix perpendicular to the direction of propagation, as illustrated in fig. 2.19. For many crystals the indicatrix is an ellipsoid of revolution and these crystals are called *uniaxial crystals*.

Assume that a crystal is placed so that an axis of the indicatrix is vertical and that incident light propagates parallel to this axis from below the crystal. Let the incident light pass through a polariser below the crystal and let there be a *crossed* polariser (*analyser*) above the crystal. If the orientation of the crystal around the vertical axis is such that the incident light is polarised parallel to either of the principal axes of the

Fig. 2.19 The indicatrix, or refractive-index ellipsoid, for a general anisotropic medium. $Ox_1x_2x_3$ are the axes of the ellipsoid and PO represents the direction of propagation (wave-normal) of light through the medium. OA and OB are the principal axes of the section of the ellipsoid normal to OP, shown shaded. The possible D vectors for the light are parallel to these axes and their lengths represent the corresponding values of the refractive indices if the ellipsoid is drawn correctly to scale. (Reproduced by permission of Oxford University Press.)

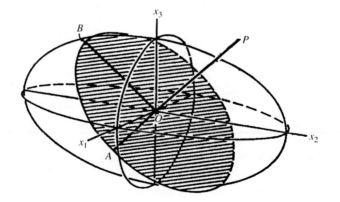

indicatrix lying horizontally, it passes through the crystal polarised in the original direction and will not pass through the analyser. The crystal will appear dark. If, however, the orientation of the crystal is such that the light is polarised in a general direction, the electric vector is resolved into components parallel to the principal axes of the indicatrix and two perpendicularly polarised beams travel through the crystal with different values of n, i.e. with different speeds. They are unlikely to be in phase when they emerge at the other side of the crystal and they thus generally combine to give *elliptically polarised light*. Elliptically polarised light has components of polarisation in all directions; thus some light now passes through the analyser and the crystal no longer appears dark. The amount of light that passes depends on how far the axes of the crystal depart from the polarisation direction.

The conclusion is that if different parts of the structure contain crystallites with their axes in different directions from each other the structure will be visible in the polarising microscope. In addition to all the usual measurements of size and shape that can be made with the optical microscope, it is possible to deduce the relative alignments of the axes of the crystallites within the structure from the relative brightness of its various parts.

Other methods for making the various parts of transparent objects visible in the optical microscope are staining techniques and dark-ground and phase-contrast microscopies. These techniques are explained in the standard optics textbooks and similar methods for use in electron microscopy are described in the next section.

In addition to making structures visible that would otherwise be invisible, the polarising microscope can also be used, generally at much lower power, for the measurement of the *birefringence* of a transparent oriented polymer sample. *Orientation* in polymers is the subject of chapters 10 and 11; it is sufficient to indicate here that stretching or other processes applied to a polymer sample may cause its molecules to become partially aligned, or oriented. In the simplest form of stretching, the sample then has different refractive indices for light polarised parallel and perpendicular to the directions of stretching and the difference is called the birefringence, Δn. If the orientation is uniform throughout the sample and if the sample is uniform in thickness, it will appear equally bright everywhere when it is viewed in the polarising microscope and the brightness will vary as the sample is rotated, going from bright to completely dark four times in a complete revolution.

The simplest method for measuring the birefringence for a moderately thick sample involves machining a tapered region on it, so that its thickness varies within this region. If the axes of the indicatrix are set at 45° to the directions of transmission of the crossed polariser and analyser of a polarising microscope, a series of light and dark bands or *fringes* is seen in the tapered region, because the path difference for the two polarised beams through the sample now varies with position. When the beams are in phase after passing through the sample, the original polarisation is restored and the light does not pass through the analyser. When the phase difference is $180°$ the two beams combine to give light polarised in a direction perpendicular to the original direction and the maximum amount of light passes through the analyser. Measurement of the thickness t of the sample and the number m of fringes in the tapered region then gives the birefringence as $\Delta n = m\lambda/t$, where λ is the wavelength of the light used.

This method is applicable only to rather special types of sample and other optical methods are more usually used. They are not discussed in this book because they are rather more complicated.

2.8.2 Electron microscopy

In this type of microscopy the wave properties of electrons are used to obtain resolution that far exceeds that of the optical microscope. According to de Broglie's relationship, the wavelength λ associated with a beam of electrons of momentum p is given by $\lambda = h/p$. If the electrons are accelerated through a potential difference V they acquire an energy $eV = p^2/(2m)$, where m is the mass of the electron and e is the electronic charge, so that $\lambda = \sqrt{1/(2meV)}\,h$. Substituting $V = 100\,000$ V leads to $\lambda = 3.9 \times 10^{-12}$ m, which is about a hundredth of the separation of the atoms in molecules or crystals. Resolution limits actually attainable are, however much greater than half this wavelength, because of the imperfections of the focusing 'lenses' of electron microscopes, but values about equal to the separations of atoms in molecules or crystals are actually attainable in some forms of electron microscopy.

There are several forms of electron microscopy in general use. *Transmission electron microscopy* (TEM) is often used with thin samples of materials in which different regions within the sample absorb electrons differently. The various different structures present in polymers tend to have similar low absorption coefficients, so that some means must be used to render them 'visible'. This is similar to the problem of looking at non-absorbing samples in ordinary optical transmission microscopy. One solution, described in the previous section, is to use the polarising microscope. Other solutions are to use suitable materials to stain various regions preferentially, or to make use of the fact that the various regions have different refractive indices and so affect the phases of the waves passing through the various regions differently. Similar methods to these can be used for TEM studies of polymers.

The staining technique for TEM uses a material that absorbs electrons and preferentially attaches itself to or reacts with certain regions of the polymer rather than other regions. Materials frequently used are uranyl acetate and osmium tetroxide. For polyethylene the technique of *chloro-sulphonation* can be used. In this method, which involves immersing the sample in chlorosulphonic acid, the electron-absorbing material becomes attached to lamellar surfaces, so that lamellae (see section 3.4.2) with their planes parallel to the direction of the electron beam become outlined in black in the micrographs.

The phase techniques rely on the fact that a plane electron wave imping-ing on the upper surface of a thin polymer sample emerges from the lower face as a distorted wave because of the different phase lags caused by various regions of the sample. This distorted wave may be regarded as the sum of an attenuated plane wave travelling in the original direction and new plane waves of low amplitude travelling in various directions at angles to the original direction, i.e. diffracted waves. If all these waves are collected by the electron lens and used to form the image of the exit surface of the sample, the image will be 'invisible', because it will only have varia-tions of phase across it, just like the wave emerging from the sample.

If, however, the wave that is not diffracted is removed and no longer contributes to making the amplitude of the wave at the image plane the same everywhere, a 'visible' image results. It is obvious that outside the sample the incident beam travels unperturbed; there is no diffraction and removal of the undiffracted wave leads to 'darkness' in the image of this region, so that the method is called *dark-field microscopy*. *Bright-field microscopy* is obtained when some of the diffracted waves are blocked rather than the undiffracted wave. The blocking is done in the back focal plane of the objective, where each different plane wave is brought to a separate focus; this plane contains the *electron diffraction pattern* of the sample.

Another method for making the sample visible by using its effects on the phase of the transmitted wave comes from the realisation that it is only in the plane immediately behind the sample, or in the image of this plane, that the amplitude is constant. In any plane just outside the sample, interference between the undiffracted and diffracted waves gives rise to an intensity distribution that is not uniform and depends on the structure of the sam-ple. It is possible to show that, if a particular plane just outside the sample is imaged, rather than the sample itself, a 'visible' image showing the structures within the sample is formed (see fig. 2.20(a)). This method is related to, but is not identical to, the method of *phase-contrast microscopy* used with optical microscopes and is sometimes known by that name.

Fig. 2.20 Examples of electron micrographs of polymers. (a) A defocussed bright-field image of a thin film of isotactic polystyrene annealed and crystallised at about 170 °C: (b) An image of a fracture surface replica from a sample of linear polyethylene crystallised from the melt at 4.95 kbar. ((a) Adapted by permission of Masaki Tsuji and (b) adapted from *Principles of Polymer Morphology* by D. C. Bassett. © Cambridge University Press 1981.)

(a)   (b)

A further technique that can be used to make electron-transparent samples visible is the *shadowing* technique. However a sample is prepared, its surface is not smooth and the various projections on the surface are related to the underlying structures. If a very thin metallic film is evaporated onto the sample in such a way that the atoms of the metal all strike the surface at the same angle well away from the normal, any region that protrudes above the surrounding surface will receive a thicker coating on the side facing the incident metal atoms than it will on the other side, which is partially or totally in shadow. (It is the similar shadowing of the sun's parallel rays that makes the craters on the moon visible.) The structures of the sample will become visible in TEM through the varying absorption of the differing thicknesses of the metal coating.

The light- or dark-ground techniques and the staining and shadowing techniques can give resolutions of about 0.5 nm, whereas the out-of-focus phase technique can give somewhat higher resolution and permits the study of polymer crystal lattices under suitable conditions, particularly when it is combined with image-processing techniques. It is in fact partially limited by the sensitivity of polymers to damage by electrons. Exposures must be kept short and a small number of electrons, like a small number of photons, cannot give rise to high resolution. All TEM studies of polymers in fact suffer from restrictions caused by electron damage, although staining and shadowing reduce the sensitivity of a sample. An important way of avoiding damage to a sample is to use *replicas* of polymer surfaces.

If a solvent is available for the polymer, the replica is made by shadowing the surface with a metal and then coating it with carbon. The sample itself is then dissolved away to leave a replica of its surface. If no solvent is available a slightly more complicated two-stage process is used. The procedures are outlined in fig. 2.21. The replica produced by either

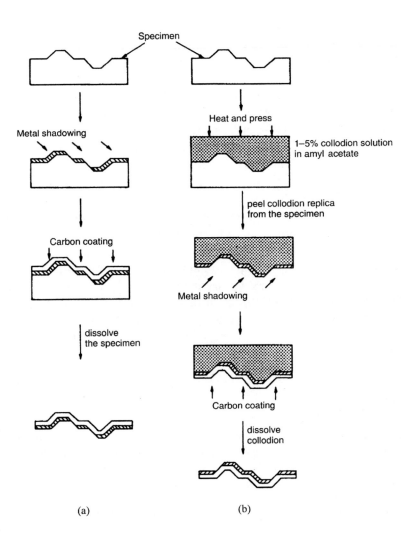

Fig. 2.21 Replication of polymer surfaces for electron microscopy: (a) the single-stage process and (b) the two-stage process. (Adapted by permission of Masaki Tsuji.)

method can be examined in the electron microscope instead of the sample. An example of a micrograph taken from a replica is shown in fig. 2.20(b).

All the methods described so far suffer from the disadvantage that they allow only the study of very thin films of polymer, by direct TEM, or of polymer surfaces, by replication. Thin enough films are difficult to make and neither they nor surfaces produced by casting or by fracture are necessarily typical of bulk material. It is therefore desirable to have a technique that can be used for any surface, including one cut from the interior of a larger sample. A technique that allows any type of surface to be prepared in a way suitable for the examination of the underlying structure is the use of *permanganic etchants*. These etchants are based on potassium permanganate dissolved in sulphuric acid, or a mixture of sulphuric and orthophosphoric acids, and they selectively etch different components of structure.

In a typical treatment of polyethylene, 1–2 μm is removed from the surface, with the reagent attacking preferentially the non-crystalline material. The surface can then be studied either by two-stage replication or by high-resolution scanning electron microscopy (SEM). The latter gives somewhat lower resolution than TEM, but is much easier to use than the production of replicas. In this technique scattered or secondary electrons emitted from the surface are collected as an electron beam is scanned across it in a raster like that used to produce a television picture and the image is built up in a similar way from the intensity detected.

It has already been mentioned above that the back focal plane of the objective lens of the microscope contains the diffraction pattern of the sample; if this plane is imaged by the imaging lens, the diffraction pattern of the sample is obtained rather than its image. Such diffraction patterns can be used to study the structure of crystallites in a similar way to X-ray diffraction patterns or to study the orientations of crystallites within a sample.

2.9   Further reading

(1) *Chemical Crystallography* by C. W Bunn, 2nd Edn, Clarendon Press, Oxford, 1961. This book gives a very good introduction to the theory of X-ray crystallography, with a considerable amount of discussion of polymers. Some of the experimental techniques described are, however, out of date.

(2) *The Vibrational Spectroscopy of Polymers* by D. I. Bower and W. F. Maddams, Cambridge Solid State Science Series, Cambridge University Press, 1989/1992. This book describes the fundamental principles of infrared and Raman spectroscopies and their application to polymers, with many examples of their use.

(3) *Nuclear Magnetic Resonance in Solid Polymers* by V. J. McBrierty and K. J. Packer, Cambridge Solid State Science Series, Cambridge University Press, 1993, and *NMR of Polymers* by F. A. Bovey and P. A. Mirau, Academic Press, 1996. These books contain a general introduction to NMR, with particular reference to the methods used in studying polymers, together with numerous examples of applications.

(4) *Polymer Microscopy* by L. C. Sawyer and D. T. Grubb, 2nd Edn, Chapman and Hall, London, 1996. This book covers all aspects of the electron and optical microscopy of polymers, with many examples of its application.

Chapter 3
Molecular sizes and shapes and ordered structures

3.1 Introduction

This chapter is concerned primarily, in section 3.3, with factors that determine the shapes that individual polymer molecules can take up, both on the local scale of a few repeat units and on the scale of the complete molecule. It is shown that, if a polymer molecule is in a particular kind of solution or in a melt of like molecules, its most likely state is a so-called *random coil* with a statistically well-defined size. This coil has an open structure that can be penetrated by chain segments belonging to other coils, so that the chains in a molten polymer are likely to be highly overlapping, or entangled, if the molar mass is high. In addition, as discussed in the following section, the molecules do not all have the same length. At first sight these two factors appear to make it unlikely that a polymer could crystallise on cooling, so that it might be expected that polymer solids would be rather structureless. However, in section 3.4 experimental evidence showing that this is not necessarily so is presented and the various ordered structures are discussed further in subsequent chapters.

3.2 Distributions of molar mass and their determination

In section 1.3.1 it is pointed out that polymers, unlike other chemical compounds, do not have fixed molar masses. The molar masses are very high and there is a distribution of molar masses, or chain lengths, that depends on the polymerisation conditions. The distribution of molar masses can have a profound influence on the physical properties of the polymer, so it is important to have methods of specifying and determining the distribution.

3.2.1 Number-average and weight-average molar masses

Figure 3.1 shows the ideal molar-mass distribution for a polymer produced by the self-condensation reaction of a pure di-functional monomer. The

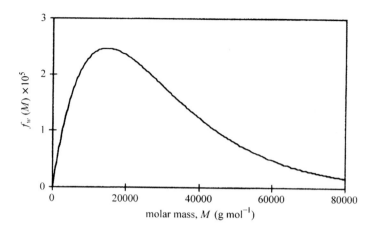

Fig. 3.1 The theoretical molar-mass distribution for a self-condensation polymer with a chemical repeat of $M = 150$ g mol^{-1} assuming that 99% of all end groups have reacted. $f_w (M)\,dM$ is the fraction by weight of the polymer that consists of chains with molar mass in the interval dM at M.

mass fraction of the polymer that consists of chains with molar mass M is plotted against M. The distributions obtained for various types of polymerisation can be very different and plots such as this one give complete information about the distribution. It is often convenient, however, simply to specify certain average values of M, partly because this allows a simple specification or comparison of different grades of polymer, and partly because some methods of obtaining information about molar masses can give only one of these averages.

If there are N_i chains with molar mass M_i then the *number-average molar mass M_n* and the *weight-average molar mass M_w* are given by

$$M_n = \frac{\sum N_i M_i}{\sum N_i} \tag{3.1a}$$

and

$$M_w = \frac{\sum W_i M_i}{\sum W_i} = \frac{\sum (N_i M_i) M_i}{\sum (N_i M_i)} = \frac{\sum N_i M_i^2}{\sum N_i M_i} \tag{3.1b}$$

where W_i is the total weight of polymer chains with molar mass equal to M_i.

The quantities M_n and M_w are equal only if all the polymer chains are of the same length, i.e. the polymer is *mono-disperse*. For any other distribution of molar masses, M_w always exceeds M_n (see example 3.1 and problems 3.1 and 3.2). The ratio M_w/M_n is thus a measure of how far the polymer departs from being mono-disperse, being larger the greater the departure, and it is called the *polydispersity index* (or *heterogeneity index*). Real distributions of molar mass are generally continuous and much wider than that suggested by example 3.1. The summations in equations (3.1) can

Example 3.1

A polymer sample consists of a mixture of three mono-disperse polymers with molar masses 250 000, 300 000 and 350 000 g mol^{-1} in the ratio $1 : 2 : 1$ by number of chains. Calculate M_n, M_w and M_w/M_n.

Solution

Let the total number of chains with molar mass 250 000 g mol^{-1} be N. Then M_n is equal to

$(N \times 2.5 \times 10^5 + 2N \times 3.0 \times 10^5 + N \times 3.5 \times 10^5)/(N + 2N + N) =$
3.000×10^5 g mol^{-1} (this result should be obvious) and M_w is equal to

$$\frac{N \times (2.5 \times 10^5)^2 + 2N \times (3.0 \times 10^5)^2 + N \times (3.5 \times 10^5)^2}{N \times 2.5 \times 10^5 + 2N \times 3.0 \times 10^5 + N \times 3.5 \times 10^5}$$
$= 3.042 \times 10^5$ g mol^{-1}.

Thus $M_w/M_n = 1.014$.

then be approximated by integrals over M. The polydispersity index can then differ substantially from unity. If p is the fraction of each of the two types of end group that has reacted for the ideal polycondensation polymer referred to above, the index is $1 + p$, so for high conversion in this type of reaction the index thus approaches 2.

3.2.2 Determination of molar masses and distributions

There are several methods for obtaining information about molar masses or their distributions. The most important method for determining M_w is the *light-scattering* method. The polymer is dissolved in a solvent and the dependence of the intensity of the scattered light on the concentration of the solution and the angle of scattering is determined. From these data it is possible to obtain not only the value of M_w but also information about the size of the molecular coils in the solution. It is more difficult to determine M_n. One method used when M_n is not too large is to determine the concentration of end groups in the polymer by means of infrared or NMR spectroscopy or by chemical analysis. A second method involves the determination of the osmotic pressure of a dilute solution of the polymer.

Because of its simplicity, a particularly important method for determining molar masses is the measurement of the *intrinsic viscosity* $[\eta]$ for a polymer solution. This quantity is defined as the limit of $(\eta - \eta_s)/(c\eta_s)$ as the concentration c of the solution tends to zero, where η and η_s are the viscosities of the solution and of the pure solvent, respectively. For a mono-disperse polymer the relationship between $[\eta]$ and the molar mass

M is $[\eta] = KM^a$, where K and a depend on the polymer, the solvent and the temperature. Values of M calculated from this expression for a polydisperse polymer lie between M_w and M_n and are usually closer to M_w.

An important method for determining the whole molar-mass distribution is *size exclusion chromatography*, often called *gel permeation chromatography*. A dilute solution of the polymer is injected into the top of a column containing a special gel that contains pores comparable in size to the molecular coils in the solution. Large coils, i.e. coils corresponding to chains of high molar mass, cannot enter the pores as easily as can small coils, corresponding to chains of low molar mass, so that the chains of higher molar mass pass out of the bottom of the column before chains of lower molar mass do. By measuring the concentration of polymer passing out of the column as a function of time the molar-mass distribution can be found. It is necessary to calibrate the column with mono-disperse samples of the polymer being studied with known molar masses in order to relate the time of emergence from the column to the molar mass.

3.3   The shapes of polymer molecules

In this section consideration is given to the factors that determine the possible shapes and sizes of individual polymer molecules. Fundamentally, these are controlled by the nature of the covalent bonds that bind the atoms of the molecule together and by the particular groupings of atoms that occur within a particular type of polymer chain.

3.3.1   Bonding and the shapes of molecules

The simplest models of covalently bonded chemical substances are the so-called 'ball-and-stick' models, in which the atoms are represented by balls and the sticks represent the bonds. In constructing such models it is usual to make use of so-called 'standard bond lengths' and 'standard bond angles'. In addition, certain orientations of groups of atoms around bonds are also assumed, i.e. 'preferred torsional angles'. These standards correspond approximately to properties of the real molecules and control the possible shapes that the molecules can take. What are the reasons for these values and what are they for various types of molecule?

(i) 'Standard' bond angles

The simplest covalent bonds consist of pairs of electrons, one electron from each of the two bonded atoms, contributed by its outermost shell of electrons. The two electrons of each pair have the same spatial wave function, but have opposite spins to comply with the Pauli exclusion principle, which says that no two spin $\frac{1}{2}$ particles in the same system can have the same total wave function. Consider now a carbon atom bonded to four other atoms of the same kind, say hydrogen atoms or chlorine atoms. The outer shell of the carbon atom now effectively has eight electrons, the maximum number that it can hold for the Pauli principle to be obeyed. These four bonds are clearly completely equivalent and, because there are no other electrons in the outer shell of the carbon atom, the mutual repulsion between the bonds and the requirements of symmetry ensure that they are all separated from each other by the same angle. This means that the hydrogen or chlorine atoms must lie at the vertices of a regular tetrahedron, as shown in fig. 3.2(a). Another way of thinking of the arrangement is shown in fig. 3.2(b). For this *tetrahedral bonding*, the angle between any two bonds is approximately 109.5°.

Another important atom for polymers is the oxygen atom. This atom has six electrons in its outer shell and this shell can be effectively completed if the oxygen atom is bonded to two other atoms. It might at first be expected that the two bonds would be 180° apart, but this is to neglect the fact that the other electrons in the outer shell are not involved in bonding. The simplest bonding theory suggests that the angle between the bonds should be 90°, but this neglects various effects such as the Coulomb repulsion between the non-bonded atoms, i.e. the two atoms bonded to the oxygen atom, and the repulsion between the bonds themselves. These effects cause the angle to open out slightly and the typical angle found in various compounds is about 110°. Similarly, the bonds formed by the triply bonded nitrogen atom that frequently occurs in polymers lie in a plane at approximately 120° to each other.

Table 3.1 shows the commonly used standard angles for various groupings of atoms that occur in polymer molecules. It must be remembered that the exact values of the angles in any particular molecule do depend on the nature of the other atoms or groups bonded to the grouping shown, for

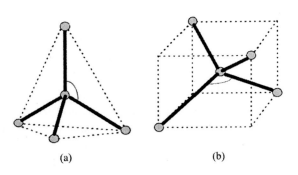

(a)   (b)

Fig. 3.2 Two ways of visualising the tetrahedral angle, (a) by means of a regular tetrahedron and (b) by means of a cube. The angles marked (and all similar angles) are equal to the tetrahedral angle, which is approximately 109.5°.

Table 3.1. *Some typical bond lengths and angles*

Bond	Length (nm)	Bond angle	
C—C	0.154	CCC	109.5°
C=C	0.134	COC	108°
C—H	0.109	OCC	110°
C—O	0.143		
C=O	0.123		
C—N	0.147		
C—Cl	0.177		
N—H	0.101		
O—H	0.096		

similar reasons to those causing the angles to deviate from those predicted from the very simplest models.

(ii) 'Standard' bond lengths

Particular types of bond, e.g. C—C, C—O, etc., generally have lengths that are approximately independent of their position in a molecule. The reason for this is that the pair of electrons forming the bond is fairly well localised to the region of the bond, so the bond is unaffected, to a first approximation, by the nature of the other bonds formed with the atom. Atomic bond lengths are always specified as the distances between the corresponding atomic nuclei. Table 3.1 shows the 'standard' lengths for the bonds most commonly encountered in polymer molecules.

(iii) 'Preferred' orientations around bonds

If three single bonds in an all-carbon backbone lie in a plane and are arranged as shown in figs 3.3(a) and (b), the middle bond is said to be a *trans bond*. This is usually the *conformation* of lowest energy. If three single bonds in an all-carbon backbone are arranged as shown in figs 3.3(c) and (d), the middle bond is said to be a *gauche bond*. Figure 3.4 shows how the energy of the n-butane molecule CH_3—CH_2—CH_2—CH_3 varies with the torsional angle around the central C—C bond. This

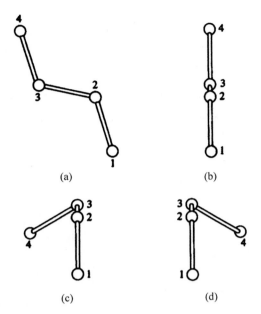

(a)   (b)

(c)   (d)

Fig. 3.3 *Trans/gauche* isomerism: (a) a *trans* bond viewed normally to the plane of the three bonds required to define it; (b) a view of (a) from the right-hand side, looking almost end-on to the *trans* bond; (c) and (d) nearly end-on views of left- and right-handed *gauche* bonds, respectively. (Reproduced from *The Vibrational Spectroscopy of Polymers* by D. I. Bower and W. F. Maddams. © Cambridge University Press 1989.)

angular variation is due partly to the inherent lowering of the energy of the bonds when the shape, or conformation, of the molecule corresponds to any of the three angles 60°, 180° and 300° in fig. 3.4 and partly due to the repulsive interaction of the groups of atoms attached to the two inner C atoms.

The energy in the *trans* conformation is lower than that in the *gauche* conformations, which are the conformations of next lowest energy, because

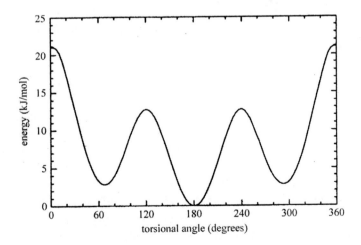

Fig. 3.4 The variation of the energy of the n-butane molecule CH_3—CH_2—CH_2—CH_3 as a function of the torsion angle of the central C—C bond measured from the *eclipsed* conformation. (Reproduced from *The Science of Polymer Molecules* by R. H. Boyd and P. J. Phillips. © Cambridge University Press 1993.)

in this conformation the two large CH_3 groups are as far away from each other as possible. The two *gauche* conformations can be imagined to be obtained by rotation of one end of the molecule with respect to the other by ±120° around a central bond that is originally *trans*. The highest-energy conformation is the *eclipsed* arrangement which brings the CH_3 groups most closely together and is obtained by a rotation of ±180° around a central bond that is originally *trans*. Note that, although the word *gauche* means 'left' in French, it also means 'awkward' in both English and French; *gauche* bonds can be either right- or left-handed. Butane molecules with right- and left-handed *gauche* bonds are mirror images of each other. As has already been implied, similar energy considerations apply to each of the single C—C bonds in a polymer backbone, but the lowest-energy states (and hence the conformation of the backbone) may be modified by the factors considered below.

(iv) *Steric hindrance*

Bulky side groups may interfere with each other, so that certain conformations of the molecule are impossible. Imagine the butane molecule in the *trans* conformation and now imagine that one of the hydrogen atoms on each of the two central carbon atoms is replaced by a large atom, or group of atoms. If the replacements are for the two hydrogen atoms originally on opposite sides of the plane of the carbon atoms, these large groups will be as far apart as they could possibly be and the molecule will probably be able to remain planar.

If, however, the two hydrogen atoms on the same side of the plane of the carbon atoms were replaced, the large groups would be much closer together and might even overlap if the molecule remained *trans*. Non-bonded atoms cannot penetrate each other because there are strong repulsive forces between them when they begin to do so. Rotation of the molecule into a *gauche* conformation would then put the two large groups as far apart as in the previous case and would probably provide the lowest-energy state of the molecule (see problems 3.4 and 4.1).

(v) *Hydrogen bonding (H-bonding)*

In some molecules *intramolecular H-bonding*, i.e. H-bonding within one molecule, can take place, which usually leads to the formation of *helical structures*. Such structures are particularly important in biopolymers e.g. the α-helical conformation of *polypeptides* (fig. 3.5(a)). The hydrogen bond is a loose kind of bond that is largely electrostatic in nature and its strength lies somewhere between that of the covalent bond and the weak *van der Waals attractive forces* that are exerted by different neutral molecules when they come within about a molecular radius of each other.

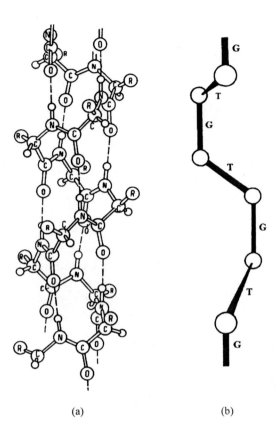

(a)   (b)

Fig. 3.5 (a) The α-helix of a polypeptide, with intramolecular hydrogen bonds shown by dashed lines. (b) The carbon backbone of a 3_1 TGTGTG helix. ((a) Reproduced by permission of Springer-Verlag.)

For the formation of a hydrogen bond it is usually necessary to have an —O—H or an $>$N—H group, i.e. a hydrogen atom attached to a small electronegative atom. These bonds are strongly polar, with the H atom positive, so that it can be attracted to a negatively polarised O or N atom in the vicinity:

$$> N^{-\delta_1}—H^{+\delta_1} \cdots \cdots O^{-\delta_2} = C^{+\delta_2} <$$

The symbols δ_1 and δ_2 indicate effective fractional charges and the dotted line represents the H-bond formed by the attraction of the positively charged H atom and the negatively charged O atom.

The result of the factors above is that the most likely (lowest-energy) conformation for many polymers that have all-carbon backbones is the *all-trans planar zigzag*, whereas for many others it is some form of *helix*. The simplest helix is the 3_1 helix for which *trans* and *gauche* bonds alternate along the backbone (fig. 3.5(b)). This helix is described in more detail in section 4.1.5.

3.3.2 Conformations and chain statistics

In the previous section the idea of the conformation of a small molecule, n-butane, was discussed, as was the idea of the local conformation of a polymer chain. It was tacitly assumed that such states were in fact different from each other, i.e. that they each had a certain degree of permanence. The plot in fig. 3.4 shows that the difference in energy between the *trans* and *gauche* states is about 3 kJ mol^{-1}, which is quite close to thermal energies even at room temperature, at which $N_A kT$ is about 2.5 kJ mol^{-1}. This means that, even though the *trans* state has lower energy, there will be a considerable fraction of molecules in the *gauche* state.

What matters, however, for ensuring that the states are essentially discrete, is that the energy barrier between the two states is about five times thermal energies at room temperature, so that most of the time the molecule will simply perform small oscillations around a discrete *trans* or *gauche* state. The various conformations of a molecule are often called *rotational isomers* and the model of a polymer chain introduced by Flory in which the chain is imagined to take up only discrete conformational states is called the *rotational isomeric-state approximation*.

Polymers typically have a very large number of single bonds around which various conformational states can exist, and a polymer molecule as a whole therefore has a very large number of conformational states. Even for a molecule with ten C-C bonds in the backbone the number is $3^8 = 6561$, assuming that all possible *trans* and *gauche* states could be reached, and the corresponding number for a typical polyethylene chain with about 20 000 C—C bonds is about 10^{9540}. It is clear from this that even though the number of conformational states that could potentially be reached is reduced somewhat by steric hindrance or overlapping of the chain with itself, only statistical methods can be used in discussing the conformations of whole polymer chains. In this discussion it is usual to start with a simplified model, that of the *freely jointed chain*, and this forms the topic of the following section. More realistic chains are considered in section 3.3.4.

3.3.3 The single freely jointed chain

Rotation around a single bond in a molecular chain generally gives a cone of positions of one part of the chain with respect to the other. If the rotation is not totally free, as for example when *trans* or *gauche* bonds have lower energy than other conformations for an all-carbon backbone, certain positions on this cone are favoured. Nevertheless, in order to develop the simplest form of the statistical theory of polymer-chain conformations the following simplified model, called the *freely jointed random-link model,* is used.

(a) The real chain is replaced by a set of points joined by n equal one-dimensional links of length l. The *contour* or *fully extended length* of the chain is then nl.
(b) It is assumed that there is *no* restriction on the angles between the links; the angles are *not* restricted to lie on cones.
(c) It is assumed that *no energy* is required to change the angles.

An important quantity in the theory is the *root-mean-square* (RMS) *length of an unperturbed randomly coiled chain*, which can be calculated as follows (see fig. 3.6):

Let r be the end-to-end vector of the chain. Then

$$r_x = \sum_i l \cos \theta_i \tag{3.2}$$

$$r_x^2 = l^2 \left(\sum_i \cos^2 \theta_i + \sum_{i \neq j} \cos \theta_i \cos \theta_j \right) \tag{3.3}$$

If r_x^2 is averaged over a large number of chains, the last term averages to zero, because there is no restriction on the angles between the links. Thus

$$\langle r_x^2 \rangle = nl^2 \langle \cos^2 \theta_i \rangle = \tfrac{1}{3}nl^2 \tag{3.4}$$

where the angle brackets $\langle \ \rangle$ denote the average over all chains. However,

$$\langle r^2 \rangle = \langle r_x^2 \rangle + \langle r_y^2 \rangle + \langle r_z^2 \rangle \qquad \text{and} \qquad \langle r_x^2 \rangle = \langle r_y^2 \rangle = \langle r_z^2 \rangle$$

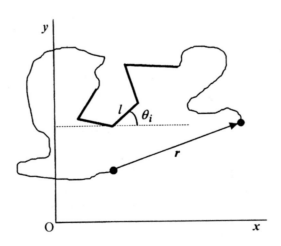

Fig. 3.6. Calculation of the RMS chain length. The thick lines represent five actual links somewhere in the chain, the rest of which is represented by the curved lines. The angle θ_i is the angle between the ith link and the Ox axis.

Thus the RMS length r_{rms} is given by

$$r_{rms} = \sqrt{\langle r^2 \rangle} = n^{\frac{1}{2}} l \tag{3.5}$$

The fully extended length of the chain is equal to nl so that the *maximum extensibility* of a random coil with $r = r_{rms}$ is from $n^{\frac{1}{2}}l$ to nl, i.e. a factor of $n^{\frac{1}{2}}$. The value of r_{rms} also gives a measure of the spatial extent of a chain. A second useful measure of this is the *radius of gyration*, r_g, which is the RMS distance of the atoms of the chain from the centre of gravity of the chain. Debye showed that, provided that n is very large, $r_g = r_{rms}/\sqrt{6}$.

In order to simplify the theory further it is usual to assume that, for all chains that need to be considered, the actual end-to-end distance is very much less than the fully extended length, i.e. $r \ll nl$, which becomes true for all chains when n is sufficiently large. A chain for which the assumption is valid is called a *Gaussian chain*. Consider such a chain with one end fixed at the origin and let the other end, P, be free to move (see fig. 3.7). With OP $\ll nl$ it can then be shown that the probability $p(x, y, z)$ that P lies in the small element of volume $dx\, dy\, dz$ at (x, y, z) is

$$p(x, y, z)\, dx\, dy\, dz = (b^3/\pi^{3/2}) \exp(-b^2 r^2)\, dx\, dy\, dz \tag{3.6}$$

where

$$b^2 = 3/(2nl^2) \tag{3.7}$$

and $r^2 = x^2 + y^2 + z^2$.

The function $p(x, y, z)$ is the *Gaussian error function* or *normal distribution*. The maximum value of $p(x, y, z)$ is at the origin, corresponding to $r^2 = x^2 + y^2 + z^2 = 0$. The most probable value of r is not, however, zero.

Fig. 3.7 The Gaussian chain with one end coincident with the origin.

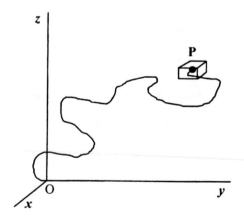

Example 3.2

Calculate the ratio of the probability of the end-to-end separation of a Gaussian chain being within a small range dr near (a) $2r_{rms}$ and (b) $3r_{rms}$ to that for it being within a small range of the same size near r_{rms}.

Solution

For any two values r_1 and r_2 of the end-to-end separation the ratio of the probabilities, from equation (3.8), is

$(r_1/r_2)^2 \exp[-b^2(r_1^2 - r_2^2)] = (r_1/r_2)^2 \exp[-3(r_1^2 - r_2^2)/(2r_{rms}^2)]$.

Setting $r_2 = r_{rms}$ gives the ratio as $x^2 \exp[-3(x^2 - 1)/2)]$, where $x = r_1/r_{rms}$.

Substituting $x = 2$ or 3 gives the ratios 0.044 and 5.5×10^{-5}.

The probability $P(r)\,dr$ that the end P lies somewhere in the spherical shell of radius r and thickness dr is given by

$$P(r)\,dr = 4\pi r^2 p(r)\,dr = 4\pi r^2 (b^3/\pi^{3/2}) \exp(-b^2 r^2)\,dr \qquad (3.8)$$

For $r = 0$ this probability is zero, and the probability peaks at $r = 1/b = \sqrt{2n/3}\,l = \sqrt{\frac{2}{3}}\,r_{rms}$. By rearrangement, equation (3.8) can be written

$$P(br)\,d(br) = (4/\pi^{1/2})(br)^2 \exp[-(br)^2]\,d(br) \qquad (3.9)$$

where br is dimensionless. $P(br)$ is plotted against br in fig. 3.8.

As discussed further in the following section, it can be shown that the statistical distribution of end-to-end distances for any real chain reduces to the Gaussian form if the number of rotatable links is sufficiently large. By suitably choosing n and l for the freely jointed random-link model, both r_{rms} and the fully extended length can be made equal to the corresponding values for the real chain. These values define the *equivalent freely jointed random chain*. For example, if it is assumed that in a real polyethylene chain (i) the bonds are fixed at the tetrahedral angle and (ii) there is free

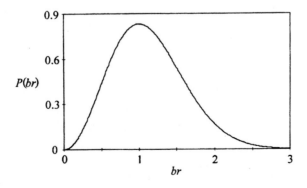

Fig. 3.8 The probability for a chain of length r in the Gaussian approximation. See the text for discussion.

rotation around the bonds, it can be shown (see problem 3.7) that one random link is equivalent to three C—C bonds in the real chain. Evidence based on stress–optical measurements (see section 11.3.3) suggests that, for real polyethylene, the number of C—C bonds in the equivalent random link is much greater than this, as might be expected, because the rotation around the bonds is not free.

The following section is concerned with more realistic chains, but it is shown that the random-link model is still applicable for polymers in molten and certain other states.

3.3.4   More realistic chains – the excluded-volume effect

The freely jointed chain model can be adapted fairly easily to somewhat more realistic chains. It can, for instance, be shown that if the polymer chain has a sufficiently long backbone consisting entirely of n_o single bonds of length l_o with a fixed angle of deviation θ_d between the direction of one bond and that of the next, but with free rotation around any of the single bonds, the RMS length, $r_{rms}(\theta_d)$, of the chain is given by

$$r_{rms}(\theta_d) = n_o^{1/2} l_o \sqrt{(1 + \cos\theta_d)/(1 - \cos\theta_d)} \tag{3.10}$$

By making a suitable choice of n and l, the value of r_{rms} for a totally freely jointed chain can be made equal to $r_{rms}(\theta_d)$ at the same time as its fully extended length is equal to that of the more realistic chain (see problem 3.7). It can more generally be shown that for any real chain an *equivalent freely jointed chain*, often called the *Kuhn chain*, can be chosen that has both the same fully extended (contour) length and the same r_{rms} as the real chain whatever the nature of the backbone bonds in the real chain and whatever the degree of rotational restriction around them, provided that interaction between chain segments separated along the chain by distances greater than the length of the equivalent random link, called the *Kuhn link length*, is neglected.

The fact that real chains cannot cross over themselves and therefore do interact at distances greater than the Kuhn link length is often called the *excluded-volume effect* and it leads to an expected increase in the RMS chain length and to a change in the way that it depends on n. Later it will be seen that the effect is not important in changing r_{rms} or its dependence on n in homopolymer melts or in solids, but the fact that the chains cannot interpenetrate each other nevertheless has important effects on their mechanical properties.

Two segments of a chain cannot interpenetrate because there is a repulsive force between them when they approach closer than a certain distance apart, and this force becomes effectively infinite when the segments 'touch'. It is easy to see why the excluded-volume effect increases r_{rms} for an isolated chain by realising that, if the chain is coiled very tightly, there will be very many close approaches of segments, and the repulsive force between them will then cause the chain to expand to a size greater than it would have in the absence of the repulsive force. If, however, the chain is in a very extended conformation there will be very few close approaches and the chain will not be expanded significantly by the repulsive forces. Averaging over all possible conformations will therefore give a larger value of r_{rms} than that for an imaginary infinitely thin chain with no repulsive forces between segments. An argument originally due to Flory, which is not quite correct but does show just how significant this effect may be, is based on minimising the Helmholz function $A = U - TS$, where U is the internal energy, T the temperature and S the entropy of the chain.

According to Boltzmann, the entropy of a system is given by $S = k \ln W$, where W is the number of configurations available to the system. The word configurations is used in the statistical-mechanics sense and corresponds here to the conformations of the chain. Consider a Gaussian chain with a particular end-to-end vector r. The probability of finding a chain with this value of r is proportional to the number of conformations that it can take up and equation (3.6) shows that $W = C \exp(-b^2 r^2)$, where C is independent of r. The entropy is then given by $S = k \ln W = k \ln C - kb^2 r^2$.

The internal energy U can be estimated by supposing that two chain segments must be within the same small volume dv if they are to interact significantly. The mean density of chain segments in the randomly coiled chain is n/V, where V is the effective volume in which the chain segments are to be found. Because both dv and the density are very small, the chance of finding one segment in any specific small volume dv is $n\,dv/V$ and the chance of finding two segments in this volume is $q = (n\,dv/V)^2$. There are V/dv volumes dv within the volume V, so that the total number of interacting pairs of segments is $qV/dv = n^2\,dv/V$.

Suppose that the mean energy of interaction is akT, where a is a dimensionless constant. The total internal energy U that the interacting pairs contribute to the free energy of the chain is then given by $U = akTn^2\,dv/V$. Assuming that the Gaussian formula can be applied approximately to the self-interacting chain and that V can be set equal to cr^3, where c is a constant, leads finally to

$$A = U - TS = kT[an^2\,dv/(cr^3) - \ln C + b^2 r^2] \tag{3.11}$$

Setting $dA/dr = 0$ and assuming that the resulting value of r is equal to r_{rms} leads to $r_{rms}^5 = 3an^2 v/(2b^2 c)$ and substituting for b from equation (3.7) leads to

$$r_{rms}^5 = al^2 n^3 \, dv/c \tag{3.12}$$

which shows that r_{rms} is proportional to $n^{3/5}$, rather than to $n^{1/2}$, which applies for any non-self-interacting chain, as shown by equation (3.5) and the fact that any non-self-interacting chain can be represented by an equivalent freely jointed chain. According to equation (3.12), quadrupling the length of a chain increases r_{rms} by a factor of $4^{3/5} = 2.3$, whereas for a non-self-interacting chain the factor would be $4^{1/2} = 2$. Although the argument is not rigorous, experimental studies of polymers in dilute solution, numerical simulations and more refined theoretical calculations support an exponent of n close to $\frac{3}{5}$.

Now consider the more realistic situation, a single chain in a very dilute solution where there is no interaction between different chains, but where each chain can interact with the solvent molecules and with itself. Suppose that the solvent is a good solvent, i.e. there is a very strong attractive interaction between the solvent molecules and the polymer chain. This will clearly cause the chain to expand, or *swell*, in a way similar to that caused by repulsion between the various segments of the chain itself, because a more extended chain will have a greater number of contacts between polymer segments and solvent molecules. The Flory argument given above now applies, with a accounting for both intra-chain and polymer–solvent interactions.

Suppose, on the contrary, that there is a much weaker attractive interaction between the solvent molecules and the polymer than there is between segments of the polymer molecules themselves, i.e. the solvent is a poor one. In order to lower the energy of the system the segments of the chain will be forced closer together so that there are fewer interactions between the chain and the solvent molecules. This is equivalent to an attractive interaction between the segments of the chain. It should now be clear that there is the possibility of a solvent that causes the chain neither to swell nor to contract.

In order to formalise this argument it is necessary to carry through a full statistical-mechanical calculation, which takes account of the entropy as well as the energy of mixing. This shows that the small volume dv should be equal to $v_0(1 - \Theta/T)$, where v_0 is a small volume that is independent of temperature and both it and Θ depend on the form of the effective interaction between the chain segments. At the *theta-temperature*, or *Flory temperature*, $T = \Theta$ and $dv = 0$, so that the chain behaves as if it did not interact with itself.

There is, however, a difficulty here; if $d\upsilon$ is zero, so is r_{rms} according to equation (3.12), which clearly does not make sense. A theoretical treatment that overcomes this difficulty makes the assumption that the expansion of a Gaussian chain in a good solvent may be well represented by simply increasing the bond length l to a new value l'. It can then be shown that if the new Gaussian chain is to be statistically representative of the real non-Gaussian excluded-volume chain and, in particular, if r_{rms} is to be equal to $n^{1/2}l'$, as it should be for a Gaussian chain, then the *swelling coefficient* $\alpha = l'/l$ is given by

$$\alpha^5 - \alpha^3 = \sqrt{\frac{6}{\pi^3}} \frac{n^{1/2}}{l^3} \frac{d\upsilon}{} \qquad (3.13)$$

The Gaussian chain expanded in this way is then the best Gaussian approximation to the real chain. This equation shows that as $d\upsilon$ tends to zero, α tends to 1, so that the chain retains its unperturbed value of r_{rms}, i.e. the best Gaussian approximation is then the unperturbed Gaussian. On the other hand, if $d\upsilon$ becomes large, α becomes proportional to $n^{1/10}$ and r_{rms} becomes proportional to $n^{3/5}$, in agreement with equation (3.12). A solvent at a temperature $T = \Theta$ is called a *theta-solvent*.

The following argument shows that any polymer chain in a homogeneous polymer melt is effectively in a Θ-solvent, so that it has the same value of r_{rms} as a chain with no interactions between segments. First consider again an isolated chain. Replace the *average* density of segments n/V used in the argument leading to equation (3.11) by the *local* density of segments ρ. The chance of finding a segment in $d\upsilon$ is then $\rho\, d\upsilon$ and the effective potential experienced by any segment in this local region of the chain is $akT\rho\, d\upsilon$. It can be shown that, on average, the value of ρ is high near the centre of mass of an isolated ideal non-self-interacting chain and falls off with distance from it. The potential thus falls with distance from the centre and gives rise to an outward force that causes the expansion of the chain. If, however, the chain is in a melt, the local density of segments ρ is constant and there is no force to cause expansion.

The idea that a chain in a melt is effectively a free chain without self-interaction was first clearly expressed by Flory and is often called the *Flory theorem*. At first sight it is paradoxical: how can the presence of other chains allow a chosen chain to take up conformations that it could not take up if the other chains were absent? The answer is that it cannot. The unperturbed chain is only equivalent to the real chain in terms of its Gaussian statistics. This is nevertheless very important, because it is the statistical properties of the chains and their link with the entropy that largely determine some of the properties of the corresponding materials.

This will become clear in chapter 6 where the elasticity of rubbers is considered.

3.3.5 Chain flexibility and the persistence length

The freely jointed chain is the most flexible type of chain that can be envisaged and, in general, real chains with many single bonds in the backbone around which rotations can take place will be highly flexible provided that the energy required to overcome the barriers to rotation is small compared with, or comparable to, thermal energies. On the other hand, a chain such as the polyparaphenylene chain shown in section 1.3.1 has no flexibility due to rotation around bonds. The only flexibility that this molecule has is due to the small changes of bond angles that take place due to thermal vibrations.

Examples of polymers with very flexible backbones are given in section 6.4, where it is shown that such flexibility can lead to the very low elastic moduli of rubbers. Examples of molecules with very low degrees of flexibility are discussed in section 12.4, where it is seen that the low flexibility can lead to liquid-crystalline behaviour, which in turn can lead to the production of materials with high elastic moduli. Because of the importance of flexibility in determining properties, it is important to have some way of defining it more precisely.

It should be clear that the Kuhn length is actually a measure of flexibility, because the relative orientations of sections of the molecule separated by more than this length are certainly random, whereas sections of the molecule that are somewhat closer together have non-random relative orientations. Slightly more sophisticated considerations lead to the definition of a *persistence length* that is closely related to the Kuhn length and, for an infinitely long chain, is equal to half of it. Flexible chains have low persistence lengths and inflexible, or rigid, chains have high persistence lengths.

Persistence lengths can be measured in dilute solution by means of light-scattering or viscosity measurements, but it is increasingly common to calculate them using computerised molecular-modelling techniques. The flexibility of a molecule changes with temperature because at higher temperatures larger fluctuations of bond angle take place and more high-energy conformations occur. The modelling techniques have the advantage that the effective temperature of the polymer can be varied over a much greater range than is available with experimental measurements and knowledge of the variation of the persistence length with temperature is important in understanding the onset of liquid crystallinity (see section 12.4.4).

3.4   Evidence for ordered structures in solid polymers

This chapter has up to now been concerned largely with the idea that polymer chains can take up essentially random structures. The rest of the chapter surveys briefly some of the evidence that various kinds of ordered structures can appear in the solid state. It thus provides an introduction to the discussion of *polymer crystallinity* in chapter 4 and of *polymer morphology* in chapter 5. The evidence is considered in the sequence leading from small- to large-scale structures and the principal techniques considered for showing that these structures exist and for indicating their sizes are (i) wide-angle X-ray scattering (WAXS), (ii) small-angle X-ray scattering (SAXS), (iii) light scattering and (iv) microscopy, but other techniques are also referred to.

3.4.1   Wide-angle X-ray scattering – WAXS

If a monochromatic X-ray beam is incident on a polymer sample, scattering is observed at angles α of about 10–50° from the direction of the incident beam. Taking a typical angle of $23° \cong \frac{2}{5}$ radian and assuming an X-ray wavelength λ of about 2 nm, standard diffraction theory shows that this implies that there are units present in the polymer with dimensions d given approximately by $d = \lambda/\alpha = 2/(2/5) = 0.5$ nm, which is of the order of the separation of atoms or of the length of the repeat units in the polymer chain. Some polymers give diffuse halos like that shown in fig. 3.9(a), which is also the type of pattern observed from a liquid and implies only short-range order within the polymer, i.e. no crystallinity. These polymers are glassy or rubbery. Other polymers give sharp rings (fig. 3.9(b)) similar to the rings observed for powders of non-polymeric crystalline materials (powder rings), which implies that there is long-

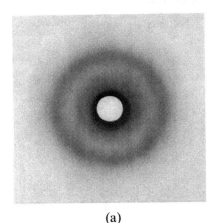
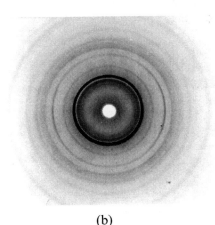

(a)   (b)

Fig. 3.9 Wide-angle X-ray scattering from (a) polystyrene, showing a diffuse halo from an amorphous sample; and (b) highly crystalline polyethylene, showing sharp 'powder' rings. (Courtesy of Dr A. P. Unwin.)

range order within the polymer, i.e. that crystallites are present. Other simple evidence for the existence of crystalline material in polymers comes from the presence of exothermic or endothermic peaks in DSC scans (see e.g. fig. 2.3).

If a stretched fibre of a crystalline polymer is placed in the X-ray machine, a so-called *fibre pattern* is observed (fig. 3.10). As discussed in chapter 4, this pattern contains information about the crystal structure of the polymer. It also contains information about the size of the crystallites and about their degree of alignment, topics discussed further in chapters 5 and 10, respectively.

3.4.2   Small-angle X-ray scattering – SAXS

When a monochromatic X-ray beam passes through some polymers peaks of scattering are observed at angles θ of the order of $1°$, or about a fiftieth of a radian, when the X-ray wavelength used is about 0.2 nm (see fig. 3.11). This implies that the linear dimension d of the structures responsible for the scattering is given approximately by $d = \lambda/\theta = \text{———} = 10$ nm, which is much greater than the chain repeat length, which is typically about 0.5–1 nm. The scattering is in fact due to regular *stacks of crystal lamellae* (thin, flat, plate-like crystals) of the form shown in fig. 3.12. Further evidence for the presence of these lamellae is provided by optical and electron microscopy and by the observation of certain low-frequency modes in the Raman spectrum. Lamellar crystals are discussed in detail in section 5.3 and lamellar stacks in section 5.4.3.

The X-ray scattering is similar to Bragg reflection (see section 2.5.1) from planes with spacing l_s. For small scattering angles θ, where the Bragg angle is $\theta/2$, first-order scattering is thus observed when $l_s\theta = \lambda$, so that l_s can be determined by measuring θ.

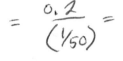

$$= \frac{0.2}{(1/50)} =$$

Fig. 3.10 A fibre pattern from oriented syndiotactic polypropylene, drawn to a draw ratio of about 5 at 109 °C. (Courtesy of Dr A. P. Unwin.)

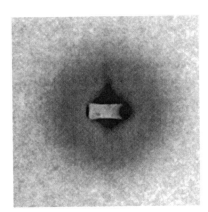

Fig. 3.11 Small-angle X-ray scattering from unoriented polypropylene. The SAXS scattering is the dark halo of diameter at its darkest equal to about half the side of the outer square. This diameter corresponds to about 1°. (Courtesy of Dr A. P. Unwin.)

3.4.3 Light scattering

If a polymer is homogeneous at least on a scale greater than about 0.1 μm, a parallel laser beam is scattered by it as if from a liquid. There are no sharp maxima, merely a smooth distribution of intensity with angle. This *Rayleigh scattering* is discussed in many texts on optics. In contrast, however, many polymers scatter strongly near specific angles θ from the direction of the incident beam, where θ is typically about 5° (about a tenth of a radian), although larger or smaller angles may be observed. Such scattering is evidence for the presence of structures with linear dimensions d given approximately by $d = \lambda/\theta$, where λ is the wavelength of the laser light. Putting $\lambda = 500$ nm, a typical value, gives $d \cong 500/(1/10) = 5\,000$ nm $= 5$ μm. Structures of this size can be studied by the conventional optical microscope, since they are several times larger than the wavelength of

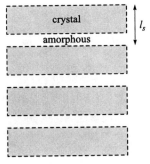

Fig. 3.12 A stack of crystal lamellae, shown schematically. The separation l_s determined by X-ray scattering is the repeat distance shown.

light. The advantage of the light scattering method is that it provides evidence for average sizes, as is discussed more fully in section 5.5.2, whereas optical microscopy allows individual structures and the variety of structures to be examined, as described in the next section.

3.4.4   Optical microscopy

Some polymers, when they are suitably prepared in thin slices or as thin films, exhibit circular features when they are viewed in the optical micro-scope (fig. 3.13), whereas others show less regular patterns, depending on the polymer and the method of preparation of the sample. In order to see these features the polarising microscope with crossed polarisers (see section 2.8.1) is used. The circular features shown in fig. 3.13 are caused by sphe-rical structures called *spherulites* which are a very important feature of polymer morphology, the subject of much of chapter 5, where the *Maltese cross* appearance seen in fig. 3.13 is explained. Each spherulite consists of an aggregate of crystallites arranged in a quite complicated but regular way.

Crystal aggregates can, of course, be examined in more detail in the electron microscope, which provides much higher resolution than does the optical microscope. Information obtained by this and other methods is also described in chapter 5.

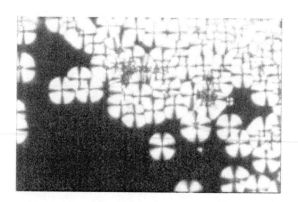

Fig. 3.13 An optical micrograph of spherulites of poly(propylene terephthalate) grown from the melt. (Courtesy of Dr M. A. Wilding.)

3.5 Further reading

(1) The general texts cited in section 1.6.1, particularly the one by Strobl.

(2) *The Science of Polymer Molecules*, by R. H. Boyd and P. J. Phillips, CUP, 1993. This book focuses on the synthesis, structure and properties of polymer molecules, covering such topics as molar masses, chain statistics, rubber elasticity and polymer solutions. The level is generally more advanced than that of the present book, but some sections are relatively straightforward.

3.6 Problems

3.1. Mono-disperse fractions of molar masses 200 000 and 400 000 g mol^{-1} are added to the polymer of example 3.1 so that the ratios of the numbers of chains, in order of increasing M, become $\frac{1}{2} : 1 : 2 : 1 : \frac{1}{2}$. Calculate the number-average and weight-average molar masses of the resulting polymer and hence show that the polydispersity index has increased.

3.2. Assuming that a polymer has a molar-mass distribution defined by $n(m) = ae^{-bm}$, where $n(m)\,dm$ represents the fraction of molecules with molar masses between m and $m + dm$, show that $a = b$ and calculate (i) the number- and weight-average molar masses, and hence the polydispersity index and (ii) the molar mass at the mode (maximum) of the distribution function.

3.3. Explain why figs 3.2(a) and (b) both show correctly the tetrahedral angle and calculate its value using fig. 3.2(b).

3.4. Calculate the smallest separations between hydrogen atoms connected to the two different carbon atoms 2 and 3 of the butane molecule when the central C—C bond, between atoms 2 and 3, is in the *trans*, *gauche* and *eclipsed* states.

3.5. Calculate the root-mean-square length of a polyethylene chain of $M = 250\,000$ g mol^{-1} assuming that the equivalent random link corresponds to 18.5 C—C bonds.

3.6. Estimate the fraction of all chains having end-to-end lengths between half and twice the most probable length for a mono-disperse polymer that obeys Gaussian statistics. Hint: look carefully at fig. 3.8 and then make sensible approximations.

3.7. An imaginary polymer has a backbone consisting entirely of n_o single bonds of length l_o with a fixed angle of deviation θ between the direction of one bond and the next. Assuming that there is free rotation about the backbone bonds, derive an expression for the fully extended length of the chain. Hence, using equation (3.10), derive an expression for the ratio of the number of equivalent random links to the number of real links (single bonds) for this polymer. Evaluate the ratio when the angle *between* bonds is the tetrahedral angle and deduce the length of the equivalent random link.

3.8. The table shows values of the swelling coefficient α for four different molar masses of polyisobutylene in cyclohexane obtained at 30°C. Show that these data are consistent with equation (3.13).

$M/10^3$ (g mol^{-1})	α
9.5	1.12
50.2	1.25
558	1.46
2720	1.65

3.9. A typical lamella in polyethylene is about 12.5 nm thick. Using the appropriate 'standard' bond angle and bond length, estimate the number of carbon atoms in a straight chain that just crosses the lamella.

3.10. If the polymer in problem 3.9 had a mean molar mass of 160 000 g mol^{-1}, how many times could a typical chain cross such a lamella, assuming that the chains fold tightly back and forth across the lamella?

Chapter 4
Regular chains and crystallinity

Ideal crystals are regularly repeating three-dimensional arrays of atoms. It follows that, if a polymer is to be able to crystallise, even if only to a limited extent, the minimum requirement is that its chains must themselves have regularly repeating units along their lengths and must be able to take up a straight conformation. This chapter is concerned with the types of regularity and irregularity that polymer chains can have, with how regular chains pack together and with the experimental study of the crystal structure at the level of the unit cell, the smallest unit from which the crystal may be imagined to be built by regular repetition of its structure in space, as described in section 2.5.1.

4.1 Regular and irregular chains

4.1.1 Introduction

It is a *geometrical* requirement that if a polymer is to be potentially capable of crystallising, its chains must be able, by undergoing suitable rotations around single bonds, to take up an arrangement with *translational symmetry*. If a polymer is actually to crystallise there is also a *physical* requirement: the crystal structure must usually have a lower *Gibbs free energy* than that of the non-crystalline structure at the same temperature. Consideration of this physical requirement is deferred to later sections; the present section deals only with the geometrical requirement.

Geometrical regularity can be considered to require

(i)   *chemical regularity* of the chain and
(ii) *stereoregularity* of the chain.

The second requirement cannot be met unless the first is met, but chemical regularity does not guarantee that the second requirement can be met. Chemical regularity means that the chain must be made up of identical *chemical repeat units*, i.e. units with the same chemical and structural formula, and the units must be connected together in the same way. A homopolymer of the type

$$—A—A—A—A—A—A—A—A—A—A—A—A—A—$$

where each A unit consists of the same group of atoms bound together in the same way, satisfies this minimum criterion. A random copolymer of the type

$$—B—A—B—B—A—B—A—A—B—A—A—A—B—B—$$

does not, so that random copolymers are unlikely to be able to crystallise. In contrast, block copolymers consisting of sequences of the type $—A_nB_m—$, where A_n and B_m represent long sequences of A or B units, may be able to crystallise; this topic is discussed in section 12.3.4. Chain branching also tends to inhibit crystallisation. An example is provided by the various types of polyethylene described in section 1.3.3: linear high-density polyethylene is more crystalline than is branched low-density polyethylene.

Stereoregularity means that the chain must be capable (by rotations about single bonds) of taking up a structure with translational symmetry. Translational symmetry means that the chain is straight on a scale large compared with the size of the chemical repeat unit and consists of a regular series of *translational repeat units*, so that the chain can be superimposed upon itself by translation parallel to the axis of the chain by the length of one translational repeat unit (see fig. 4.1). The translational repeat unit may be identical to the chemical repeat unit, but often there is more than one chemical repeat unit in a translational repeat unit. The most important

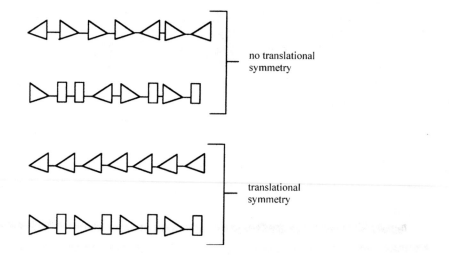

Fig. 4.1 Schematic representations of chains with and without translational symmetry. The various shapes represent different chemical groups. True translational symmetry would require the two lower chains to extend to infinity, repeating regularly as shown.

no translational symmetry

translational symmetry

example of this is where the polymer chains form a *helix*. For the simplest
type of helix, the translational repeat unit consists of a small number n of
chemical repeat units arranged in such a way that the chain can be superposed upon itself by translation through l_t/n along the axis of the chain and
rotation through $2\pi/n$ around the axis, where l_t is the length of the translational repeat unit. Helices may, however, be more complicated, as
described in section 4.1.5. Some types of polymerisation automatically
lead to potential translational symmetry, but others do not. In the next
sections examples of these types of polymers are considered in turn.

4.1.2 Polymers with 'automatic' regularity

Several classes of polymer are regular, in the sense that they potentially
have translational symmetry, as a simple consequence of the particular
polymerisation process used in their preparation. Important examples
are the condensation polymers. For example, there is only one possible
kind of chemical repeat unit in poly(ethylene terephthalate) (PET), viz.

The important consequence of the polycondensation reaction is that
the chain can only consist of $-(OCH_2CH_2O)-$ units joined to
$-(CO)-\bigcirc-(CO)-$ units in an alternating sequence. The single bonds
along the backbone of the molecule are not collinear as shown in the
chemical formula above, but it is nevertheless possible for the chain to
acquire translational symmetry by suitable rotations around them and,
for this molecule, the chemical and translational repeat units are the
same. Another example is a polyamide such as nylon-6,6 (Bri-nylon)

Again, the important thing is that the condensation process ensures that
this unit repeats itself always the same way round as the chain is traversed
from one end to the other.

Each of these condensation polymers can have only one possible *configuration*, i.e. arrangement in which the atoms are joined together, but can
take up many different *conformations*. To change the configuration would
require the breaking and reforming of bonds. To change the conformation

merely requires rotation around single bonds. Addition polymers can have types of irregularity that are not possible for these simple condensation polymers produced from symmetric divalent precursors. The irregularities that can occur for vinyl and diene polymers are considered in the next two sections.

4.1.3 Vinyl polymers and tacticity

Vinyl polymers are addition polymers that have the general formula $(CHX-CH_2)_n$, where X may be any group and is, for example, a chlorine atom for poly(vinyl chloride) (PVC). Slightly more complicated polymers are those such as poly(methyl methacrylate) (PMMA), which has the formula $(CXY-CH_2)_n$, where X is a methyl group (CH_3) and Y is the methacrylate group (($CO)OCH_3$) (see fig. 1.2). These polymers are sometimes also classed as vinyl polymers. Both types of polymer may have two principal kinds of irregularity or *configurational isomerism*:

(i)   the monomer units may add *head-to-head* or *tail-to-tail* and
(ii) the chains need not exhibit a regular *tacticity*.

(i) *Head-to-head, etc., arrangements*
The three possible ways for adjacent monomer units of a vinyl polymer to join are

$$-(CHX-CH_2)-(CHX-CH_2)- \qquad \text{head-to-tail}$$

$$-(CHX-CH_2)-(CH_2-CHX)- \qquad \text{tail-to-tail}$$

$$-(CH_2-CHX)-(CHX-CH_2)- \qquad \text{head-to-head}$$

The head-to-tail arrangement predominates for most vinyl polymers prepared by the normal methods, so it is assumed in the following sections that only head-to-tail structures are present unless stated otherwise. This leads to chains that are chemically regular in the sense discussed in section 4.1.1.

(ii) *Tacticity*
Figure 4.2 illustrates vinyl chains that are *isotactic, syndiotactic* and *atactic*. When considering the tacticity of a vinyl polymer, imagine the carbon backbone in the *planar zigzag* conformation and imagine looking at this plane edge on, as shown in figs 4.2(b)–(d). Then if all X units are

on the *same side* of the plane the polymer is *isotactic*;
alternately on opposite sides the polymer is *syndiotactic*;
randomly on either side of the plane the polymer is *atactic*.

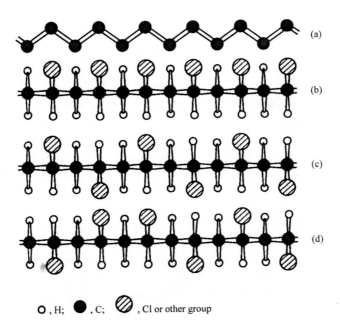

Fig. 4.2 Imagined planar zigzag conformations of vinyl polymers: (a) the carbon backbone, (b) a regular isotactic configuration, (c) a regular syndiotactic configuration and (d) a random atactic configuration. The planar zigzag conformation might not be possible for the real chain because of steric hindrance. (© Cambridge University Press 1989.)

○, H; ●, C; ⊘, Cl or other group

More generally, isotactic means that all repeat units have the same *handedness* and syndiotactic means that the handedness alternates along the chain. Alternate *chemical* repeat units along the syndiotactic chain are mirror images of each other. In the translationally symmetric conformations shown in fig. 4.2 the translational repeat unit for the syndiotactic chain is two chemical repeat units, whereas it is only one for the isotactic chain.

A particular state of tacticity is a particular *configuration* of the molecule and cannot be changed without breaking and reforming bonds and, at ordinary temperatures, there is not enough thermal energy for this to happen. Rotations around bonds produce only different *conformations*. A vinyl polymer is therefore unlikely to be appreciably crystalline unless it is substantially either isotactic or syndiotactic; the atactic chain cannot get into a state in which it has translational symmetry.

For example, commercial PVC is usually of rather poor crystallinity ($\sim 10\%$) because the probability of syndiotactic placement of any one —$CHCl$—CH_2— group with respect to the preceding one along the chain is only about 0.53; the polymer is almost atactic. PMMA, which has bulky side groups and is atactic in the commercial form, is non-crystalline, which is why it forms a very good polymeric glass; there are no crystallites to scatter the light. In poly(vinyl alcohol), $(CHOH—CH_2)_n$, the side group, —OH, is very small and the all-*trans* planar zigzag can exist, so the chains can crystallise, even though they are atactic. It is often possible to produce vinyl polymers of fairly regular tacticity, either experimentally or commercially, and these can crystallise, which gives them very different physical properties from those of the non-crystalline types.

The following techniques (and others) can give information about tacticity.

(a) *Measurement of (i) crystallinity (ii) melting point.*
Both increase with increasing regularity, because this leads not only to higher crystallinity but generally also to the existence of larger crystallites, which have a higher melting point than do smaller ones (see section 5.6.1). Both are indirect measures of tacticity and need to be calibrated against samples of known tacticity prepared physically in the same way.

(b) *Infrared (IR) and Raman spectroscopy.*
The principle behind these techniques is that certain modes of vibration can occur only for either the syndiotactic or the isotactic version of a particular polymer. Figure 4.3 shows the IR spectra of isotactic and two forms of syndiotactic polypropylene which illustrate this. The spectra are of oriented samples and the reason for the differences between the spectra for the two directions of polarisation is explained in chapter 11. They are shown here merely to illustrate how different the spectra for the different tacticities can be.

For a simple unoriented mixture of two tactically pure samples of a vinyl polymer, measurement of the ratio of the IR absorbances or Raman intensities of two modes, one specific for isotactic and one for syndiotactic material, would be a measure of the ratio of the isotactic and syndiotactic contents. Even in this ideal case the method would require calibration with known mixtures of the two types of polymer, because it is not possible to predict how the absolute values of the absorbances or Raman intensities depend on concentration, but only that they are proportional to the concentration.

The problem is further complicated by two factors. The first is that it is not usually a mixture of the two types of polymer that is of interest, but rather a polymer in which there is a certain probability of a syndiotactic or isotactic placement, so that not only may there be junction points between syndiotactic and isotactic stretches of chains but also there may be essentially atactic stretches.

These factors lead to broadening of the absorption or scattering peaks and to the presence of extra peaks. The configurational disorder also leads to conformational disorder, which introduces further new vibrational modes and takes intensity away from others. The problem then becomes very complicated, so that essentially empirical methods are often used. For instance the ratio of the absorbances or intensities at two points in the spectrum that do not necessarily coincide with obvious peaks may be found to correlate with the tacticity.

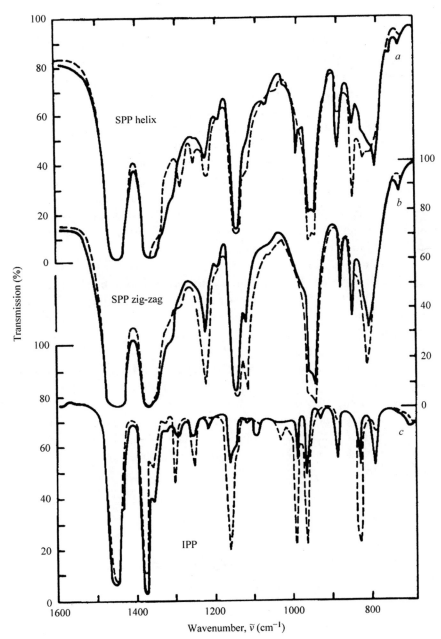

Fig. 4.3 The polarised infrared spectra of various forms of oriented polypropylene: (a) syndiotactic helix, (b) syndiotactic planar zigzag and (c) isotactic; ——— indicates the perpendicular and – – – the parallel polarised spectra. (Reproduced by permission from Tadokoro, H. *et al.*, *Reps. Prog. Polymer Phys. Jap.* **9**, 181 (1966).)

(c) *High-resolution NMR spectroscopy.*

In principle this is a method of determining tacticity absolutely. As indicated in section 2.7, under suitable conditions the strengths of the signals in the various peaks of an NMR spectrum are directly proportional to the numbers of ^{13}C, ^1H or other nuclei in particular local chain environments. In a vinyl polymer there are two kinds of carbon nuclei that are chemically distinct, those of the *α-carbon atoms* of the CHX group and those of the *β-carbon atoms* of the CH_2 group. As an example of the method, the use of the spectrum of the α-carbon atoms is considered.

Each α-carbon may be considered to be at the centre of a *triad* of CHX groups and there are three possible types of triad, as shown in fig. 4.4. Each triad may itself be considered to be made up of two *dyads*, each of which may be *mesic* (m) or *racemic* (r), where a mesic dyad consists of two —CHX—CH_2— units with the same handedness and a racemic dyad consists of two —CHX—CH_2— units with opposite handednesses. The resonance frequencies for ^{13}C nuclei in the three different types of location are slightly different, so that three peaks are observed in the region of the spectrum due to the α-carbon nuclei.

An example is shown in fig. 4.5, in which each of the three peaks is in fact further broadened or split due to the sensitivity of the chemical shifts to *pentad* structure. The area under each peak is proportional to the number of carbon atoms in the corresponding location, so that the tacticity, i.e. the probability P_r of a syndiotactic placement, can be calculated from the ratios of the peak areas (see example 4.1).

NMR can only be used to determine tacticity routinely for polymers that are totally soluble. If a polymer is highly crystalline it may be difficult to dissolve it completely and, because the crystalline regions are the more highly ordered regions and also the regions most difficult to dissolve, any measurement of tacticity in solution is likely to underestimate the true degree of order in the polymer. Some information can, however, be obtained from solid-state NMR.

Fig. 4.4 α-carbon triads for a vinyl polymer: (a) isotactic, or mm; (b) syndiotactic, or rr, and (c) mr, where m = mesic and r = racemic.

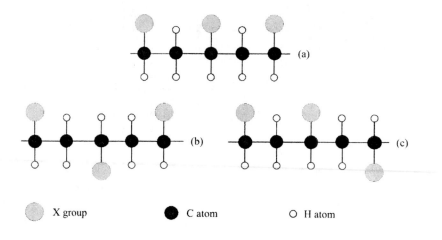

X group   ● C atom   ○ H atom

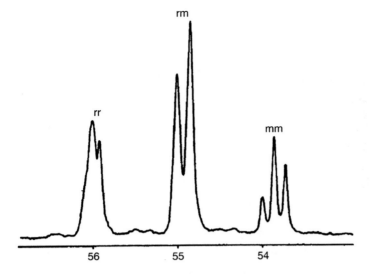

Fig. 4.5 The ^{13}C NMR spectrum in the α-carbon region for a PVC sample dissolved in *o*-dichlorobenzene. The frequencies are expressed in ppm from a standard reference frequency. (Reproduced from King, J., Bower, D. I., Maddams, W. F. and Pyszora, H., *Makromol. Chemie* **184**, 879 (1983).)

Polymers of the form $(CHX—CHY)_n$ have more possible kinds of regular tacticities than do vinyl polymers and polymers of the type $(CH_2—CXY)_n$, as discussed, for instance, in reference (1) of section 4.5.

Example 4.1

If P_r is the probability of a racemic placement for a growing vinyl polymer chain, deduce an expression for the ratios of the intensities of the ^{13}C NMR peaks for the rr, mr and mm α-carbon triads and evaluate the ratios for $P_r = 0.6$. Assume that *Bernoullian statistics* apply, i.e. that the probability of a racemic placement is independent of the nature of the previous placements. (This is not necessarily true for a real system.)

Solution

Every α-C atom at the centre of an rr triad arises as the result of at least two sequential r placements and every two sequential r placements produce one α-C atom. The probability P_{rr} that any α-C atom lies at the centre of an rr triad is therefore P_r^2. Similarly, the probability P_{mm} for an mm triad is $(1 - P_r)^2$ and the probability P_{mr} for an mr triad appears to be $P_r(1 - P_r)$. However, mr and rm triads give rise to the same frequency and are really the same, so that $P_{mr} = 2P_r(1 - P_r)$. The ratios of the intensities are in the ratios of the probabilities (which, of course, add up to 1) and are thus given by $P_{rr} : P_{mr} : P_{mm} \equiv P_r^2 : 2P_r(1 - P_r) : (1 - P_r)^2$, which for $P_r = 0.6$ become $0.36 : 0.48 : 0.16$.

4.1.4 Polydienes

A *diene* contains two double bonds and during polymerisation only one of them opens, which leads to the formation of polymers of three distinctly different kinds. The three forms for polybutadiene, obtained from butadiene, $H_2C\!=\!CH\!-\!CH\!=\!CH_2$, are shown in fig. 4.6. The vinyl *1,2* form, for which the double bond is in the side group X, can have any of the types of tacticity discussed above. The two *1,4* types, for which the double bond is in the chain backbone, illustrate a form of *configurational isomerism*. Rotation around a double bond is not possible, so these two forms are distinct. In the *cis* form the two bonds that join the unit shown to the rest of the chain are on the same side of a line passing through the doubly bonded carbon atoms of the unit, whereas for the *trans* form they are on the opposite side of that line.

When the *trans* form of a polydiene is stretched out into its straightest form the chain is rather compact laterally and this form tends to be more crystalline than the *cis* form, which has a bulkier chain. The *cis* forms of the polydienes are in fact useful as synthetic rubbers. *Natural rubber*, which comes from the sap of the tree *Hevea braziliensis*, is the *cis* form of *1,4-*polyisoprene, which differs from polybutadiene only by having a CH_3 group on the $_2C$ carbon atom of every repeat unit instead of a hydrogen atom. The corresponding *trans* form is *gutta percha*, which comes from the sap of a different tree. Natural rubber has a crystal melting point of 28 °C, but supercools and is amorphous at room temperature, and it is elastic and strong when it is vulcanised (see section 6.4.1). In contrast, gutta percha has a crystal melting point of 75 °C and is hard and tough when it is vulcanised, so that it was at one time used to form the core of golf balls. The lower melting point of the *cis* form is due to the less compact nature of its chains, which reduces the attractive forces between them because they can make fewer close contacts.

4.1.5 Helical molecules

A regular helical chain may be defined as a chain that possesses a *screw axis*, $C_n^{(k)}$, where $n > 1$. The operation corresponding to this is a rotation

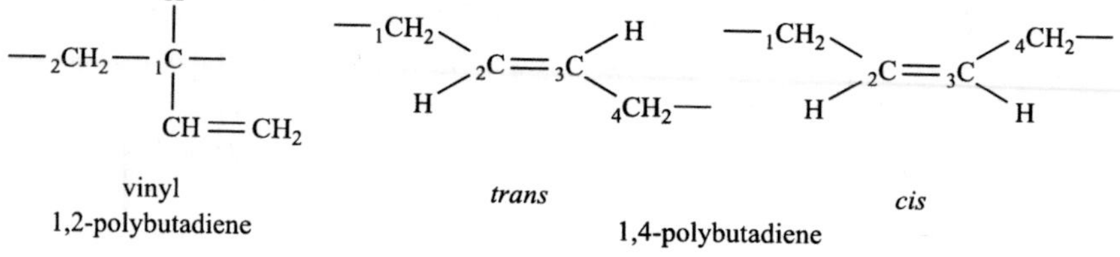

Fig. 4.6 The three possible forms of polybutadiene.

vinyl
1,2-polybutadiene

trans

cis

1,4-polybutadiene

through $2\pi k/n$ around the chain axis followed by translation through a distance l_t/n parallel to the axis, where l_t is the translational repeat length of the chain. The helix can thus be described as consisting of n *physical repeat units* in k turns, which constitute one translational repeat, and it is called an n_k *helix*. The physical repeat unit is often the same as the chemical repeat unit but need not be. The planar zigzag polyethylene chain possesses a $C_2^{(1)}$ axis and may thus be considered to be a 2_1 helix, whereas the conformation often assumed by isotactic vinyl polymers $(CHX\text{---}CH_2)_n$ is the 3_1 helix illustrated in fig. 4.7(a). This structure is formed by alternating *trans* and *gauche* bonds in the backbone, with all the *gauche* bonds having the same handedness. It places the X groups far away from each other and avoids the steric hindrance to the formation of the planar zigzag conformation (see problem 4.1).

A rather more complicated helix is formed by the polytetrafluoroethylene (PTFE) chain, $(CF_2\text{---}CF_2)_n$. At first sight a planar zigzag chain similar to that of polyethylene would be expected, but the F atom is larger than the H atom and there is slight steric hindrance to that form. The 'plane' of the zigzag actually twists slightly along its length and the CCC bond angle opens slightly to give rise to a helical form that, below $19\,^\circ C$, contains 13

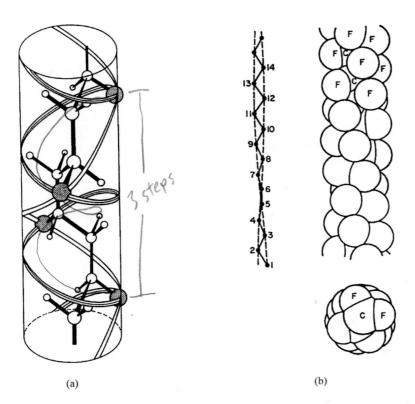

(a)   (b)

Fig. 4.7 (a) The threefold helix for an isotactic vinyl polymer. The X group of the structural unit $(CHX\text{---}CH_2)$ is shown shaded. The carbon backbone forms a right-handed helix and the right-handed 3_1 and the left-handed 3_2 helices are indicated by the 'ribbons'. (b) The 13_6 helix of PTFE, showing the twisted backbone, and a space-filling model viewed from the side and viewed along the chain axis. ((a) reproduced from *The Vibrational Spectroscopy of Polymers* by D. I. Bower and W. F. Maddams. © Cambridge University Press 1989; (b) reprinted by permission of Macmillan Magazines Ltd.)

CF_2 units in one translational repeat unit, as shown in fig. 4.7(b). This repeat unit corresponds to a twist of the backbone through $180°$. In polyethylene a rotation of $180°$ is required to bring one CH_2 group into coincidence with the previous one but, because of the twist, the angle required for PTFE to bring one CF_2 group into coincidence with the next is $180°$ less $180°/13$, or $12\pi/13$, so that the structure is a 13_6 helix.

In specifying a helix, n and k should have no common factor, otherwise more than one translational repeat unit is being referred to. The value of k is also arbitrary to the extent that any integral multiple pn of n may be added to it or subtracted from it, because this will simply insert p extra whole turns per physical repeat unit to give an identical molecule. Thus k should always be less than n to exclude such redundant turns. For greatest simplicity of description $k < n/2$ (or $k = n/2$ for n even), because a helix consisting of n physical repeat units in k clockwise turns can equally well be described as consisting of n physical repeat units in $n - k$ anticlockwise turns. This is illustrated for $n = 3$ in fig. 4.7(a). The helix with $k < n/2$ is not, however, always of the same handedness, or *chirality* as the helix described by the bonds in the backbone of the molecule, as it is in fig. 4.7(a), and the description with the higher value of k may then be preferred. The two helices can, however, differ in chirality only if there is more than one backbone bond per physical repeat unit.

4.2 The determination of crystal structures by X-ray diffraction

4.2.1 Introduction

The determination of the crystal structure of a substance with small molecules is nowadays almost a mechanical procedure; for a polymer the problem is more difficult. The reason for the difficulty lies in the fact that large (a few tenths of a millimetre) fairly perfect single crystals can readily be obtained for most small-molecule compounds. As discussed in more detail in the next chapter, the best that can be obtained for a polymer, apart from a few very special cases, is a piece of material in which a mass of crystallites is embedded in a matrix of amorphous material.

In order to obtain the maximum amount of information about the crystal structure it is necessary to align the crystallites, which can be done by methods described in detail in chapter 10. It is sufficient to note here that suitable orientation is often produced by stretching a fibre of the polymer. In the simplest cases the chain axis of each crystallite, which is designated the c-axis, becomes aligned towards the fibre axis, but there is no preferred orientation of the other two axes around the c-axis. From such a sample a *fibre pattern* can be obtained, of the type shown in fig. 3.10.

In order to determine the crystal structure it is necessary to determine the positions and intensities of all the spots in the fibre pattern. The positions of the spots provide information from which the shape and dimensions of the unit cell can be calculated and the intensities provide information about the contents of the unit cell. The following section considers the relationship between the fibre pattern and the unit cell.

4.2.2 Fibre patterns and the unit cell

Before considering fibre patterns it is useful to consider the less informative pattern obtained from a randomly oriented crystalline polymer, such as that shown in fig. 3.9(b). For every possible set of crystal planes, some crystallites are oriented so that the angle θ between the incident X-ray beam and the planes satisfies Bragg's law, $2d \sin \theta = n\lambda$ (see section 2.5.1). Because the crystallites are randomly oriented, the normals to such sets of correctly oriented planes are randomly distributed around the direction of the incident X-ray beam, so that the scattering at angle 2θ will form a cone of scattered rays that intersects a plane placed normal to the incident X-ray beam in a circle, as shown in fig. 4.8. This is the origin of the 'powder' rings seen. The angles 2θ are easily measured, so the values of d for all planes can be calculated, because the value of λ is known. If the randomly oriented crystallites were replaced by a single crystal, it would be unlikely that any set of planes would be correctly oriented to satisfy Bragg's law, so no diffraction would be seen.

A highly oriented fibre consists of a very large number of crystallites, which all have one particular crystallographic direction oriented almost parallel to the fibre axis (usually the chain axis, the c-axis) and the remaining directions are oriented randomly around this direction. Assuming that the axis of the fibre is normal to the incident X-ray beam, the scattering expected is therefore almost exactly the same as that which would be observed from a single crystal with its c-axis parallel to the fibre axis if this crystal were rotated continuously around the c-axis during the exposure of the X-rays.

Imagine a particular set of planes in such a single crystal and consider what happens as the crystal is rotated about the c-axis, which is assumed to be vertical. During the rotation the angle θ changes continuously. Imagine

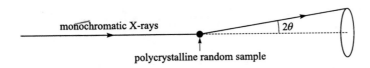

monochromatic X-rays 2θ

polycrystalline random sample

Fig. 4.8 Formation of a powder ring from a particular set of planes.

starting from the position where the normal to the set of planes lies in the plane containing the incident X-ray beam and the fibre axis. As the rotation takes place continuously in one direction from this position the angle θ changes until at some angle of rotation, say α, Bragg's law is satisfied. If the crystal is held in this position a diffraction spot will be seen. The important point is that this spot must lie on the corresponding circle that would have been seen if a powder sample had been used, because the single crystal can be considered as just one of all the possible crystallites that would be present in the powder. Now imagine rotating the crystal around the fibre axis in the opposite direction from the starting point. It is obvious that, at the same angle of rotation α in this direction, Bragg's law must be satisfied, so a crystal in this orientation will give another spot on the powder ring. It should also be clear that the line joining the two spots will be horizontal. Assume that this horizontal line lies above the X-ray beam. Now imagine returning to the starting position and first turning the crystal by 180° around the vertical axis from this position. The same argument can be used from this new starting position as from the old one, so that two more spots will appear, this time lower than the X-ray beam.

The conclusion is that, for a rotating crystal, most sets of crystal planes give rise to four spots lying at particular points on the imaginary circle where the corresponding powder-pattern circle would have been. These points are symmetrically placed with respect to the plane that contains the incident X-ray beam and the rotation axis and to the plane that contains the incident X-ray beam and is normal to the rotation axis, as shown in fig. 4.9. A similar conclusion can be drawn for a stationary, highly oriented polymer fibre. Planes parallel to the rotation or fibre axis give rise to only two diffraction spots.

The radius of the (imaginary) powder circle and the positions of the four spots on it for a particular type of crystal plane depend on the indices of the planes, which are also the diffraction indices for the spots, as explained in section 2.5.1. The diffraction indices h, k and l are the numbers of

Fig. 4.9 The production of the four diffraction spots corresponding to a given set of planes for a rotation pattern: (a) the starting position, where θ_0 is greater than the Bragg angle; and (b) the location of the four diffraction spots corresponding to a given set of planes.

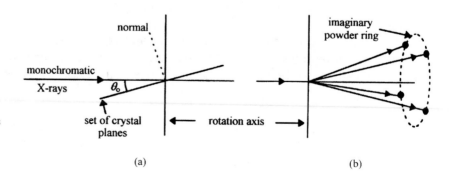

(a)   (b)

wavelengths path difference between waves scattered from adjacent points along the a-, b- and c- axes, respectively, of the crystal. The value of 2θ depends not only on the indices but also on the unit-cell dimensions a, b and c and on the angles between the axes. As explained in section 2.5.1, this relationship is in general rather complicated. It is, however, possible to see a very simple relationship between the value of c and the location in the fibre pattern of the diffraction spots for all sets of planes with a common value of the diffraction index l.

In an ideal highly oriented fibre the c-axes of all the crystallites are parallel to the fibre axis. Figure 4.10 illustrates the scattering from a single column of unit cells separated by the distance c along the fibre axis, each unit cell being represented by a large dot. Strong scattering will be observed at the angle ϕ shown only if

$$c \sin \phi = l\lambda \quad \text{(l an integer)}$$

where λ is the wavelength of the X-rays, because the path difference for scattering from adjacent unit cells along the c-axis must be an integral number of wavelengths. The symbol l has been used for this integer because it is clear from the definitions of the diffraction indices that this must be the value of the third index of any set of planes that scatter at this angle ϕ. A single column of unit cells would, however, scatter everywhere on a vertical cone of semi-angle $\pi/2 - \phi$, with the column of cells as its axis. If the column were surrounded by a cylindrical film the cone would intersect the film in a horizontal circle This intersection corresponds to the lth *layer line*. On a flat film normal to the incident X-ray beam the layer lines become hyperbolae, whereas on a cylindrical film the powder rings are no longer circular, but flattened in the horizontal direction (see problem 4.4).

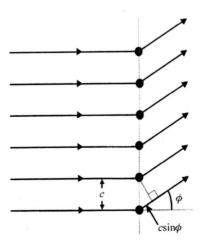

Fig. 4.10 Scattering from a single column of unit cells parallel to the fibre axis.

The reason why scattering is not observed at all points on a layer line is that equation (4.1) expresses only one of the three conditions necessary for strong diffraction, namely that there should be whole numbers h, k and l of wavelengths path difference for scattering from adjacent unit cells along each of the three axes a, b and c, respectively. Each powder ring corresponds to a particular set of values h, k and l and it follows that diffraction spots can be seen only at those places where the lth layer line crosses the position where a powder circle corresponding to the same value of l would have been seen, as shown schematically in fig. 4.11. Each such circle and pair of layer lines for $\pm l$ thus gives rise to the four spots previously shown to arise from any particular set of planes. Because there can be various sets of planes with different values of h and k but the same value of l there will be several pairs of spots on each layer line. The layer line for $l = 0$ is called the *equator* and the normal to this through the point where the incident X-ray beam would strike the film is called the *meridian*.

Figure 4.12(a) shows a fibre pattern obtained from a sample of poly-(vinylidene chloride) and fig. 4.12(b) shows a rotation photograph obtained from a single crystal of gypsum. All the orientations of crystallites obtained sequentially by rotating a single crystal are present simultaneously for the fibre. The spots on the rotation photograph are much sharper than those in the fibre pattern. The reason for this is that the orientation of the c-axes within the fibre is not perfect, so that the spots are broadened into small arcs along the directions of the powder-pattern rings. The crystallites in the polymer are also rather small, which leads to broadening in the radial direction.

There are no spots on the meridian for perfect fibre alignment; spots occur in pairs symmetrically on either side, but it is easy to determine the c-axis repeat distance c from a knowledge of where the layer lines cut the meridian, using equation (4.1). For imperfect alignment, *meridional spots*

Fig. 4.11 A schematic diagram showing the relationship among layer lines, powder rings and diffraction spots in a fibre diagram. For simplicity the layer lines are shown straight and the powder rings as circles, but see the discussion in the text.

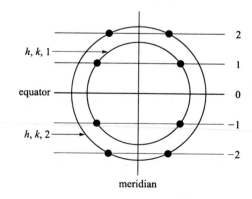

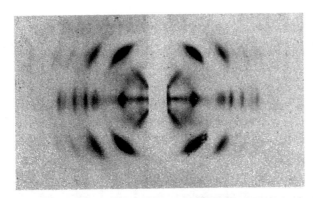

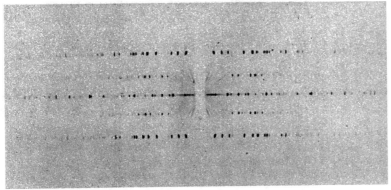

Fig. 4.12 Upper: a fibre pattern for PVDC; and lower: a rotation photograph for gypsum. (Adapted by permission of Oxford University Press.)

may occur as the remnants of powder rings and they correspond to planes normal or nearly normal to the axis which preferentially aligns (usually the *c*-axis). Thus (00*l*) spots are often seen.

It can be shown (see problem 4.5) that, if the *ab* plane is perpendicular to the *c*-axis, $\phi > 2\theta$ unless $\phi = \theta = 0$, where ϕ is the angle of the *l*th layer line and 2θ is the angle of the (00*l*) powder ring. When there is sufficient misalignment, the layer lines are effectively broadened and can intersect the (imaginary) powder rings on the meridian If meridional spots of this type occur, Bragg's equation can be used to find d_{001}:

$$2d_{001} \sin \theta = l\lambda \qquad \text{for a meridional spot of order } l \qquad (4.2)$$

The difference between this equation and equation (4.1), which refers to the *l*th layer line, should be noted carefully. If meridional spots do not occur, the sample can be tilted to observe them.

It can be concluded that, if layer lines or appropriate meridional spots occur, the *c*-axis repeat distance *c* or d_{001} can be determined by the use of equation (4.1) or equation (4.2). If the *c*-axis is normal to the *ab* plane, $c = d_{001}$. The other dimensions and the angles of the unit cell can also be determined from measurements of spot positions only. If the unit cell is orthorhombic, i.e. if $\alpha = \beta = \gamma = 90°$, the determination is straight-

---

Example 4.2

An X-ray diffraction pattern was obtained from a very highly oriented fibre of a polymer using a cylindrical camera of radius 5.00 cm and X-rays of wavelength 0.154 nm, with the fibre coincident with the axis of the camera. First-order layer lines were observed at distances 3.80 cm above and below the equator. Determine the c-axis spacing.

Solution

The angle of the first-order layer line ($l = 1$) is given by $\phi = \tan^{-1}(r/h)$, where r is the radius of the camera and h is the height of the layer line above or below the equator. Equation (4.1) shows that the c-axis spacing is given by $c = l\lambda/\sin\phi$. Thus $c = l\lambda/\sin[\tan^{-1}(r/h)]$. Substitution of the given data yields $c = 0.255$ nm.

---

forward. For the simplest type of orthorhombic unit cell the spots on the equator nearest to the meridian will be the (100) or (010) spots, from which the corresponding value of a or b can be determined by the use of Bragg's law. The next closest spot will generally be the (110) spot and the corresponding plane spacing deduced from Bragg's law can be used with an already known value of a or b to find the other or as an extra check on the accuracy of the values of a and b. For some types of crystal there are *missing equatorial spots*, which implies that the dimensions in one or both directions perpendicular to the c-axis are twice those suggested by the lowest equatorial reflections (see problem 4.6).

As pointed out in section 2.5.1, where the reciprocal lattice is defined, it is possible to determine the shape and dimensions of any real lattice from the corresponding information about the reciprocal lattice. The spots on the lth layer line correspond to reciprocal-lattice points lying on the reciprocal-lattice plane containing all points (hkl) for different h and k. The fibre diagram is in fact similar to what would be obtained if the following imaginary experiment were performed. Place the origin of a suitably scaled version of the reciprocal-lattice at the point where the incident X-ray beam would strike the film, with the normal to the planes of constant l parallel to the fibre axis and therefore to the meridian of the fibre pattern. Rotate the reciprocal lattice around the meridian and mark a point on the film every time a lattice point intersects it. To understand the difference between this pattern and a real fibre pattern it is useful to consider fig. 4.13.

In this figure O represents the point of intersection of the incident X-ray beam and the fibre and R is the radius of a sphere, the *Ewald sphere*, drawn around O, where R is equal to the radius of the camera if a cylindrical camera is used or to the fibre–film distance if a flat film is used. The point C is the centre of the fibre pattern. Imagine that a set of planes (hkl) is at the

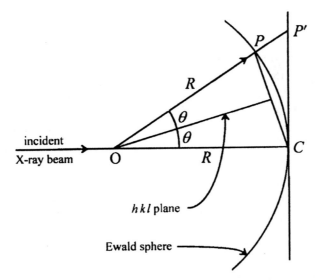

Bragg angle θ to the incident X-ray beam and that the normal to the planes lies in the plane of the diagram. OP is then the direction of the scattered beam and the length CP is equal to $2R\sin\theta = R\lambda/d_{hkl}$, i.e. to $R\lambda$ times the length of the reciprocal-lattice vector corresponding to the (*hkl*) planes. It is also clearly parallel to the normal to these planes; therefore, P is the point at which the reciprocal-lattice point (*hkl*) would have intersected the sphere if a reciprocal lattice scaled by the factor R had been rotated around the meridian. The intersection P' of the extended line OP with the film gives the position of the corresponding spot in the fibre pattern. In the diagram a flat film is assumed. It is a simple matter of geometry to relate P' to P and hence to calculate the length of the corresponding reciprocal-lattice vector. For small angles θ the distances measured directly on the film are approximately proportional to the lengths of the reciprocal-lattice vectors, but this is clearly not so for larger angles.

It is possible from measurements of a limited number of reciprocal-lattice vectors and observations of systematic absences of 'reflections' to deduce not only the shape and dimensions of the unit cell, but also the complete symmetry of the crystal structure. The determination of the complete structure now requires the contents of the unit cell to be deduced from the intensities of the 'reflections'. These are usually determined by using diffractometers rather than film to record the diffraction. A diffractometer is usually a device that allows the recording of the intensity of scattering in any particular direction in space. Modern types, using CCD arrays, can determine the intensity over a range of directions for one setting of the instrument. This greatly speeds up the collection of data but leads to some complication in terms of the need to calibrate the different pixels of the CCD array and to perform more complex calculations to relate the positions on the array for any setting to the reciprocal lattice.

For non-polymeric materials it is possible to get very good X-ray diffraction patterns and the intensities can be determined for many well-defined spots. The determination of the contents of the unit cell is then essentially a mechanical process using modern computational techniques. Because of the often rather poor quality of the fibre diffraction patterns obtained for polymers, the number of spots observed may be rather small and the spots are often quite broad and overlapping. It is thus difficult or impossible to determine the individual intensities accurately. Fortunately, the nature of polymer molecules themselves leads to some simple methods for deducing the likely contents of the unit cell, as discussed in the next section.

4.2.3 Actual chain conformations and crystal structures

If the crystal structure of a polymer is known, it is possible to calculate the expected intensities of all the diffraction spots by using the known scattering powers of its various atoms. If, therefore, it is possible to predict the structure from the measured unit-cell dimensions and some assumed form for the polymer chains within the unit cell, it is possible to compare a predicted set of intensities with the observed diffraction pattern. If the structure assumed is close to the actual one, only minor adjustments will be needed in order to get a more perfect fit between the predicted and observed scattering patterns and thus to determine the structure. Otherwise a new prediction must be made.

Crystallisation takes place only if the Gibbs free energy $U + PV - TS$ is lowered. This is more likely if the chains can pack closely together to give many van der Waals interactions to lower the internal energy, U, or if specific intermolecular bonding, usually hydrogen-bonding, can take place. A prediction of the likely shape of individual polymer molecules within the crystal can, however, be made by realising that, in the absence of intermolecular hydrogen-bonding, or other strong intermolecular bonding, the shape of the molecule in the crystal is likely to be very similar to one that would give the isolated molecule a low energy.

Two important types of polymer chain are the planar zigzag and the helix, but whatever form the chains take in crystallites, they must be straight on a crystallite-size scale and the straight chains must pack side by side parallel to each other in the crystal. The ideas of 'standard' bond lengths, angles and orientations around bonds discussed in chapter 3 can be used to predict likely possible low-energy model chain conformations that exhibit translational symmetry (see example 4.3). As discussed in the

Example 4.3

By assuming the 'standard' bond lengths, angles and preferred orientations around bonds, estimate the most likely chain repeat length for polyethylene.

Solution

The 'standard' C—C bond length l is 0.154 nm, the 'standard' CCC bond angle is tetrahedral, 109.5°, and the planar zigzag structure is the most likely conformation. The

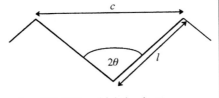

value of c is given by $c = 2l \sin\theta$, where $\theta = 109.5°/2$, which leads to $c = 0.251$ nm.

This value is in fact <2% different from the observed value. The X-ray patterns can be well accounted for by a bond angle of 112° and a C—C bond length of 0.153 nm.

previous section, one of the easiest things to determine experimentally is the dimension of the crystal unit cell in the chain-axis direction; only the intensities predicted on the basis of models that correspond approximately to the correct value for this length need be compared with the X-ray data.

For helical molecules the layer lines vary in intensity; those that correspond to scattering for which waves from adjacent physical repeat units are all in phase are generally stronger than the others, so that the length of the physical repeat unit and the number n of these in the helix repeat length can be deduced. The precise form of the variation of layer-line intensity gives information about the way that the units are arranged in the helix, i.e. in how many turns the n units occur. As discussed in section 4.1.5, a helix with n physical repeat units in k turns is called an n_k helix.

Information about the lateral packing is contained in the equatorial reflections, as has already been considered for the orthorhombic cell, and the complete structure can often be postulated on the basis of this information plus the symmetry of the crystal structure and the information about the conformation derived in the way just described. This postulated structure can then be used as the starting point for an energy-minimisation program in a computerised modelling system which takes account of both intermolecular and intramolecular interactions in finding the structure corresponding to the energy minimum. The intensities of all the spots can then be predicted and compared with the observed X-ray patterns. The comparison is often done not with the fibre pattern, but with a section drawn through a powder pattern obtained from a random sample, which contains information from all lattice planes. The refined model then

Fig. 4.14 A comparison of
the observed powder ring
profile for isotactic 1,4-*cis*-
poly(2-methyl-1,3-
pentadiene) with that
calculated from the
structure with minimum
energy: (a) the observed
profile and (b) the
calculated profile, with
peaks broadened to give
the best fit. (Reprinted with
permission from the
American Chemical
Society.)

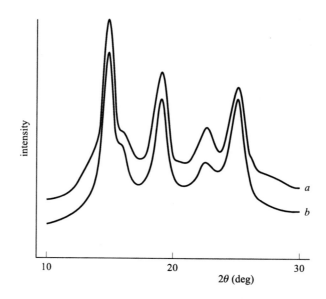

predicts the radius and intensity for each ring. Figure 4.14 shows such a
comparison for the β crystal form of isotactic 1,4-*cis*-poly(2-methyl-1,3-
pentadiene) (β-iPMPD), $-(C(CH_3)CH=C(CH_3)CH)_n$, for which peak
broadening has been applied to the calculated diffraction peaks in order
to obtain a best fit.

The lateral packing is often approximately regular hexagonal, because
the polymer chains are often rod-like, particularly when the chains are
helical. This can sometimes be seen fairly directly in the equatorial reflections, where two or three strong reflections occur at almost the same angle.
Polyethylene provides a good example of pseudo-hexagonal packing (see
problem 4.7) and a fibre pattern for a very highly oriented sample is shown
in fig. 4.15.

Fig. 4.15 A fibre pattern
from a highly oriented
sample of polyethylene.
Note the two strong
equatorial reflections very
close together just over a
third of the way out from
the centre of the pattern
shown. They are, from the
centre outwards, the (110)
and (200) reflections,
respectively. (Courtesy of
Dr A. P. Unwin.)

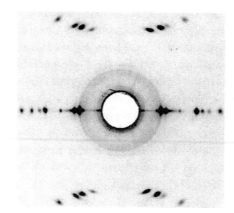

4.3 Information about crystal structures from other methods

Although X-ray scattering is the principal method used for obtaining information about crystal structures, other methods can provide information that may either confirm conclusions derived from X-ray studies or help to resolve ambiguities that arise from the difficulties of obtaining well-defined fibre patterns. NMR spectroscopy can give information about the strength of dipole–dipole coupling between nuclear spins, which depends on the distance between the spins and is thus capable of providing information about the separations of nuclei and hence about molecular conformation. Some two-dimensional studies have been performed, for instance, on highly oriented polyethylene to give $^{13}C/^1H$ dipolar coupling along one axis and chemical shift along the other, but the method has not been used widely. It is limited by the fact that the dipole–dipole interaction is of very short range, so that only very local information can be obtained. Infrared or Raman spectroscopy, on the other hand, can provide information not only about local conformational structure but also about the longer range conformation of the chain and the number of chains per unit cell.

The selection rules for vibrational spectroscopy state that, for a crystal, the only modes of vibration that can be active in either spectrum are those in which corresponding atoms in all unit cells vibrate in phase. If there is only one chain passing through each unit cell, corresponding atoms in every chain in the crystal must vibrate in phase for spectral activity to be observed. If, however, there is more than one chain passing through the unit cell, two new possibilities arise: (i) corresponding atoms in different chains within one unit cell can vibrate with different phases and (ii) the chains can vibrate bodily with respect to each other in various ways. The first of these possibilities leads to *correlation splitting* and the second to the observation of additional *lattice modes*.

A simple example of correlation splitting is provided by the vibrational spectrum of polyethylene. Figure 4.16 shows the splitting of the Raman-active CH$_2$ twisting mode as the temperature of the sample is cooled. In the twisting modes the H atoms move parallel to the chain axis and the two H atoms on any one C atom move in opposite directions. In the Raman-active mode all the H atoms on one side of the plane of the backbone move in phase. Figure 4.16(d) shows that, even when each chain vibrates in this way, there are two possible phase relationships between the vibrations of the two chains in the unit cell. The frequencies of these two vibrations are slightly different because of the van der Waals forces between the two chains; the difference becomes larger as the temperature of the sample is reduced because the chains move closer together. Correlation splitting can also be observed in the IR spectrum, in particular for the IR-active CH$_2$

Example 4.4

Obtain a general expression for the number of lattice modes for a polymer with m chains passing through the unit cell and hence derive the numbers for $m = 1$ and 2.

Solution

The number of degrees of freedom for m isolated rods is $6m$. $3m$ of these correspond to translational motion and $3m$ to rotational motion. In a crystal the rods cannot rotate around any direction perpendicular to their lengths without the crystal rotating as a whole. This removes $2m$ degrees of freedom. In addition, three of the degrees of freedom correspond to translation of the crystal as a whole. There remain $4m - 3$ degrees of freedom and therefore $4m - 3$ lattice modes; i.e. one for a single chain per unit cell and five for two chains per unit cell. Note that the single mode for one chain is a rotatory vibration, or *libration*, around the chain axis and that this does *not* correspond to a rotation of the crystal as a whole.

Fig. 4.16 The effect of temperature on the correlation splitting of the Raman-active CH₂-twisting doublet of polyethylene: (a) the sample at room temperature; (b) the sample at −196 °C; (c) the sample at −196 °C with increased spectral resolution and expanded wavenumber scale (⊢⊣ indicates 5 cm⁻¹ in each spectrum); and (d) forms of the vibrations. The + and − signs represent the simultaneous displacements up and down with respect to the plane of the diagram for one phase of the vibration. Only the two chains in one unit cell are shown. (Reproduced from *The Vibrational Spectroscopy of Polymers* by D. I. Bower and W. F. Maddams. © Cambridge University Press 1989.)

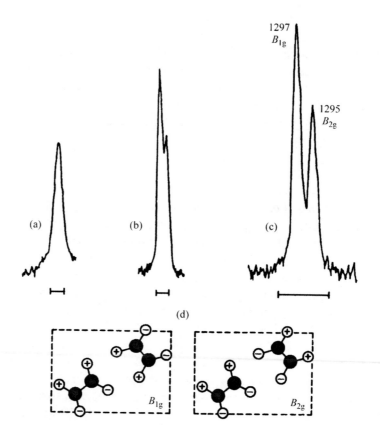

bending mode, which is observed at 720 and 731 cm^{-1}. The bending mode involves the periodic variation of the HCH angles.

For polyethylene, X-ray studies alone were sufficient to establish the crystal structure, but the gradual elucidation of the structure and spectral assignments of the low-temperature form of PTFE provides a good example of the complementary role to conventional crystallographic studies of polymers that vibrational spectroscopy can play. Early X-ray studies suggested that this form had a structure with only one chain per unit cell, whereas the interpretation of splittings observed for several peaks in the Raman spectrum suggested that there were two chains per unit cell and that there would be five lattice modes, all of which would be infrared-active. In addition, studies of the IR spectrum revealed a number of absorption peaks in the 30–90 cm^{-1} region, four of which could be assigned as lattice modes. More recent X-ray results have confirmed that there are indeed two chains per unit cell in the low-temperature form.

In the following section examples of the actual crystal structures of some common polymers are given.

4.4   Crystal structures of some common polymers

4.4.1   Polyethylene

Polyethylene is orthorhombic, with two chains per unit cell. The cell dimensions are $a = 0.742$, $b = 0.495$ and $c = 0.255$ nm. The angle between the plane of the zigzag backbone and the b-axis is approximately $45°$ and the planes of the two chains passing through the unit cell are at right angles to each other. The chains are almost close-packed, as discussed in section 4.2.3. The structure is shown in fig. 4.17.

4.4.2   Syndiotactic poly(vinyl chloride) (PVC)

Commercial PVC is very nearly atactic and therefore crystallises only to a very small extent. Syndiotactic PVC, which can be made only by a rather special polymerisation technique, crystallises with an orthorhombic unit cell. Like the unit cell of polyethylene, there are two chains passing through it, but this time they are oriented at $180°$ to each other, with the planes of both backbones parallel to the b-axis. The cell dimensions are $a = 1.026$, $b = 0.524$ and $c = 0.507$ nm. The structure is shown in fig. 4.18.

4.4.3   Poly(ethylene terephthalate) (PET)

PET is a polyester; indeed, the word polyester is often used in the textile trade to mean PET. The crystal structure is triclinic, i.e. none of the cell angles is a right angle, and there is only one chain per unit cell. The cell

Fig. 4.17 The crystal structure of polyethylene: (a) a perspective view of the chains and the unit cell and (b) a schematic plan view looking down the *c*-axis, with only the directions of the interatomic bonds shown. ((a) Reproduced by permission of Oxford University Press.)

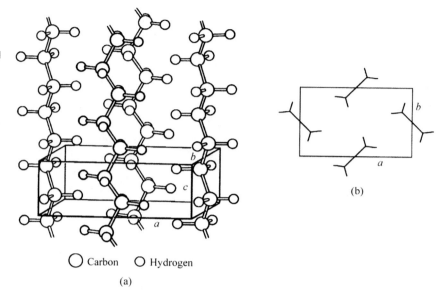

○ Carbon   ○ Hydrogen

(a)

Fig. 4.18 The crystal structure of syndiotactic poly(vinyl chloride). The unit cell contains two chains. For clarity the hydrogen atoms are not shown.

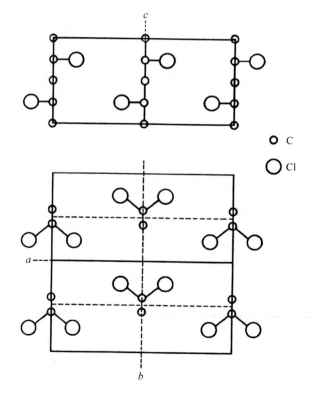

○ C

○ Cl

parameters are $a = 0.456$ nm, $b = 0.594$ nm, $c = 1.075$ nm, $\alpha = 98.5°$, $\beta = 118°$ and $\gamma = 112°$. The chain is nearly fully extended and the direction of *para*-disubstitution in the benzene ring makes an angle of about 19° with the c-axis. The planes of the rings are almost stacked on top of each other and the packing is good; the projections on one molecule fit hollows on the next. The structure is shown in fig. 4.19.

4.4.4 The nylons (polyamides)

The general formula for the repeat unit of a polyamide described as nylon-*n,m* is $-(NH(CH_2)_nNH.CO(CH_2)_{m-2}CO)-$. Particularly important examples are nylon-6,6, or poly(hexamethylene adipamide), and nylon-6,10. Related polyamides of slightly simpler structure have repeat units of the type $-(NH(CH_2)_5CO)-$, which is polycaprolactam, or nylon-6.

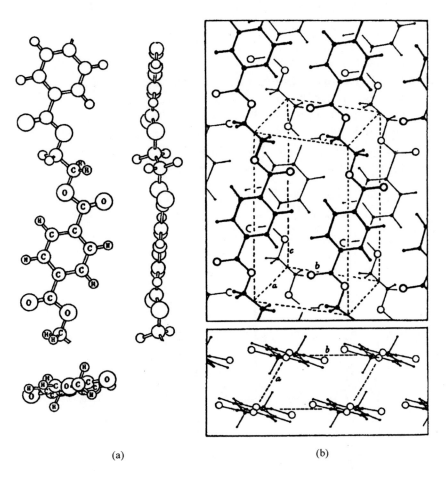

(a)   (b)

Fig. 4.19 The crystal structure of poly(ethylene terephthalate) (PET): (a) views of the chain conformation taken up in the crystal and (b) the arrangement of the chains in the crystal, with one chain passing through the triclinic unit cell. (Reproduced by permission of the Royal Society.)

Fig. 4.20 Hydrogen-bonded sheets in (a) nylon-6,6 and (b) nylon-6. The unit-cell faces are shown by the dashed lines. (Reprinted by Permission of John Wiley & Sons, Inc.)

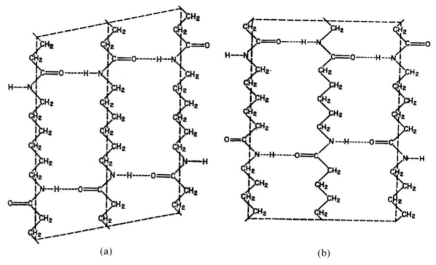

(a)   (b)

The crystalline regions of nylons contain H-bonded sheets of fully extended chains, as illustrated schematically in fig. 4.20 for nylon-6,6 and nylon-6. The chains of nylon-6,6 are centro-symmetric, but the chains in sheets of nylon-6 are arranged alternately up and down. The H-bonding is clearly established by IR spectroscopy, since the —N—H stretching frequency is lowered on crystallisation. The H-bonded sheets are stacked to form a three-dimensional structure, which is shown schematically in fig. 4.21 for the α and β forms of nylon-6,6. The α form is more stable, but both forms can occur together in one crystallite.

Fig. 4.21 The stacking of the hydrogen-bonded sheets in the α and β crystalline forms of nylon-6,6. The lines represent the chains and the circles the oxygen atoms. Full and broken circles represent atoms on the near and far sides of the chain, respectively. The dashed lines show the unit cell. (Reproduced by permission of the Royal Society.)

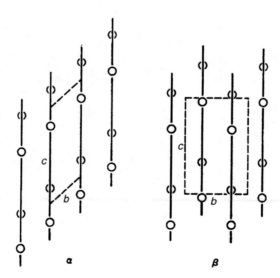

4.5 Further reading

(1) *Precisely constructed polymers*, by G. Natta, *Scientific American*, Aug. 1961. This article gives a description of the possible ordered forms of vinyl and allied polymers.
(2) *Chemical Crystallography* by C. W. Bunn. See section 2.9.
(3) *The Science of Polymer Molecules*, by R. H. Boyd and P. J. Phillips. See section 3.5.

4.6 Problems

All required data not given in any question are to be found within chapter 3 or 4.

4.1. Calculate the separation of the centres of nearest-neighbour chlorine atoms in the planar zigzag syndiotactic and isotactic PVC chains. The van der Waals diameter of the Cl atom is 0.362 nm. From this information and the result of the above calculation, what do you conclude about the isotactic chain?

4.2. A particular vinyl polymer obeys Bernoullian statistics and has a syndiotacticity $P_r = 0.55$. Calculate the ratios of the intensities of the six β-carbon tetrad peaks in its ^{13}C NMR spectrum.

4.3. A certain isotactic vinyl polymer crystallises with the chains in a simple 3_1 helical conformation, with only one chain passing through each unit cell. At approximately what angle will the second layer line be observed in an X-ray fibre photograph obtained using radiation of wavelength 0.194 nm?

4.4. Derive an expression for the ratio of the 'radius' of a powder ring observed in the equatorial direction to the 'radius' observed in the meridional direction for scattering at the angle 2θ in a cylindrical X-ray camera. Evaluate the ratio for $2\theta = 30°$. (Assume that the sample is on the axis of the camera.)

4.5. Assuming that $\alpha = \beta = 90°$, so that the *ab* plane is normal to the *c*-axis, show that, unless $\phi = \theta = 0$, $\phi > 2\theta$, where ϕ is the angle of the *l*th layer line and θ is the Bragg angle for the (00*l*) powder ring or meridional spot.

4.6. The structure of polyethylene, shown in fig. 4.17, has glide planes perpendicular to the *b*- and *a*-axes, which means that reflection in these planes followed by translation by $a/2$ or $b/2$, respectively, leaves the structure unchanged. Explain why, for any crystal with such glide planes, it follows that the (*h0l*) and (0*kl*) 'reflections' have zero intensity, i.e. the (*h0l*) and (0*kl*) spots are absent, unless *h* and *k* are even.

4.7. Show that, for an orthorhombic unit cell, $d_{110} = b\sin[\tan^{-1}(a/b)] = a\sin[\tan^{-1}(b/a)] = ab/\sqrt{a^2 + b^2}$. Calculate the spacings of the d_{110} and d_{200} planes for polyethylene and hence show that the chain packing is pseudo-hexagonal.

4.8. Calculate the diffraction angles corresponding to the (110), (200) and (020) equatorial 'reflections' for a highly oriented polyethylene fibre if the fibre photograph is obtained using radiation of wavelength 0.194 nm. Calculate also the angle ϕ at which the first layer line is observed.

4.9. For a certain highly crystalline polymer a fibre pattern obtained using X-rays of wavelength 0.154 nm had spots on the equator at angles of diffraction of 21.6° and 24.0° and the first layer line was at the angle 37.2°. Assuming that the crystal structure is orthorhombic and that the equatorial spots correspond to the (110) and (200) reflections, respectively, calculate the cell dimensions. The Raman and infrared spectra of the polymer show that several of the vibrational modes of the molecules occur as doublets. If the density of the polymer is 1.01×10^3 kg m^{-3}, calculate the molar mass per repeat unit of the chain and suggest what the polymer might be.

4.10. The fibre pattern obtained for a particular sample of polyoxymethylene $-(CH_2-O)_n$ using X-rays of wavelength 0.154 nm exhibits equatorial reflections at angles of 21.9°, 37.7° and 23.2°. Show that these spots may be consistently indexed as the (110), (200) and (020) reflections, respectively, of an orthorhombic unit cell. The first layer line corresponds to the angle 25.6° and the unit cell contains four repeat units. Calculate the density of the crystalline material.

4.11. Calculate the separation of the (221) spots along the layer line for $l = 1$ for a polymer that crystallises in an orthorhombic unit cell with $a = 1.03$, $b = 0.54$, $c = 0.50$ nm if the fibre pattern is obtained using X-rays of wavelength 0.154 nm in a flat-plate camera with separation 8.0 cm between fibre and film. (Approximate by assuming that the curvature of the layer line can be neglected.)

Chapter 5
Morphology and motion

5.1 Introduction

Chapter 4 is principally concerned with the crystal structure of polymers, that is to say the shape and size of the smallest unit of a crystal—the unit cell—and the arrangement of the atoms within it. The present chapter considers the structure of polymers on a scale much larger than the unit cell. Not all polymers crystallise and, even in those that do, there is always some remaining non-crystalline material, as shown for instance by the presence of broad halos in WAXS patterns from unoriented polymers in addition to any sharp rings due to crystalline material. The widths of the rings due to the crystallites indicate that, in some polymers, the crystallite dimensions are only of the order of tens of nanometres, which is very small compared with the lengths of polymer chains, which may be of order 3000 nm measured along the chain, i.e. about 100 times the dimensions of crystallites. The following questions therefore arise.

(i)   How can long molecules give rise to small crystallites?
(ii) What are the sizes and shapes of polymer crystallites?
(iii) How are the crystallites disposed with respect to each other and to the non-crystalline material?
(iv) What is the nature of the non-crystalline material?

These questions are the basis of *polymer morphology*, which may be defined as the study of the structure and relationships of polymer chains on a scale large compared with that of the individual repeat unit or the unit cell, i.e. on the scale at which the polymer chains are often represented simply by lines to indicate the path of the backbone through various structures. In addition to the four questions above, morphology is concerned with such matters as the directions of the chain axes with respect to the crystallite faces and with the relationship between the crystallites and the non-crystalline material, a particular aspect of which is the nature of the crystalline–'amorphous' interface. Sections 5.2–5.5 are concerned mainly with describing observed structures and methods of investigating them, whereas section 5.6 is concerned with how the structures arise.

The morphology of a polymer plays an important role in determining its properties, but the molecular motions that take place within the polymer play an equally important role. The later part of this chapter deals with the types of motion that can take place in solid polymers and the evidence for these motions. This topic of motion is taken up again in subsequent chapters, particularly in chapters 7 and 9, where the effects of motion on the mechanical and dielectric properties of solid polymers are discussed.

5.2   The degree of crystallinity

5.2.1   Introduction

Before considering the details of how the chains are arranged in the crystalline and non-crystalline regions of a polymer, it is useful to consider how the amount of material contained within the two types of region can be determined. It is important to realise also that the simple two-phase model, in which there are only two types of region, crystalline and non-crystalline, is an approximation that applies to some polymer samples better than it does to others. For the moment it will be assumed that it is a sufficiently good approximation. In principle, almost any property that is different in the crystalline and non-crystalline regions could be used as the basis for a method of determining the *degree of crystallinity*, χ, or, as it is usually more simply put, *the crystallinity*, of a polymer sample. In practice the most commonly used methods involve density measurements, DSC measurements and X-ray diffraction measurements.

However the crystallinity is determined, it is important to distinguish between two slightly different measures, the *volume crystallinity*, χ_v, and the *mass crystallinity*, χ_m. If V_c is the volume of crystalline material and V_a is the volume of non-crystalline (amorphous) material within a sample, then $\chi_v = V_c/(V_a + V_c)$ and, in a similar way, $\chi_m = M_c/(M_a + M_c)$, where M_a and M_c are the masses of crystalline and non-crystalline material within the sample. From these definitions it is easy to show (see problem 5.1) that

$$\chi_m = \frac{\rho_c}{\rho_s} \chi_v \qquad\qquad (5.1)$$

where ρ_c and ρ_s are the densities of the crystalline regions and of the sample, respectively.

The values obtained for χ_m and χ_v by the methods described below vary considerably from polymer to polymer. For a given polymer they can depend markedly on the method of solidification and subsequent heat treatment (annealing). For a nearly atactic vinyl polymer like commercial

PVC the range of values is about 0%–10%, whereas for polyethylenes of various types χ_m and χ_v can range from about 40% to 70% or even wider for specially prepared samples.

5.2.2 Experimental determination of crystallinity

It is easy to show (see problem 5.2) that, if ρ_a is the density of the amorphous material, then

$$\chi_v = \frac{\rho_s - \rho_a}{\rho_c - \rho_a} \tag{5.2}$$

With the assumption that the densities of crystalline and amorphous material are known, determination of the density of the sample easily provides a value for the crystallinity.

The density of the crystalline material can be obtained from the chemical formula of the polymer and the lattice parameters, which can be found in the literature. A difficulty is that the lattice parameters of polymer crystals often depend on the precise conditions under which crystallisation took place, so a value obtained for a sample prepared in as similar a way as possible to the one whose crystallinity is to be determined should be chosen. The density of the amorphous material is quite difficult to determine because it too depends on the method of preparation. When fully amorphous material cannot be produced the densities of a series of samples whose crystallinities have been determined by another method can be extrapolated to zero crystallinity. The density of the sample is usually obtained by means of a *density-gradient column*, the principles of which are explained in section 2.3.

In principle the density method is readily applicable to oriented samples as well as random samples, although care must be taken in choosing appropriate values for ρ_c and ρ_a. Another method that is applicable both to random and to oriented samples is the use of DSC. As explained in section 2.2, this method allows the change in enthalpy due to melting or crystallisation to be determined for any sample. If the change in enthalpy per unit mass of crystals is known from measurements on samples for which the crystallinity is known, the crystallinity of the sample can be determined. Care must be taken that the value used for the enthalpy per unit mass is appropriate for material crystallised in a similar way to the sample.

A method that is less easy to apply to oriented material is the X-ray method. This method uses the fact that, for a given mass, the total scattering integrated over all scattering angles is independent of the crystalline or amorphous nature of the material. The scattered intensity plotted against the scattering angle 2θ generally looks something like the full curve in fig. 5.1. The sharp peaks are due to the crystalline material and the broad background to the amorphous material. Various methods can be used to estimate the shape of the amorphous contribution under the peaks, shown by the dotted line. The ratio of the total scattered intensity within the sharp rings to that within the amorphous halos is then easily calculated, provided that the sample is not oriented, and in principle gives the ratio of crystalline to amorphous material. In practice a number of corrections must be made, one of which is that the observed intensity must be multiplied by an angular factor that takes account of the polarisation of the scattered X-rays at different angles 2θ. This is important because the crystalline and amorphous contributions are different at different angles. Other corrections that must be made include a correction for incoherent scattering, which is independent of the crystalline or amorphous nature of the material, and a correction for crystal imperfections, including thermal vibrations. For oriented samples it is necessary to perform three-dimensional scans.

5.3 Crystallites

The most obvious question that needs to be answered about polymer crystallites is question (i) of section 5.1, 'How can long molecules give rise to small crystallites?'. Two principal types of answer have been given; they lead to the *fringed-micelle* model and the *chain-folded* model for polymer crystallites. A further type of crystallite, the *chain-extended* crystal, can also occur when samples are prepared in special ways. These three types of crystallite are considered in the following sections.

Fig. 5.1 Determination of crystallinity from X-ray scattering. The full curve shows the observed X-ray-scattering intensity as a function of 2θ, where θ is the Bragg angle and 2θ is the angle between the incident and scattered X-ray beams. The dashed curve indicates the estimated contribution of the amorphous material. (Adapted by permission of IUCr.)

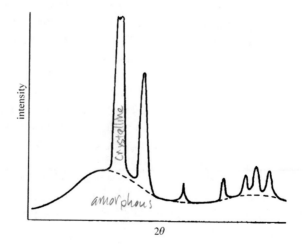

Example 5.1

Calculate the volume and mass crystallinities of a sample of polypropylene of density 910 kg m^{-3} assuming that the densities of the crystalline and amorphous regions are 936 and 853 kg m^{-3}, respectively.

Solution
Equation (5.2) shows that

$$\chi_v = \frac{\rho_s - \rho_a}{\rho_c - \rho_a} = \frac{910 - 853}{936 - 853} = \frac{57}{83} = 0.69$$

and equation (5.1) shows that

$$\chi_m = \frac{\rho_c}{\rho_s}\chi_v = \frac{936}{910} \times \frac{57}{83} = 0.71$$

5.3.1 The fringed-micelle model

The fringed-micelle model was an early attempt to inter-relate long molecules, small crystals and a 'sea' of amorphous material. It was proposed in 1930, by Hermans and others, to explain the structure of gelatin and was subsequently applied to natural rubber. It is now believed to be incorrect as the basic model for polymer crystallites, but it is worth describing for historical reasons and because it may be a good approximation to the true structure in special cases. The essentials of the model are illustrated in fig. 5.2.

The model suggests the occurrence of *fibrillar crystallites*, which can grow both parallel and perpendicular to the chain axes. Each chain

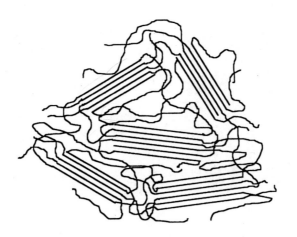

Fig. 5.2 A schematic diagram showing fringed-micelle crystallites.

typically passes through several crystallites and small crystallites can thus be reconciled with long chains.

5.3.2 Chain-folded crystallites

This type of crystallisation was first suggested by Storks in 1938. He made films of gutta percha 27 nm thick by evaporation from solution. Electron diffraction showed that the films were composed of large crystallites with the chain axes *normal* to the plane of the film. The only possibility was that the chains folded back and forth upon themselves, so that adjacent segments were parallel and in crystal register.

This idea lay dormant until the early 1950s, but then very small *single crystals* of polymers were produced from dilute solutions. These crystals were regularly shaped and of lateral dimensions up to 0.3 mm, but their thicknesses were only about 12 nm. When they were rotated between crossed polarisers in the polarising microscope, they gave uniform extinction, i.e. darkness, at a certain angle, which proved that they were indeed single crystals. Studies by Keller and his group using electron diffraction showed that the chain axes were parallel to the thickness direction of these *lamellar crystals* and, once again, the only possibility was chain folding. Figure 5.3 shows examples of solution-grown crystals of polyethylene.

Shortly afterwards (in 1957) Fischer showed by electron microscopy that the crystallites in melt-grown spherulites of polyethylene and nylon were most likely to be lamellar rather than fibrillar, as would be expected from the fringed-micelle model. It is now accepted that *chain-folded lamellar crystallites* play an important part in the structure of most ordinary

Fig. 5.3 Electron micrographs of single crystals of polyethylene crystallised from dilute solution in xylene: (a) diamond-shaped crystals and (b) truncated crystals. (Reprinted by permission of John Wiley & Sons, Inc.)

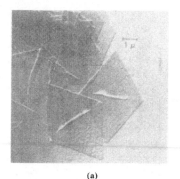

(a)   (b)

crystalline polymers. Figure 5.4(a) shows schematically how the chains fold in a perfect chain-folded crystallite.

The following topics are now considered briefly:

(i)   the production of single crystals from solution,
(ii) the description of solution-grown crystals,
(iii) the nature of the chain folds and the crystal 'surfaces' and
(iv) the reason for the observed form of solution-grown crystals of poly-
      ethylene.

Discussion of the reason for chain folding is deferred until section 5.6.3.

(i) *Production of single crystals from solution*
They form in super-cooled dilute solution, e.g. 0.01% of linear polyethyl-ene in *p*-xylene at 70 °C, and appear as a suspension, which can be sedi-mented and examined by optical or electron microscopy. The first systematic observations were on linear polyethylene, but many other poly-mer single crystals have now been obtained in a similar way.

(ii) *Description of solution-grown crystals of polyethylene*
When they are grown at sufficient dilution, the crystallites approximate to *lamellae* with a uniform thickness of about 12 nm, the precise value depending on the temperature of growth. Electron diffraction shows that the chain axes are approximately perpendicular to the planes of the lamel-lae. The crystals are not exactly flat, but have a *hollow-pyramidal structure*, with the chain axes parallel to the pyramid axis. This pyramidal structure is seen clearly in fig. 5.5, which shows a single crystal of polyethylene floating in solution. This should be compared with fig. 5.3(b), which shows similar crystals flattened on an electron-microscope grid. The dark lines on the crystals in fig. 5.3(b) show where the pyramid has broken when the crystal flattened.

When crystals are grown at low temperature their outlines are bounded by straight (planar) growth faces, in the simplest cases (110)-type faces, as shown in fig. 5.3. When the growth temperature is higher, the faces can

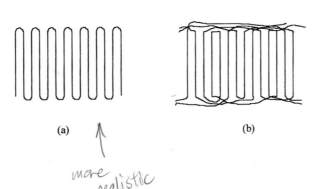

(a)

more realistic

(b)

Fig. 5.4 Schematic diagrams showing (a) regular chain folding with adjacent re-entry, as envisaged for a perfect chain-folded crystallite; and (b) the switchboard model.

Fig. 5.5 A single crystal of
polyethylene floating in
solution, showing the
pyramidal form. (Reprinted
with permission from
Elsevier Science.)

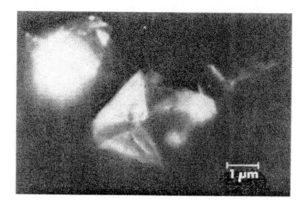

become curved, leading to a shape rather like an ellipse but pointed at the
ends of the long axis.

(iii) *The nature of chain folds and of crystal surfaces*

Three questions arise. (a) Are the folds loose or sharp? (b) Is adjacent re-
entry the main mechanism for folding? (c) Are all fold lengths the same, i.e.
is the fold surface smooth?

Keller originally suggested that solution-grown crystals exhibited regular sharp folding with adjacent re-entry (see fig. 5.4(a)). A later model with
irregular re-entry, the *switchboard model*, was proposed by Flory and is
illustrated in fig. 5.4(b). (The name arises from the similarity in the appearance of the crystal surface in this model to the criss-crossing of the wires on
an old-fashioned manual telephone switchboard.) Evidence for the nature
of the chain folding has been obtained from experiments using selective
degradation of the crystalline material by nitric acid, from infrared-absorption measurements and from neutron-scattering experiments.

On treatment of the crystalline polymer with fuming nitric acid the
chains are cut at the crystal surface, because the acid first dissolves the
amorphous material at the surface and then attacks the surface itself but
cannot readily penetrate the bulk of the crystallites. Gel-permeation chromatography of the degraded material gives the resultant distribution of
chain lengths. The results showed that there were several well-defined
lengths for solution-grown crystals, which could be interpreted as showing
that they were formed by sharp folding with adjacent re-entry. Melt-grown
crystallites gave much less clear results.

The infrared-absorption spectra of mixed crystals of ordinary polyethylene, $-(CH_2-)_n$, and fully deuterated polyethylene, $-(CD_2-)_n$, yielded information about adjacent re-entry through *correlation splitting*, which is due
to the in- and out-of-phase vibration of the two chains in the unit cell, as

discussed in section 4.3. The results are difficult to analyse, but the principle is that correlation splitting is seen only if the two chains in the unit cell are identical. This will happen more frequently for mixed crystals if adjacent re-entry is the rule, giving adjacent chains of the same type, rather than random re-entry, for which adjacent chains are likely to be of different types.

For solution-grown crystals adjacent re-entry was found to predominate, with a probability of about 0.75 for folding along (110) planes (the predominant growth faces), whereas for melt-grown crystals adjacent re-entry probably occurs quite frequently, but much less regularly. Mixtures of short-chain paraffins exhibit a random arrangement of the molecules containing hydrogen or deuterium. These results are consistent with those from nitric-acid-etching experiments and data from neutron-scattering experiments also support them. Figure 5.6 shows photographs of a model of a possible sharp chain fold in polyethylene involving four *gauche* bonds.

Various pieces of evidence suggest that the *fold surfaces* are not perfectly regular, particularly for melt-crystallised materials. *Buried folds* exist up to 2.5 nm below the surface, but the number of sharp folds increases as the overall surface is reached and there are many folds near the mean lamellar

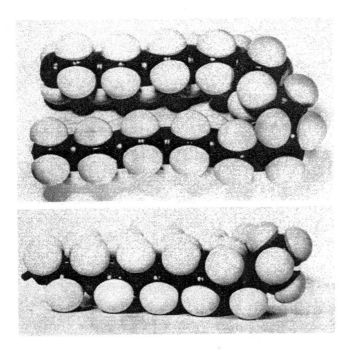

Fig. 5.6 A model of a polyethylene chain folded using four *gauche* bonds, as suggested by Frank. Upper, view normal to the plane of folding; lower, view along the plane of folding. (Reprinted by permission from John Wiley & Sons Limited.)

surface. Outside this there are loose chain ends and long loops (non-adjacent re-entry). In melt-crystallised material *tie molecules* pass from one lamella to adjacent ones in the *multi-layer stacks* that are generally found in such material. There is thus an *amorphous layer* between the crystal lamellae. As the molar mass increases chain folding becomes less regular and the degree of crystallinity also falls, because the freedom of the chains to rearrange themselves on crystallisation decreases.

There is evidence from neutron-scattering experiments that lamellar crystallisation from the melt can occur without significant change of overall molecular dimensions, such as the radius of gyration. A model that fits this evidence and incorporates the ideas described above is shown in fig. 5.7. Lamellar stacks are discussed further in section 5.4.3.

(iv) *The reason for the pyramidal form of solution-grown crystals of polyethylene*

Figure 5.8 shows schematically the arrangement of the chains within a solution-grown single crystal. When the chains fold back and forth parallel to the (110)-type faces, they can be imagined to produce flat ribbons of folded chains, which in the various sectors of the crystal are at angles to one another determined by the angles between the different sets of (110)-type planes. If the fold lies slightly out of the plane of the ribbon, as suggested in the lower part of fig. 5.6, the close packing of these folded ribbons within a sector will require an offset of one ribbon with respect to the previous ribbon in the direction of the chain axes. This accounts for the hollow-pyramidal form of the crystallites and shows that the chain axes will lie parallel to the axis of the pyramid. Packing of adjacent folds within one folded ribbon may also require a displacement within the plane of the ribbon in order to obtain close packing, so that the length of the ribbon is not normal to the chain axes. In this case the base of the pyramid will not be flat, unlike that of the pyramid shown in fig. 5.8.

Fig. 5.7 The 'solidification model' of the crystallisation process, showing how a chain can be incorporated into a lamellar structure without significant change of overall shape. (Reproduced by permission of IUPAC.)

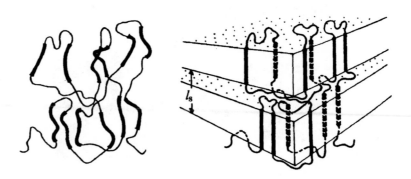

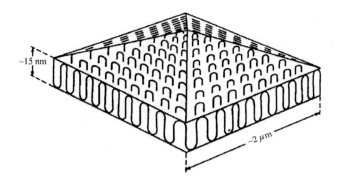

Fig. 5.8 A schematic diagram of chain folding in a solution-grown single crystal of polyethylene. (Reproduced from *The Vibrational Spectroscopy of Polymers* by D. I. Bower and W. F. Maddams. © Cambridge University Press 1989.)

As the chain folds it must twist through 90° so that the chains are in correct crystal register (see fig. 4.17(b)) and, although the various sectors of the crystal grow by chain folding in different planes, there is no discontinuity of crystal structure across the sector boundaries, as is shown most simply by uniform extinction in the polarising microscope.

5.3.3   Extended-chain crystallites

Extended, or fully extended, chain crystallites contain straight chains at least 200 nm long and have been obtained for only a few polymers, such as polytetrafluoroethylene (PTFE), polyethylene and polychlorotrifluoroethylene, using special crystallisation techniques. Extended-chain polytetrafluoroethylene can be obtained by slow crystallisation from the melt; the other two are obtained by crystallisation from the melt under elevated pressure. Solution crystallisation has so far not been shown to give rise to extended-chain crystals.

These materials tend to be very brittle because, although they are highly crystalline, there are few inter-crystalline linking molecules. Figure 5.9 shows an electron micrograph of a replica of a fracture surface of extended-chain crystalline material of polychlorotrifluoroethylene. The lamellae are about 1–2 μm thick and there appears to be no limit on the lateral dimensions of the crystallites except space and availability of material.

5.4   Non-crystalline regions and polymer macro-conformations

5.4.1   Non-crystalline regions

The non-crystalline regions of polymers are usually called amorphous, but they may sometimes have some kind of organisation and they may certainly be oriented, as discussed in chapters 10 and 11. They are the

Fig. 5.9 A transmission electron micrograph of a replica of a fracture surface of extended-chain polychlorotrifluoro-ethylene. The sample was crystallised under 100 MPa pressure at 250 °C for 16 h. The sample was first heated above 295 °C. After crystallisation it was cooled rapidly to room temperature, followed by release of the pressure. (Reproduced by permission of the Society of Polymer Science, Japan.)

component of polymer morphology that iş least well understood and they are often represented by a tangled mass of chains rather like a bowl of spaghetti. Such regions can exist in molten polymers, totally amorphous polymers in the glassy or rubbery states and as a component of semicrystalline polymers. Figure 5.10 illustrates schematically three models that have been put forward as possible structures for these regions in non-crystalline polymers or polymers of low crystallinity.

Initially it was thought that there would have to be some non-random organisation in order to account for the high packing density of chains in non-crystalline material, which is usually 85%–95% that of the crystalline material. The difference in electron density between the high-density bundles postulated in the model shown in fig. 5.10(a) and the surrounding material should show up in small-angle X-ray scattering, but does not. Neutron-scattering data show that the radius of gyration of the molecules in a glass or melt is the same as that in a Θ-solvent, whereas the meander model of fig. 5.10(b) would predict greater values. On the other hand, computer modelling has shown that randomly coiled chains can pack

Fig. 5.10 Proposed structures for the non-crystalline regions of polymers: (a) the bundle model, (b) the meander model and (c) the random-coil model. ((a) and (c) reprinted by permission of Kluwer Academic Publishers; (b) reprinted by permission of John Wiley & Sons, Inc.)

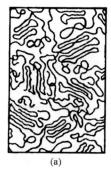

(a)

(b)

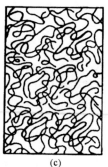

(c)

together at the required densities, in agreement with the Flory theorem (see section 3.3.4), so the random-coil model for unoriented non-crystalline material, illustrated schematically in fig. 5.10(c), is favoured at present.

5.4.2 Polymer macro-conformations

The possible *macro-conformations* of polymers (as opposed to the *micro-conformations* in the unit cell) can be illustrated schematically by means of *Wunderlich's triangle diagram* (fig. 5.11). Notice that the fringed micelle is modified to include some chain folding. This diagram is rather like the three-colour triangle diagram for colour mixing. Any particular overall type of morphology can be represented as a point on the diagram and the concentration of each of the three principal components is represented by how close it is to each of the vertices corresponding to them. It should, however, be noted that this representation does not take account of a very important feature of polymer morphology, namely, how the crystallites are arranged with respect to each other and to the amorphous material. This relationship is rarely random and one of the most important types of organisation is the multi-layer stack of alternating crystalline and non-crystalline material described in sections 3.4.2 and 5.3.2 and now considered further.

5.4.3 Lamellar stacks

Figure 3.12 shows an important characteristic of the multi-layer stack, i.e. the repeat period l_s, which consists of the combined thickness of one crystalline lamella and one non-crystalline region between lamellae and is usually of the order of tens of nanometers. Other important questions about the stacks are the following. (a) What are the individual thicknesses of the crystalline and non-crystalline layers? (b) What is the nature of the non-crystalline material? (c) Can the crystalline and non-crystalline layers

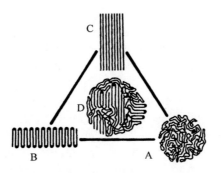

Fig. 5.11 A schematic representation of the macro-conformations of polymer chains. The vertices indicate the limiting cases. A, amorphous; B, chain-folded; C, chain-extended. The area indicates intermediate structures: D, fringed micelle. (Reproduced by permission of Academic Press.)

really be considered to be totally distinct or is there a third, intermediate type of structure at the interfaces? (d) Are the chain axes in the crystalline lamellae normal to the lamellar plane? The answers to these questions vary not only from one type and molar mass of polymer to another but also according to the conditions of preparation of any individual polymer sample.

If the stacking is regular, as is often the case, l_s can be determined by SAXS, as described in section 3.4.2. If the stacking is very irregular, electron microscopy must be used to sample different regions. Electron microscopy shows that the lamellae are not always flat but may be corrugated, particularly in melt-crystallised material. Electron diffraction shows that the chain axes are then usually perpendicular to the overall plane of the lamella rather than being normal to the local thickness direction. Information about the thicknesses of the crystalline layers can be obtained from Raman spectroscopy.

To a good approximation, certain low-frequency Raman-active modes of vibration of the chains within a lamellar crystallite are like the long-itudinal vibrations of elastic rods. These modes are called the *longitudinal acoustic modes* (LAMs). The LAM with the lowest frequency corresponds to a rod having a displacement node at its centre and antinodes at its ends, so that the length of the rod is half a wavelength. If the Young modulus and density of the rod are E and ρ, respectively, the frequency of this mode is given by

$$\nu = [1/(2l)]\sqrt{E/\rho} \tag{5.3}$$

where l is the length of the rod, which in the simplest case is equal to the lamellar thickness. There are complications, such as those due to uncertainties about how the frequency is affected by interactions of the chains in the crystallites with those in the non-crystalline regions, but it is generally accepted that, if l is expressed in nanometres and the frequency is expressed in cm^{-1} ($\tilde{\nu}$) then, within about 20% absolute accuracy, $l = 3000/\tilde{\nu}$ for polyethylene.

Another method that can give information about the periodicity and at the same time can help to answer questions (a) and (c) is NMR spectroscopy, which is now considered in some detail because it provides a link with the discussion of molecular motion later in the chapter.

In section 2.7.5 the idea of spin diffusion is introduced. In any kind of diffusion experiment the time taken for the diffusing species to pass through a particular thickness of material depends on the thickness and on the diffusion coefficient. If either of these is known, a measurement of the diffusion time can be used to determine the other. This idea can be applied to yield information about the average separation of the phases in lamellar structures provided that one of the phases can be preferentially populated with the diffusing species, so that the time taken for it to diffuse into the other phase can be measured. In NMR terms this means selectively exciting the spins in one of the phases. This can frequently be done by making use of the fact that the T_2 relaxation time is usually very much shorter for the crystalline regions than it is for the non-crystalline regions because of the greater mobility of the chains in the non-crystalline regions (see section 5.7.2).

A suitable pulse sequence can then set up a state in which the net magnetisation of the crystalline phase is zero and that of the non-crystalline phase is in equilibrium with the applied magnetic field B_o. Application of a 90° pulse after a time t and analysis of the subsequent FID then allows the determination of the magnetisation $M(t)$ of the crystalline phase as a function of time as it relaxes back towards equilibrium by the transference of magnetisation to it by spin diffusion from the non-crystalline phase.

A response function $R(t)$ that increases from 0 to 1 as t goes from 0 to infinity can be defined as follows:

$$R(t) = 1 - [M(t) - M(\infty)]/[M(0) - M(\infty)] \tag{5.4}$$

and for a lamellar stack structure it can be shown that

$$R(t) = 1 - \exp(z^2)\, \mathrm{erfc}(z) \tag{5.5}$$

where $z^2 = Dt/b^2$, $\mathrm{erfc}(z) = (2/\sqrt{\pi}) \int_z^\infty \exp(-x^2)\,\mathrm{d}x$ is the complementary error function, D is the spin-diffusion coefficient and b is the average thickness of the non-crystalline phase. Figure 5.12 shows how $R(t)$ depends on z.

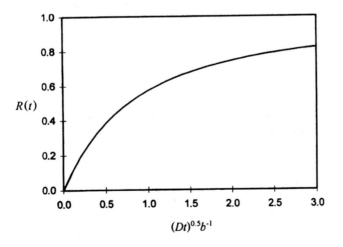

Fig. 5.12 The response function $R(t)$ for recovery of magnetisation during one-dimensional diffusion.

Provided that D is known, a measurement of $R(t)$ as a function of time t can be used to determine b. The ratio of the thicknesses of the crystalline and amorphous regions can be found from the magnitudes of the signals from the two regions in a simple $90°$ pulse experiment, and the lamellar spacing l_s can then be calculated. Unfortunately it is not easy to obtain an accurate value of D. Theoretical estimates can be made and D can also sometimes be determined experimentally from samples for which b is reliably known from other measurements.

In some cases it is not possible to fit the data to a simple model in which there are only two phases, crystalline and non-crystalline; it is necessary to assume the presence of an intermediate *interfacial phase*. For instance, in a particular set of spin-diffusion experiments on polyethylene the data could be fitted to a three-phase model in which the thickness of the interfacial region was found to be 2.2 ± 0.5 nm both for low- and for high-density polymers, compared with thicknesses of about 9 and 40 nm, respectively, for the crystallites.

If the structure is not known to be lamellar, the NMR results are more difficult to interpret. The possibility that the non-crystalline phase is present as rod-like or 'blob'-like (spherical, cubic, etc.) regions must be considered and the interpretation cannot always be unambiguous. The multi-layer lamellar stack is itself usually a component within a morphological structure of greater size, which for unoriented polymers is frequently the *spherulite*. This type of structure is discussed in the next section.

Example 5.2

In a spin-diffusion experiment on a sample of polyethylene the magnetisation of the crystalline phase was found to have risen to 0.57 of its maximum value 120 ms after the start of the FID. Assuming that the magnetisation of this phase is essentially zero at the start of the FID and that the appropriate diffusion coefficient $D = 8.3 \times 10^{-16}$ m^2 s^{-1}, calculate the mean thickness of the non-crystalline layers, assuming that the polymer consists of lamellar stacks of crystalline and non-crystalline material.

Solution

If $M(0) = 0$, equation (5.4) reduces to $R(t) = M(t)/M(\infty)$, so that $R(t) = 0.57$ for $t = 120$ ms. Figure 5.13 shows that Dt/b^2 is close to 1 for $R(t) = 0.57$. A more precise value of 0.98 can be found by, for instance, plotting R against Dt/b^2 using tables of the erfc function or a spreadsheet. Thus $Dt/b^2 = 0.98$, or $b = \sqrt{Dt/0.98} = \sqrt{8.3 \times 10^{-16} \times 120 \times 10^{-3}/0.98} = 1.01 \times 10^{-8}$ m, or 10.1 nm. (Note that, in a real determination, a fit to a large number of values of $R(t)$ would be used (see problem 5.5).)

5.5   Spherulites and other polycrystalline structures

Spherulites, which consist of aggregates of crystal lamellar stacks, are important structural features found in very many polymers crystallised from the melt. The simplest ways of observing them are directly in the optical (or electron) microscope and indirectly by light scattering. Other polycrystalline structures observed under certain circumstances are *axialites* and *shish-kebabs*, which are described briefly in section 5.5.4.

5.5.1   Optical microscopy of spherulites

Spherulites are usually observed in the optical microscope with the sample between crossed polarisers, as described in section 3.4.4 and illustrated in fig. 3.13. Figure 5.13 shows the growth of eventually very large poly(ethylene oxide) spherulites from the melt. The size of the spherulites obtained depends on the time of crystallisation and on the separation of the nuclei from which they grow. Growth is eventually limited by the meeting of spherulite boundaries (see figs 3.13 and 5.13), which leads to an irregular final structure. Before this the spherulites are indeed spherical (or, in very thin films, possibly circular).

Two important observations can be made from figs 3.13 and 5.13. The first is that a spherulite consists of *fibrils* growing out in a radial direction; the second is that each spherulite exhibits a *Maltese-cross* pattern. This pattern is shown particularly clearly in fig. 3.13.

The first observation suggests that the fibrils probably branch at fairly small angles as they grow outwards, otherwise the fibrils themselves would have to increase in lateral dimensions in order to fill all space, whereas the

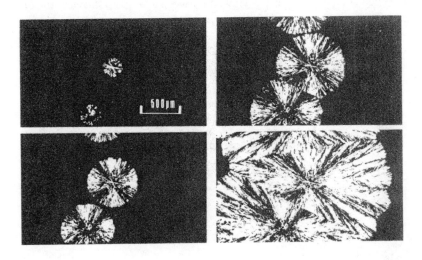

Fig. 5.13 Growth of poly(ethylene oxide) spherulites from the melt, over a period of about 1 min. The originally separate spherulites eventually impinge on each other to form an irregular matrix. (Reproduced with the permission of Nelson Thornes Ltd.)

photograph shows no evidence of any increase in the lateral dimensions. The second observation suggests that the orientation of the crystallites is related in a simple way to the radial direction within the spherulite, because the following simple argument then explains the origin of the Maltese cross.

Consider a spherulite observed between the crossed polariser and ana-lyser in a polarising microscope (see fig. 5.14). Assume that the crystallites within the spherulite have a constant orientation with respect to the radius vector. The corresponding orientations of the indicatrices are then as shown in fig. 5.14(b) and, according to the principles of the polarising microscope explained in section 2.8.1, the Maltese cross will appear in the orientation shown in fig. 5.14(c). Even if the shorter axis of the indicatrix is parallel to the radius vector the orientation of the cross will not change. All that matters for the field to appear dark is that one of the principal axes of the indicatrix should be parallel to the axis of the polariser.

Electron microscopy shows that the fibrils are in fact lamellar crystallites and detailed studies using birefringence and micro-X-ray methods show that, in polyethylene, the spherulites crystallised under conditions that produce chain folding have the c-axes of the crystallites normal to the radius vector and the b-axes parallel to the radius vector. The indicatrix for polyethylene crystals is approximately cylindrically symmetric around the c-axis, which corresponds to its longest axis. This means that the long axes of the indicatrices are normal to the radius vector of the spherulite. Under certain conditions of growth the lamellae twist around the radius vector as they grow, although this twisting is not usually uniform but changes of orientation take place almost discontinuously at periodic intervals. When such a spherulite is viewed between crossed polarisers the cross-section of the indicatrix perpendicular to the direction of propagation of

Fig. 5.14 Formation of the Maltese cross for a spherulite: (a) the directions of transmission of the polariser and analyser, (b) a schematic representation of the orientation of the indicatrices in the spherulite and (c) the appearance of the Maltese cross. In (b) the indicatrix orientations are shown for only eight radial directions, for clarity, and should be imagined for the remainder.

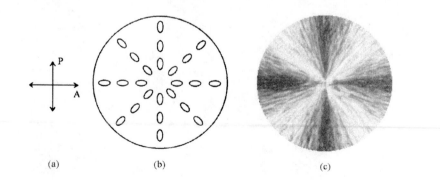

Maltese Cross

the light through the spherulite periodically becomes circular and the polarisation is unmodified as the light traverses the spherulite, so that dark circles appear at regular intervals in addition to the Maltese cross. Such spherulites are called *banded spherulites*. Examples are shown in fig. 5.15.

5.5.2   Light scattering by spherulites

The general principles of light-scattering experiments are described in section 2.4. If a simple photographic system is used with the spherulitic polymer sample between crossed polarisers, a pattern of four scattering peaks is observed, as shown schematically in fig. 5.16.

To find a value for the radius R of the spherulites the angle θ_{max} corresponding to the maxima of scattered intensity for crossed polarisers is measured. If it is assumed that the spherulite can be approximated by an isolated anisotropic sphere embedded in an isotropic matrix, a detailed theoretical treatment then shows that

$$R = 4.1\lambda/[4\pi \sin(\theta_{max}/2)] = 0.326\lambda/[\sin(\theta_{max}/2)] \qquad (5.6)$$

The value of R found in this way is heavily weighted towards larger spherulites, because the intensity scattered by a spherulite is proportional to its volume. The method is useful for comparing samples or following the growth of spherulites, since the rate of increase of radius is usually independent of the size of spherulite. The theory has also been worked out for more complicated forms of spherulites, for truncated spherulites and for other non-spherulitic types of crystalline aggregates.

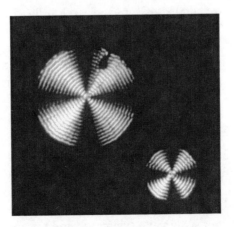

Fig. 5.15 Banded spherulites growing in a thin film of polyhydroxybutyrate. (Courtesy of Dr P. G. Klein.)

Fig. 5.16 (a) A schematic
diagram of the observed
light scattering from a
spherulitic polymer film
placed between crossed
polarisers. The straight
lines indicate the
directions of polarisation
and the closed curves
indicate one contour of
intensity for each of the
four patches of light seen.
The crosses indicate the
positions of maximum
intensity. (b) A photograph
of the corresponding
pattern for a polyethylene
film. ((b) Reproduced by
permission of the
American Institute of
Physics.)

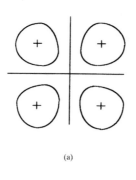

(a)

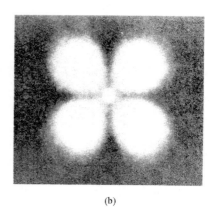

(b)

5.5.3 Other methods for observing spherulites

Spherulites can also be observed by means of electron microscopy, electron
diffraction and X-ray diffraction. These are methods for observing a single
spherulite or a few spherulites in detail, as is the optical-microscope
method, whereas light scattering, as pointed out above, gives only an
average size for many spherulites in the sample.

Particularly interesting are studies by electron microscopy of the early
stages of spherulite growth, which show that the spherulite starts as a single
lamella or small stack of lamellae. The lamellae grow in their preferred
growth direction to become ribbon-like structures that gradually branch at
small angles. This leads to sheaf-like structures in the early stages of
growth, as illustrated in fig. 5.17(a). As the growth and branching continue
the ends of the sheaf meet and the spherulite is gradually constrained to
grow radially, as shown clearly in fig. 5.17(b).

After the growth of these primary lamellar stacks, which do not occupy
the whole of the volume within the boundary of the spherulite, growth of
secondary stacks occurs. These usually consist of material of lower molar
mass that had been excluded from the primary growth.

5.5.4 Axialites and shish-kebabs

Axialites are generally found in polymeric material melt-crystallised at
temperatures only just below the melting temperature (see section 5.6.3).
They can be observed in the optical or electron microscope in a similar way
to spherulites, but their appearance is rather less distinctive because they
are not spherically symmetric and the apparent shape of any axialite there-

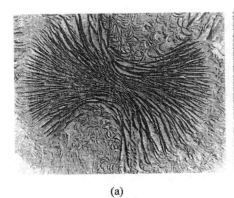

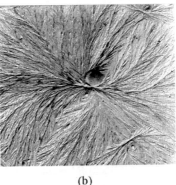

(a)   (b)

Fig. 5.17 Early stages in the growth of spherulites: (a) the sheaf-life stage in the growth of a polyethylene spherulite and (b) the beginnings of radial growth in a spherulite of poly(4-methylpentane). ((a) Reprinted by permission of Kluwer Academic Publishers; (b) © Cambridge University Press 1981.)

fore depends on the angle at which it is viewed. An axialite viewed from the appropriate direction can appear rather like the immature spherulite shown in fig. 5.17(a). When, however, it is viewed from an appropriate direction at right angles to this, it is seen that the structures within it are lamellae that are shaped rather like those observed in solution-grown single crystals, rather than fibrils as in a spherulite.

Shish-kebabs are generally produced only when crystallisation occurs under strain, for instance when a crystallising solution or melt is stirred or flows. Under certain kinds of flow molecular chains can be stretched out and tend to become aligned parallel to each other. Fibrillar crystals sometimes then grow with chain-folded lamellae attached to them at intervals along their length, with the planes of the lamellae at right angles to the axis of the fibril (see fig. 5.18). Similar structures can be produced by extruding a melt through a die under certain extreme conditions. This can lead to a structure in which highly aligned parallel fibrils are formed with chain-folded lamellae attached to them at right angles in such a way that the lamellae on adjacent fibrils interlock.

5.6 Crystallisation and melting

The previous sections of this chapter have concentrated on describing the morphological features to be found in solid polymers and, in particular, the predominance in crystalline polymers of the lamellar crystallite; they have concentrated on what is observed. The present section deals with why some of the observed features are as they are and at what rate they appear. This involves a discussion of the melting temperature, the factors that determine the overall crystallinity achievable and the rate of

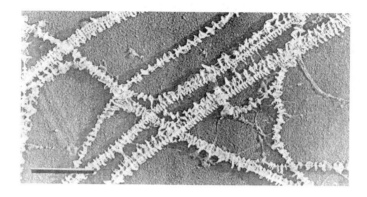

Fig. 5.18 Shish-kebab morphology produced by stirring a 5% xylene solution of polyethylene at 510 rpm and 104.5 °C; the scale bar represents 1 μm. (Adapted by permission of John Wiley & Sons, Inc.)

growth of crystallinity and what determines the fold length in lamellar crystallites.

5.6.1 The melting temperature

The general condition determining the melting point of any substance is that the change in the Gibbs free energy Δg per unit mass on melting is zero, where $\Delta g = \Delta h - T \Delta s$ and Δh and Δs are the increases in enthalpy and entropy per unit mass on melting. The melting temperature T_m is thus given by $T_m = \Delta h / \Delta s$. Polymers with high Δh and low Δs therefore have higher melting points than do those with low Δh and high Δs. For example, polyethylene, which has no specific interactions between the molecules in the crystal and therefore has a low Δh, but has a great deal of flexibility in the chain because of the single bonds in the backbone and therefore has a high Δs, has a melting point of 418 K. Nylon-6,6, on the other hand, has hydrogen-bonding in the crystallites, giving it a high enthalpy, but has similar chain flexibility to polyethylene and has a melting point of 543 K. Table 5.1 shows the melting points of some common polymers.

The melting points just quoted are the *equilibrium melting points*, T_m^o. It is generally understood that crystals of pure small-molecule compounds have well-defined melting points. However, this is true only provided that the crystals are large enough. For any crystallite, the total Gibbs free energy is the sum of a part proportional to its volume and a part due to the extra energy involved in forming its surfaces. As the crystallite becomes larger the ratio of volume to surface area increases and the contribution due to the surfaces becomes negligible compared with that due to the volume. For a small crystallite, however, the surface energy contributes significantly to the enthalpy, so that the effective enthalpy per unit mass is higher. This means that the increase in enthalpy required to melt the crystallite is lower and, assuming that Δs is temperature-independent, T_m is

Table 5.1. *Approximate equilibrium melting points of some common polymers*

Polymer	Melting point (°C)
Polyisoprene (*cis* form)	28
Poly(ethylene oxide)	66
Polyisoprene (*trans* form)	80
1,2-Polybutadiene (isotactic form)	120
Polyethylene (linear)	135
1,2-Polybutadiene (syndiotactic form)	154
Polypropylene (syndiotactic form)	163
Polyoxymethylene	183
Polypropylene (isotactic form)	187
Poly(vinyl chloride)	212
Polystyrene (isotactic form)	240
Poly(vinyl alcohol)	250
Nylon-6	260
Poly(ethylene terephthalate)	270
Nylon-6,6	270
Polytetrafluoroethylene	332
Polyacrylonitrile (syndiotactic)	341

lower than that of large crystals. For lamellar polymer crystals the fold surfaces contribute by far the larger part to the surface energy, both because they are much larger in area than the other surfaces and because the surface energy per unit area is much higher owing to the high energy required to make the folds. The lowering of T_m for a lamellar crystal therefore depends only on its thickness l, so it is easy to show (see problem 5.4) that

$$T_m = T_m^o[1 - 2\sigma/(l\rho_c \, \Delta h)] \tag{5.7}$$

where σ is the surface free energy per unit area and Δh is the increase in enthalpy per unit mass on melting for an infinitely thick crystal. Equation (5.7) is the *Thompson–Gibbs equation*.

5.6.2 The rate of crystallisation

When the temperature of a melt is lowered below the melting point, crystallisation takes place at a rate that depends on the *degree of supercooling*, $\Delta T = T_m^o - T_c$, where T_c is the temperature at which crystallisation takes

Example 5.3

Samples of polyethylene containing lamellae of thicknesses 30 nm and 15 nm are found to melt at $T_1 = 131.2\,°C$ and $T_2 = 121.2\,°C$, respectively. Assuming that the surface free energy of the fold surfaces is 93 mJ m^{-2}, that the density of the crystallites $\rho_c = 1.00 \times 10^3$ kg m^{-3} and that the increase in enthalpy per unit mass, Δh, on melting is 2.55×10^5 J kg^{-1} for an infinitely thick crystal, determine the equilibrium melting temperature.

Solution

Equation (5.7) shows that

$$T_1 - T_2 = \frac{2 T_m^o \sigma}{\rho_c \Delta h}\left(\frac{1}{l_2} - \frac{1}{l_1}\right)$$

$$= T_m^o\left(\frac{2 \times 93 \times 10^{-3}}{1 \times 10^3 \times 2.55 \times 10^5}\right)\left(\frac{10^8}{1.5} - \frac{10^8}{3}\right) = 0.0243\, T_m^o$$

Hence $T_m^o = (131.2 - 121.2)/0.0243 = 411$ K $= 138\,°C$. (In a real determination of T_m^o data from many samples with different values of l would be used.)

place. Two different quantities need to be distinguished here, which are the rate at which a given crystallite grows and the rate at which the crystallinity of the sample increases, but both depend on ΔT. Consider first the rate at which an individual crystallite grows.

The increase in free energy Δg per unit mass on melting is $\Delta g = \Delta h - T_c \Delta s$. Assuming that Δh and Δs are constant in the temperature range of interest, it follows from the fact that Δg is zero at the true melting point T_m^o that, at any temperature of crystallisation T_c,

$$\Delta g = \Delta h (T_m^o - T_c)/T_m^o = \Delta h\, \Delta T/T_m^o \tag{5.8}$$

Thus Δg is larger the further T_c is from the melting point and this tends to lead to an increasing rate of crystallisation as T_c is lowered. This tendency is opposed by the fact that the mobility of the chains becomes lower as the temperature is lowered, so that the rate at which chain segments are brought up to the growing crystals falls. There is thus a maximum in the rate of growth of any crystallite at some temperature below T_m and a corresponding maximum in the rate of increase of crystallinity. Models that attempt to predict the dependence of the growth rate of polymer lamellae on temperature are considered in the next section, 5.6.3.

Once sufficient growth has been added to the central nucleus the radius r of a spherulite increases at a rate $\dot{r}$ equal to the rate of growth of each crystallite at the growth temperature. Assuming that all spherulites in a sample start to grow at the same time $t = 0$, it is easy to show that, if volume changes on crystallisation are neglected, the volume fraction $v_s(t)$ of the sample occupied by spherulites at time t is given by the equation

$$1 - v_s(t) = \exp(-4\pi\dot{r}^3 g t^3/3) \tag{5.9}$$

More generally it can be shown that, whatever the precise nucleation and growth mechanisms, the following equation, called the *Avrami equation*, holds:

$$1 - v_s(t) = \exp(-Kt^n) \tag{5.10}$$

where K and t depend on the nucleation and growth mechanisms. An equation of this form applies to all types of crystallisation, not just to polymers. Spherulites are only partially crystalline and, if the volume-fraction crystallinity of the sample at time t is $v_c(t)$, then $v_s(t) = v_c(t)/v_c(\infty)$, so that

$$v_c(t) = v_c(\infty)[1 - \exp(-Kt^n)] \tag{5.11}$$

Taking account of volume changes during crystallisation leads to a slightly more complicated equation. In equation (5.11) $v_c(\infty)$ should be taken as the crystallinity at the end of primary growth of the spherulites before crystal thickening or growth of subsidiary lamellae has taken place. Avrami exponents n found experimentally for polymers usually range between 2 and 6.

5.6.3 Theories of chain folding and lamellar thickness

Attempts to understand the growth of lamellae have been made using a number of theoretical approaches, but the subject is rather complicated and not yet completely understood. It is therefore the intention here to give only a brief outline of the various approaches. Two principal types of theory have been put forward: *thermodynamic theories* and *kinetic theories*.

In thermal equilibrium the conformation of the molecules within the crystals of any substance and the way that they pack together in the crystal must correspond to a minimum in the Gibbs free energy. Short-chain molecules that can take up straight conformations, such as the paraffins, $CH_3\text{—}(CH_2)_n\text{—}CH_3$ with fairly small values of n, crystallise with straight chains because the introduction of a bend in the chain would significantly increase the free energy. The *thermodynamic theories* suggest, however, that, for very long molecules, the fold period observed at a particular temperature T_c of crystallisation corresponds to a true minimum in the Gibbs free energy density of the crystal at T_c, provided that all contributions to the free energy, including thermal vibrations, are properly accounted for.

These theories are usually rejected on the grounds that any minimum in the free energy at a particular fold length is likely to be broad and shallow compared with the free energy of crystallisation, so that equilibrium would not be reached within the usual crystallisation times, and also that the predicted dependence of the crystal thickness on the degree of supercooling is not in agreement with experimental observations. More recent theories of the crystal fold length are therefore *kinetic theories*, which suggest that the observed fold length is close to that for which the rate of crystallisation is greatest.

The principle behind kinetic theories is the assumption that, although the free energy of the system is reduced after additional chain stems have been added to a growing crystal face, there is a free-energy barrier to be overcome as the stems are added. The nature of this barrier is considered below. It is rather like a person jumping over a wall where the ground is lower on the landing side than it is on the take-off side; in order to end up with a lower potential energy a higher potential energy must first be acquired. Just as a person can jump back over the wall, but requires more energy to do so, molecular stems can also detach themselves from the growing crystal, but require more free energy to do so than they required to become attached to the crystal. The free energy is supplied by the random thermal motions and the probability of acquiring the higher energy required to detach is always lower than that of acquiring the energy to attach, so that the crystal grows.

The reason why the crystal grows with a preferred thickness at a given temperature is that the energy barriers to attachment and detachment depend on the length of the chain between folds. For very long fold lengths the rates of attachment and detachment become very low, because the energy barriers are high, whereas for very short fold lengths the barriers are low and the rates become high but approximately equal. Therefore, at some intermediate fold length there is a maximum rate of crystal growth. This is, of course, not the only fold length that will occur, so the theory is usually developed to give the average fold length $\bar{l}$ and the linear growth rate G of the advancing crystal face. These are found to be given by expressions of the form

$$\bar{l} = \frac{C_1}{\Delta T} + C_2 \tag{5.12a}$$

and

$$G = \beta \exp\left(\frac{-K_g}{T_c \Delta T}\right) \tag{5.12b}$$

where C_1, C_2, β and K_g depend on the polymer and are approximately independent of temperature, at least for ranges of values of ΔT small compared with T_c. The quantity β describes the rate at which chain segments are brought up to the growing crystal surface. It goes to zero as the crystallisation temperature approaches the temperature at which diffusion ceases, about 50 °C below the glass transition temperature for a melt (see section 7.5.3).

The first version of the kinetic theory was presented by Lauritzen and Hoffman in 1960. In their version of the theory the most important contributions to the free-energy barrier are the surface energies of the crystallites, including the high surface energy associated with the folds in the lamellar surfaces. The energy barrier is thus mainly enthalpic. This version of the theory has been developed further over the last 40 years and enthalpy-based kinetic theories of crystal growth are often referred to as *Lauritzen–Hoffman theories*. In 1983 Sadler introduced the idea that the free-energy barrier might actually be entropic rather than enthalpic in nature. Further developments along these lines by Sadler and Gilmer led to this type of theory being called the *Sadler–Gilmer* theory. The essential difference between the two assumptions lies in the way that the chain segments are assumed to be laid down on the growing surface.

It is assumed in the Lauritzen–Hoffman theory that, once a chain segment has been laid down on the growing surface parallel to the chains already forming the crystal growth face, the next units in the chain continue to be laid down to form a straight stem. When a lamellar face is reached the chain folds and the following units are laid down adjacent to

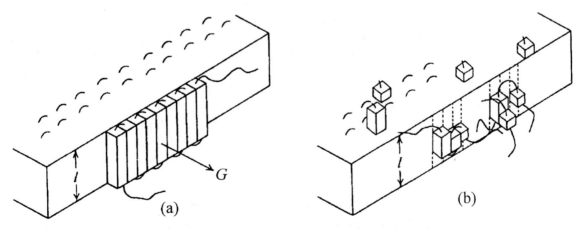

Fig. 5.19 Addition of chain segments to a growing crystallite according to (a) the Lauritzen–Hoffman theory and (b) the Sadler–Gilmer theory. Note that, in the Sadler–Gilmer theory, there are no perfectly regular fold surfaces. ((a) Adapted by permission of Kluwer Academic Publishers.)

the section previously laid down until the other face of the lamella is reached, when chain folding takes place again. The whole process is one of the orderly laying down of one length l after another to form smooth flat growth faces, as illustrated schematically in fig. 5.19(a), and in the simplest form of growth no new layer is started until the previous layer is complete. The principle is thus to assume that regular chain-folded lamellae are formed and then to predict the average fold length and the rate of growth. The Sadler–Gilmer theory was introduced in recognition of the fact, as mentioned in section 5.3.2, that curved growth faces are sometimes observed. This curvature implies steps on the growth surface and is inconsistent with the basic Lauritzen–Hoffman theory.

In the Sadler–Gilmer theory, units consisting of straight segments of chain containing only a few monomers are imagined to be laid down on the growing crystal face parallel to the pre-existing chains. The places where these units can add to the crystal are subject to certain rules that attempt to take account of the fact that the units belong to polymer chains and may thus be connected to other units, but they are otherwise chosen randomly. Several islands of partially crystallised units can exist simultaneously on the growing face and units can even add to incomplete layers (see fig. 5.19(b)). The energy of the interaction of each unit with the surface is only of order kT, so that units not overlaid by others can detach themselves again. Groups therefore attach and detach until randomly, but subject to the rules, they take up the required straight conformations that permit the close fitting of groups and the lowering of the energy of the interaction between them. There is thus an entropic barrier, a barrier due to disorder, that must be surmounted before the free energy can be lowered, and it is clear that this barrier will be higher the longer the straight segments must become. Lamellar growth is not explicitly required in this model but is predicted when the rules are incorporated into a computer simulation.

In comparing equation (5.12a) with experimental results it should be noted that the predicted value of $\bar{l}$ applies to the lamellae as they are first formed; care must be taken to ensure that the experimental value is not influenced by the crystal thickening that takes place on annealing. Provided that this is done, reasonable agreement is usually obtained. Equation (5.12b) can be compared with experiment by plotting $\ln(G/\beta)$ against $1/(T \Delta T_c)$, which should result in a straight line with slope $-K_g$. Experiment frequently produces data that can be fitted to two straight lines with a ratio of slopes of two to one, with the higher slope for the data at higher T_c (lower ΔT). This region corresponds to the growth of axialites and the region at higher supercooling corresponds to the growth of spherulites. The Lauritzen–Hoffman theory associates this transition with a transition from single nucleation on the growing crystal face at low supercooling to multiple nucleation at higher supercooling and correctly predicts the ratio of two for the growth rates.

Finally, it should be noted that, although the kinetic theories can explain much observed experimental data, other theories are also under consideration, including some that postulate that crystallisation may take place via metastable phases, i.e. types of unit cell that occur only in the early stages of crystallisation and are subsequently converted into the observed final type of unit cell. These theories may have implications for the understanding of chain folding.

5.7 Molecular motion

5.7.1 Introduction

The description of the structure of polymers given so far in this chapter has concentrated on the spatial arrangements of the molecules. Just as important for the physical properties of the materials are the various motions that the molecules can undergo; in later chapters their effects on the mechanical and dielectric properties are examined in detail. This section provides an overview of the types of motion to be found, with evidence taken mainly from NMR experiments.

The range of 'frequencies' of the motions observable in solid polymers is extremely wide, from about 10^{-3} to 10^{14} Hz. The reason for placing the word frequencies between quote marks is that, although the higher frequencies, in the range 10^{12}–10^{14} Hz, correspond to the true vibrational modes of the polymer chains that are studied by means of infrared and Raman spectroscopy, many of the lower frequencies are the frequencies of jumps between different relatively discrete states.

These jumps may be, for instance, those between the three different equivalent positions of a methyl group around the bond joining it to the rest of the molecule or the flipping of a phenylene (*para*-disubstituted benzene) ring through 180° around the two collinear bonds joining it to the rest of the molecule. These types of motion are due to *stochastic*, or *random processes* and their *average* frequencies of occurrence depend on the temperature of the polymer. At room temperature the methyl rotational jump can take place more than 10^9 times per second and the phenylene ring flip in polycarbonate can take place more than 10^6 times per second.

NMR experiments are capable of probing motions in the range from about 10^{-2} to about 10^{12} Hz and the methods employed fall into three principal groups: two-dimensional exchange measurements, line-shape measurements and relaxation-rate measurements. The three types of measurement are applicable in the approximate 'frequency' ranges of motion

10^{-2} to 3×10^5, 10^4 to 3×10^6 and 10^7 to 10^{12} Hz, respectively. In section 5.7.4 an example of each of these methods is discussed and in section 5.7.5 a brief description of some of the general findings from NMR spectroscopy with relevance to motion in polymers is given. Before these descriptions are given it will be useful to understand the relationship between the *relaxation phenomena* observed in NMR, described in section 2.7, and those observed in mechanical and electrical measurements. This is described in the next section, which is followed by a description of a simple model for the temperature dependence of the *relaxation time* observed in mechanical or electrical measurements for a particularly simple type of relaxation.

5.7.2   NMR, mechanical and electrical relaxation

The relaxation processes observed in NMR are due to the fluctuating magnetic fields produced at any particular nucleus by the motion of the surrounding nuclei. Because relaxation takes place by jumps from one energy state to another and can thus be caused only by specific frequencies, it is the strength of the corresponding frequencies in the randomly fluctuating magnetic field that determines the relaxation time.

It is a general result that the power spectrum $J(\omega)$ of any quantity $X(t)$ that fluctuates with time t is given by the Fourier transform

$$J(\omega) = \int_{-\infty}^{\infty} C(\tau)e^{-i\omega\tau}d\tau \qquad (5.13)$$

of its time-correlation function $C(\tau) = \langle X(t).X(t+\tau)\rangle$, where the angle brackets $\langle\ \rangle$ denote the average over all t and . indicates the scalar product if X is a vector. If $X(t)$ is taken to be the fluctuating magnetic field $H(t)$, it is usual to assume for simplicity that

$$C(\tau) = \langle H(t).H(t)\rangle \exp(-|\tau|/\tau_c) \qquad (5.14)$$

where τ_c is called the *correlation time*. This is equivalent to assuming an exponential decay of the 'memory' of any initial value of $H(t)$. If the relaxation is due to a single process of random jumping between two molecular states, it is clear that τ_c must be proportional to the average time between jumps. It is shown in the following section that τ_c is in fact equal to $1/\nu$, where ν is the number of jumps per unit time between the two states. Each different possible relaxation process has its own value of τ_c.

Substituting $C(\tau)$ from equation (5.14) into equation (5.13) leads to

$$J(\omega) = H_m^2\tau_c/(1 + \omega^2\tau_c^2) \qquad (5.15)$$

where H_m is the root-mean-square value of $H(t)$. The NMR relaxation rates, i.e. the reciprocals of the relaxation times T_1 and T_2, each depend linearly in different ways on the values of $J(\omega_o)$ and $J(2\omega_o)$, where ω_o is the resonance frequency (see section 2.7.1). In addition to depending on τ_c in the way described by equation (5.15), these values of $J(\omega)$ also depend, through H_m, on the inter-nuclear distances and on whether the relaxation is due to the interaction of like or different spins. This means that the relationship between τ_c and the relaxation times is complicated, but known. The relaxation times can be found using special pulse sequences and τ_c can thus be determined. Matters are somewhat simpler for mechanical and electrical relaxation.

If a polymer sample is subjected to a mechanical stress or to an electric field the structure responds, i.e. relaxes, in such a way as to reach an equilibrium under the stress or the field. This response generally involves some rearrangement of the structural units of the polymer, i.e. changes in the proportions of the various conformational or orientational states of its molecules, and it is the nature of these changes and the extent to which they take place that determine the elastic modulus or dielectric constant of the polymer. In the rest of this section and in section 5.7.3 it is assumed that the equilibrium ratio of the numbers of molecules in a particular pair of states between which jumps take place is actually influenced by the stress or electric field. This is so for many such pairs of states, but not for all pairs (see sections 7.6.1 and 9.2.5).

If there is only one mechanism by which the structure can relax and this mechanism conforms to the exponential behaviour expressed by equation (5.14), the material might be expected to relax exponentially towards a new equilibrium strain or polarisation, with time constant τ_c. This simple argument neglects the fact that the relaxing units are embedded in a medium, so that the electric field or stress they experience is not equal to the macroscopic field or stress, which can lead to a difference between the measured macroscopic relaxation times and τ_c, as discussed in section 9.2.5. Dielectric and mechanical studies of relaxation are usually made using alternating electric fields or stresses. As discussed in sections 7.3.2 and 9.2.4, if measurements are made over a range of frequencies, any relaxation mechanism gives rise to the presence of a peak in the dissipation of energy per cycle, called a *loss peak*, at an angular frequency equal to the reciprocal of the relaxation time. Alternatively, because the relaxation time for any relaxation depends on the temperature (see the following section for an example), the loss peaks can be studied by varying the temperature at a fixed frequency, high-temperature peaks corresponding to low-frequency peaks at a fixed temperature.

In general a polymer will have several different relaxation mechanisms, each of which will behave in a way similar to that just described. Many mechanical and dielectric relaxation measurements were made before the advent of the NMR techniques that increasingly allow the identification, often unambiguously, of the various mechanisms involved at the molecular level. Because of this lack of knowledge the various loss peaks were usually labelled α, β, γ, etc., starting with α at the highest temperature, or lowest frequency. The mechanisms available for relaxation in different polymers are different, however, so that relaxations labelled with the same label, say β, in different polymers need not have any fundamental similarity in their mechanism. However, the α relaxation in amorphous polymers is usually related to the glass–rubber transition.

Before describing experimental evidence for some of the specific motions that take place in solid polymers it is useful to consider the so-called *site model* which allows the relaxation time for a relaxation process due to a single mechanism of random jumping to be related to the temperature and the height of the energy barrier that must be surmounted for the corresponding jump to take place.

5.7.3 The site-model theory

In the simplest form of this theory there are two 'sites', each representing a particular local conformational state of the molecule, separated by an energy barrier, as shown in fig. 5.20. The ΔGs are the Gibbs free-energy differences per mole (constant pressure conditions are assumed).

Assume that there are n_1 molecules in the conformational state 1 represented by site 1 at a particular time and that the probability that any molecule initially in state 1 will make the transition to state 2 in any small interval of time dt is $v_{12}\,dt$. The total number of molecules 'moving from site 1 to site 2' in unit time is then $n_1 v_{12}$. A similar argument applies for transitions in the reverse direction, with v_{21} replacing v_{12} and n_2 replacing n_1. Thus, if n_1^o and n_2^o are the equilibrium values of n_1 and n_2, respectively, and v_{12}^o and v_{21}^o are the corresponding equilibrium values, then

Fig. 5.20 The site model: a representation of the energy barrier between two molecular states.

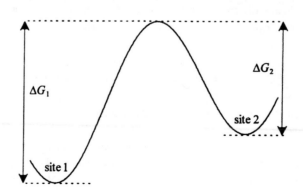

$$n_1^o v_{12}^o = n_2^o v_{21}^o \quad \text{or} \quad n_1^o v_{12}^o - n_2^o v_{21}^o = 0 \qquad (5.16)$$

Assume that an applied stress σ disturbs the equilibrium by causing *small* changes in the free energies of the sites (and hence in the transition probabilities) and that, at any subsequent time, there are n more molecules in site 1 than the original equilibrium number. Then

$$dn/dt = d(n_1^o + n)/dt = -n_1 v_{12} + n_2 v_{21} \qquad (5.17)$$

Substituting $n_1 = n_1^o + n$ and $n_2 = n_2^o - n$ and using equation (5.16) leads to

$$
\begin{aligned}
dn/dt &= -n(v_{12} + v_{21}) - n_1^o(v_{12} - v_{12}^o) + n_2^o(v_{21} - v_{21}^o) \\
&= -n(v_{12} + v_{21}) + C\sigma
\end{aligned}
\qquad (5.18)
$$

where it is assumed that both transition probabilities are changed from their equilibrium values by an amount proportional to the stress σ, so that C is a constant. The solution of this equation, as is easily checked by differentiation, is

$$n = C\sigma\tau(1 - e^{-t/\tau}) \qquad (5.19)$$

where

$$\tau = 1/(v_{12} + v_{21}) \qquad (5.20)$$

Now assume that the strain e is equal to an, where a is a constant, i.e. that each time a molecule changes site it causes the same increment of strain. It then follows from equation (5.19) that

$$e = aC\sigma\tau(1 - e^{-t/\tau}) \qquad (5.21)$$

The probabilities $v_{12}\,dt$ and $v_{21}\,dt$ must be proportional to the probabilities that a molecule gains enough energy to surmount the respective barriers between the sites. Provided that the effect of stress on ΔG_1 and ΔG_2 is small compared with their equilibrium values, it follows that

$$v_{12} = A\exp[-\Delta G_1/(RT)]; \qquad v_{21} = A\exp[-\Delta G_2/(RT)] \qquad (5.22)$$

where A is constant for a particular pair of sites. Differentiation of these equations shows that the assumption made above, namely that the changes in v_{12} and v_{21} are proportional to the stress, is equivalent to the assumption that ΔG_1 and ΔG_2 change linearly with stress for small stresses.

Substituting v_{12} and v_{21} from equation (5.22) into equation (5.20) and assuming, as is often true, that $\Delta G_1 - \Delta G_2 \gg RT$, leads to

$$\tau = (1/A)\exp[\Delta G_2/(RT)] = 1/v_{21} \qquad (5.23)$$

Equations (5.21)–(5.23) show that the strain relaxes to its equilibrium value $e = a\sigma\tau$ exponentially with a time constant, or *relaxation time*, τ that is temperature dependent and is equal to the average time between jumps over the barrier.

This process of relaxation is equivalent to the loss of memory of the initial state of occupation of the sites with a time constant τ, so that $\tau = \tau_c$, the correlation time. Because $\Delta G = \Delta H - T\Delta S$,

$$\tau = A' \exp[\Delta H_2/(RT)] \tag{5.24}$$

where $A' = (1/A)\exp(-\Delta S_2/R)$. This is the *Arrhénius equation*. It shows that the relaxation time for a particular process of the type considered here changes rapidly with temperature, which leads to a strong temperature dependence of the mechanical or dielectric properties. The enthalpy difference ΔH_2 is often written as ΔE and is called the *activation energy* for the process. A plot of the logarithm of τ against $1/T$, called an *Arrhénius plot* is a straight line when a relaxation process obeys equation (5.2.4).

Example 5.4

If the relaxation time τ of a polymer is governed by an activation energy $\Delta E = 96$ kJ mol^{-1}, calculate the increase in temperature from 300 K which will produce an order of magnitude decrease in τ, i.e. a decrease by a factor of ten.

Solution
Replacing ΔH_2 by ΔE in equation (5.24) gives $\tau = A' \exp[\Delta E/(RT)]$. Thus $\tau/\tau_o = \exp[(\Delta E/R)(1/T - 1/T_o)] = 10^{-1}$, or $(\Delta E/R)(1/T - 1/T_o) = -\ln(10)$. Hence $1/T - 1/T_o = -R\ln(10)/\Delta E$ and to $1/T = 1/T_o - R\ln(10)/\Delta E$. Thus $1/T = 1/300 - 8.31\ln(10)/(9.6 \times 10^4) = 3.134 \times 10^{-3}$ and $T = 319$ K, which corresponds to a rise of 19 °C.

5.7.4   Three NMR studies of relaxations with widely different values of τ_c

Because this book is about solid polymers and the motions of molecules in solids are generally imagined to be fairly restricted, particularly if the solid is crystalline, the first example to be described is one that demonstrates that, even in the crystalline regions of solid polymers well below their melting points, slow but significant motion of the chains can take place. The detection of these slow movements requires the use of two-dimensional NMR, the principles of which are described briefly in section 2.7.6.

For the detection of slow processes, the experiment consists essentially of the application of a 90° pulse, followed by an evolution time t_1, a mixing time t_m and an evolution time t_2. If no changes of the structure take place during t_m, the spectrum observed in the two-dimensional plot of intensity against (ω_1, ω_2) is confined entirely to the diagonal $\omega_1 = \omega_2$. If, however, a nucleus moves during t_m to a new location in which it has a different resonance frequency, intensity is correspondingly seen off the diagonal. A simple way in which such a change in resonance frequency can occur is through the anisotropy of the chemical-shift tensor.

Suppose that a ^{13}C atom in a particular group of atoms in a polymer chain has a chemical-shift tensor whose axes are in a fixed orientation with respect to the group and that, during t_m, the group rotates into a different orientation with respect to the applied field B_o. The ^{13}C atom will now have a different resonance frequency, because the chemical shift has changed. In such a process spins are exchanged between two or more frequencies and the corresponding spectrum is called a *two-dimensional (2-D) NMR exchange spectrum*.

Figure 5.21 shows the ^{13}C 2-D exchange spectrum for a sample of commercial polyoxymethylene (POM), $-(CH_2-O)_n$, at 252 K obtained with a slightly more complicated pulse sequence that ensures that the amorphous regions do not contribute to the spectrum. All the intensity lies on the diagonal and the shape of the peak gives information about the chemical-shift tensor. For this spectrum $t_m = 1$ s, which shows that at 252 K no jumps take place within a time of 1 s. Figure 5.22 in contrast shows contour maps of the spectra for a highly oriented sample of POM obtained at two higher temperatures still well below the melting point, 456 K, of the crystallites. These spectra have complicated-looking patterns of off-diagonal intensity, which means that jumps are taking place during the mixing time of 1 s.

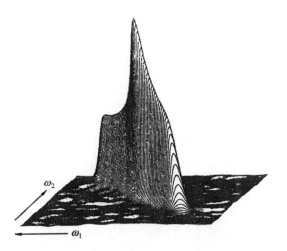

ω_2

ω_1

Fig. 5.21 The two-dimensional ^{13}C spectrum of commercial polyoxymethylene at 252 K obtained with a mixing time of 1 s. (Reproduced by permission of Academic Press.)

Fig. 5.22 Experimental and simulated contour plots of two-dimensional ^{13}C spectra of highly oriented polyoxymethylene. The bottom part of the diagram shows the sub-spectra corresponding to jumps through various angles. The main diagonal runs from bottom left to top right in each figure. (Reproduced by permission of Academic Press.)

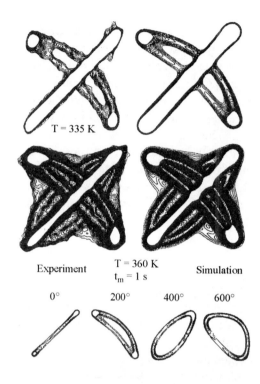

Experiment   T = 360 K   Simulation
                t_m = 1 s

0°   200°   400°   600°

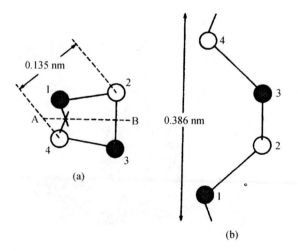

Ordinary POM crystallises in a hexagonal unit cell with one chain in a helical 9_5 conformation passing through each cell. This conformation is illustrated in fig. 5.23. In the highly oriented sample all the helix axes are very nearly parallel to the symmetry axis of the sample but the crystallites have no preferred orientation around it. This axis is placed in the spectro-

Fig. 5.23 The 9_5 helical conformation of polyoxymethylene: (a) viewed along the chain axis and (b) viewed perpendicular to the chain axis and to the line AB.

meter at an angle to B_o and it follows that there is still a distribution of orientations of $-(CH_2-O)-$ groups with respect to B_o, although the distribution is different from that for a random sample. The two-dimensional spectrum would still show only a diagonal ridge in the absence of motion during t_m and the reason for using an oriented sample is that stronger, more easily interpreted signals are observed. The spectra in fig. 5.22 show very clearly that motion is taking place on a time-scale of 1 s and that different types of motion take place at the different temperatures. The first off-diagonal component of the spectrum to appear as the temperature increases is the bow-shaped feature, which is followed by the boat-shaped feature parallel to the main diagonal and finally by the broader feature beneath the bow.

The calculated spectra shown on the right-hand side of fig. 5.22 are based on knowledge of the chemical-shift tensor and the assumption that the jumps correspond to the movement of a chain by one $-(CH_2-O)-$ unit at a time, so that the chain is displaced by the distance 0.193 nm and rotated through the angle $(5 \times 360/9)^\circ = 200^\circ$ and thus becomes superimposed upon itself. What matters in producing the spectrum is that each $-(CH_2-O)-$ group has changed its orientation during the jump. The different off-diagonal features correspond to successive jumps by 200°, so that, at 360 K, some of the chains have been displaced by three $-(CH_2-O)-$ groups in the mixing time of 1 s. Further information can be obtained from three-dimensional spectra, including the probabilities for forward and backward jumps.

It is pointed out in section 2.7.5 that studies of the line-shape of the ^2H spectrum can give information about motion in the approximate range 10^{-7} s $> \tau_c > 10^{-3}$ s. Figure 5.24 shows the observed and calculated spectra for poly(butylene terephthalate) deuterated at the middle two methylene units of the butylene group,

$-((C=O)-\circledcirc-(C=O)-O-CH_2-(CD_2)_2-CH_2-O)-$,

where $-\circledcirc-$ represents the phenylene group and D the deuterium atom, ^2H. The simulated spectra were calculated on the assumption that the C—D bond flips randomly between two directions making the angle 103° with each other. The most likely motions to give these jumps are three-bond motions involving either the relocation of a *gauche* bond from the sequence $ttg^\pm$ to $g^\pm tt$ or the creation of the sequence $g^\pm tg^\mp$ containing two *gauche* bonds from the all-*trans* sequence ttt, as shown in fig. 5.25. If the bonds were exactly at the tetrahedral angles to each other, the jump angle for each C—D bond would be $109\frac{1}{2}^\circ$. As shown in fig. 5.24, the jump rate was observed to go from $\ll 10^4$ s^{-1} at -88 °C to nearly 6×10^5 at $+85$ °C.

Jump rates greater than about 10^7 s^{-1} can be studied by determining relaxation times. It is easily shown by differentiating equation (5.15) that

Fig. 5.24 Observed and simulated ^{2}H spectra for poly(butylene terephthalate) deuterated at the middle two carbon atoms of the butylene unit for various temperatures of the sample. The jump rates used for the calculations were (a) $\ll 10^4$ s^{-1}, (b) 4.7×10^4 s^{-1}, (c) 2.1×10^5 s^{-1}, and (d) 5.7×10^5 s^{-1}. (Adapted with permission from the American Chemical Society.)

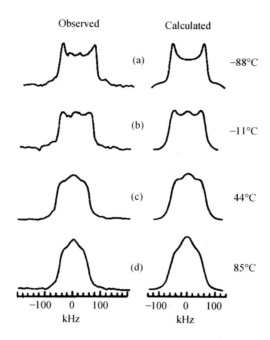

the value of $J(\omega)$ has a maximum value for a fixed value of ω when $\tau_c = 1/\omega$. If the rate of spin–lattice relaxation were proportional to $J(\omega_o)$, the intensity of the fluctuating field at the resonance frequency, then the relaxation time T_1, which is proportional to the inverse of the relaxation rate, would have a minimum when $\tau_c = 1/\omega_o$. For like spins the relaxation rate actually depends linearly on $J(2\omega_o)$ as well as on $J(\omega_o)$ and the minimum value of T_1 occurs when $\tau_c = 0.62/\omega_o$ (see problem 5.9). For unlike spins T_1 depends on the resonance frequencies ω_a and ω_b of the two types of spin and the minimum value of T_1 occurs in the region between $\tau_c = 1/\omega_a$ and $\tau_c = 1/\omega_b$. For very much higher values of τ_c, $J(\omega)$ and $J(2\omega)$ are proportional to $1/(\tau_c\omega^2)$, so that T_1 is proportional to $\tau_c\omega_o^2$. For very much lower values of τ_c, $J(\omega)$ and $J(2\omega)$ are proportional to τ_c, so that

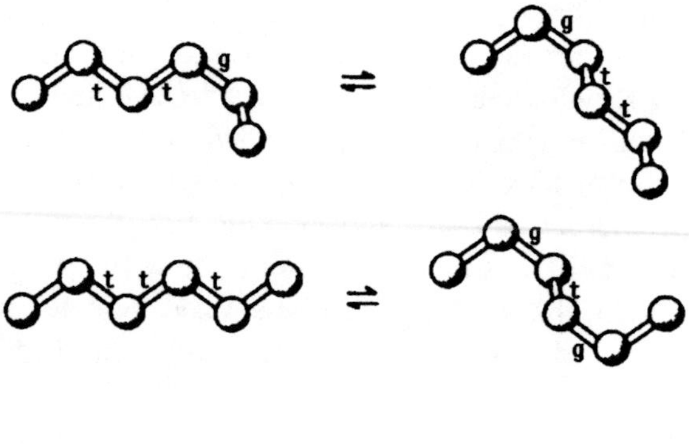

Fig. 5.25 The most likely conformational jumps of the butylene group in poly(butylene terephthalate). Each jump requires rotation around the second and fourth bonds from the left. The middle two carbon atoms are deuterated in each case. (Adapted with permission from the American Chemical Society.)

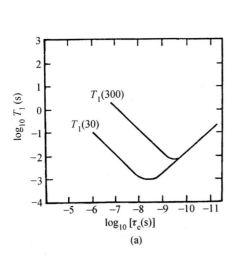

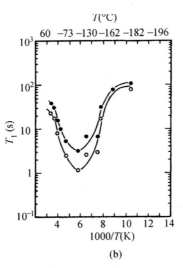

Fig. 5.26 (a) The ideal response of T_1 as a function of τ for a pair of protons 0.1 nm apart. The numbers in parentheses denote the values of $\omega_0/(2\pi)$ in megahertz. (b) The ^{13}C T_1 relaxation times for the (●) methylene and (○) methine carbon atoms of isotactic propylene. ((a) Reproduced from *Nuclear Magnetic Resonance in Solid Polymers* by V. J. McBrierty and K. J. Packer. © Cambridge University Press, 1993; (b) reproduced by permission of IBM Technical Journals.)

T_1 is proportional to $1/\tau_c$. Figure 5.26(a) shows these relationships among T_1, τ_c and ω_0 for a pair of protons 0.1 nm apart.

Remembering that, for a relaxation process that obeys equation (5.24), $\ln \tau = a + b/T$, where a and b are constants, a plot of T_1 against $1/T$ for any particular resonance should exhibit similar behaviour to the plot of T_1 against $\ln \tau_c$. Figure 5.26(b) shows the spin–lattice relaxation for the methine (CH) and methylene (CH_2) carbons of isotactic polypropylene, $-(CH_2-CH(CH_3))_n-$, as a function of temperature. The material was 90% isotactic and 70% crystalline and the data reflect the behaviour of the crystalline material. Both curves have minima in the region of $-110\,^\circ$C, which is the temperature at which the proton T_1 also has a minimum attributed to methyl-group rotations. The ^{13}C relaxation observed at low

---

Example 5.5

The jump frequency ν for a particular relaxation process in a polymer is found to be 3×10^5 at 100 K and 3×10^7 s^{-1} at 170 K. Assuming that the relaxation obeys the Arrhénius equation, calculate the activation energy ΔE for the process.

Solution

The Arrhénius equation (5.27) can be written $\nu = \nu_0 \exp[-\Delta E/(RT)]$. Thus $3 \times 10^5/3 \times 10^7 = \exp[-\Delta E/(100R)]/\exp[-\Delta E/(170R)]$, or

$$\ln(100) = \frac{\Delta E}{R}\left(\frac{1}{100} - \frac{1}{170}\right)$$

which leads immediately to $\Delta E = 9.3$ kJ mol^{-1}.

temperatures is thus due not to segmental chain motion, but rather to methyl-group (CH_3) rotation. The value of T_1 for the methine carbon is lower than that for the methylene carbon because of the closer proximity of the methine carbon atom to the methyl group. The experiments were performed at 30 MHz, so it can be concluded that, even at $-110\,°C$, the jump rate for rotation of the methyl group is of the order of 10^7 per second. The activation energy is 9.7 kJ mol^{-1}.

5.7.5   Further NMR evidence for various motions in polymers

In this section some further examples of conclusions reached from NMR studies of motion in polymers are given. The emphasis is on the results, rather than on the specific NMR techniques used. It is important to understand that, although in what follows only one type of relaxation mechanism is mentioned for any one polymer, several different relaxations can usually take place.

In the previous section the helical jumping of the molecules in the crystallites of POM was discussed. Similar motions have been observed in a number of other crystalline polymers, including poly(ethylene oxide) (PEO), isotactic polypropylene (IPP) and polyethylene. At 240 K each helical chain in PEO makes about three jumps per second. In the crystalline regions of IPP helical jumps along the 3_1 helix are observed at temperatures in the range 340–380 K, whereas in syndiotactic polypropylene, which has a more complicated 4_1 helix, no jumps are observed.

The experiment required to detect the corresponding helical jumps for the 2_1 helix of polyethylene has to be slightly different from that described for POM because the chemical-shift tensor is invariant under a rotation through 180°, so that this does not give any change of frequency. Under magic-angle spinning conditions, however, a slight difference between the frequencies of the ^{13}C signals from the crystalline and amorphous regions can be observed because of the presence of *gauche* bonds in the amorphous regions. Because the jumps lead to a gradual diffusion of sections of the chains between the two regions, off-diagonal peaks can be seen and data obtained in the region 319–377 K have successfully been modelled on the basis of a random walk of jumps within a lamellar crystallite, taking account of restrictions imposed by chain loops in the amorphous regions. The data were found to fit a straight line on an Arrhénius plot (see section 5.7.3), with an activation energy of 105 ± 5 kJ mol^{-1}, in good agreement with dielectric data for the α relaxation. The mechanism for this motion is discussed further in section 7.6.3.

A somewhat different kind of jump has been observed close to room temperature in the α-crystalline form of poly(vinylidene fluoride) (PVDF), in which the jump is a conformational change and the jump angle is 113°. By combining the NMR data with data from dielectric measurements on oriented samples the conformational change involved has been identified as TGTG′ to G′TGT, for which exact tetrahedral bonding would give a jump angle of $109\frac{1}{2}°$.

Rotational jumps not involving translations are observed in many polymers, particularly important types being the rotational jumps of methyl groups, as considered above for isotactic polypropylene and observable for a wide variety of polymers, and the 180° flips of *para*-disubstituted benzene rings (phenylene rings, —◎—). These flips have been observed in many polymers, including amorphous polymers such as bisphenol-A polycarbonate (PC), $+$◎$-C(CH_3)_2-$◎$-O-CO-O+$, semicrystalline polymers such as poly(ethylene terephthalate) (PET), $+O-CO-$◎$-CO-O-(CH_2)_2+$, and highly crystalline polymers such as poly(phenylene vinylene) (PPV), $+$◎$-CH=CH+$. Ring flips have also been observed for the monosubstituted benzene ring (phenyl ring —◎) in polystyrene (PS), $+CH_2-CH$◎$+$. Sometimes small oscillations (librations) around the ring axis are observed as well as ring flips.

So far, two classes of motion have been considered predominantly, namely the motion of whole chain segments in crystals and the motions of specific groups that do not involve conformational changes in the backbone of the molecule. An exception was the *trans–gauche* jumps in the $(CH_2)_4$ regions of poly(butylene terephthalate) and even there the rotations take place around two parallel bonds in such a way that there is no overall change in the direction of the backbone. In general, for all such motions that involve specific jumps through well-defined angles, a plot of the logarithm of the correlation time τ_c against $1/T$ (an Arrhénius plot) gives a straight line, corresponding to a well-defined activation energy.

Above the glass-transition temperature T_g of an amorphous polymer, relaxations take place that cannot be described in terms of jumps through specific well-defined angles, but correspond to a continuous range of jump angles. The correlation times for such processes generally do not give a straight line on an Arrhénius plot. An example is given in fig. 5.27. This plot shows data obtained from 2-D ^2H exchange spectra of polyisoprene deuterated at the methyl groups. The spectra show that the C—D bonds change direction in an essentially random way, with no preference for particular jump angles. This is caused by conformational changes of the backbone that do not correspond to localised jumps between two or more preferred conformations. At 260 K, about 55 °C above T_g, the correlation time τ_c for the process is $< 10^{-6}$ s, but it has

Fig. 5.27 An Arrhénius plot
of the correlation times
from two-dimensional
exchange ^{2}H spectra of
polyisoprene deuterated
at the methyl groups. The
significance of the fitted
curve is explained in
section 7.5.2. (Adapted by
permission of the
American Institute of
Physics.)

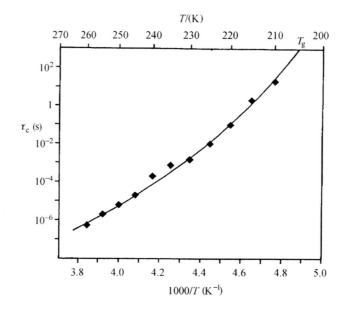

risen to about 10 s when the temperature is only 5 °C above T_g. In this
particular example, for a specific temperature well above T_g the spectrum
actually requires a small range of values τ_c to be used in its interpretation
and the range of values increases as T_g is approached.

Such spectra often require a wide range of values of τ_c for their interpretation or the assumption of a non-exponential form of the correlation
function $C(\tau)$ in equation (5.13), such as the *stretched exponential* (see
problem 5.10). Relaxations associated with the glass transition are discussed in more detail in chapter 7. Chapter 6 is concerned with the
mechanical properties of polymers well away from the glass transition
and from other relaxation regions.

Table 5.2 gives a simple classification of the possible types of
motion, but it must be recognised at the outset that, just as the classification of polymeric material into two sharply defined categories of
crystalline or amorphous is an approximation, so is the classification of
motions into discrete types. Nevertheless, this approach is useful for
developing models with which to correlate observations made by using
different techniques.

Finally, it should be noted that, although this chapter has concentrated
mainly on the use of NMR for studying motion in polymers, this is closely
allied to its use for studying morphology, because different components of
the morphology generally have different types or degrees of motion. For
instance, NMR evidence suggests that in PET the structure below the glass
transition should be considered to consist of crystalline material and two

Table 5.2. *Typical types of motion observed in polymers*[b]

Motion	Region[a]	Nature	Typical activation energy $(kJ \, mol^{-1})$	Remarks
Primary main chain	C	Hindered rotations, oscillations or translations	> 125	May be activated by defects
Primary main chain	A	Large-scale rotations and translations	< 400	Associated with glass transition and breakdown of long-range order: ΔE is usually temperature-dependent
Secondary chain motion	C/A	Localised motion of interfacial material: cilia, folds, tie molecules, chains of low molar mass	40–60	Characteristic of linear polymers: some or all chains may be involved
Side group	C/A	Highly localised motion of specific moiety	8–80	Usually in the low-temperature regime below T_g
Impurity motions	C/A	Motions induced by trace solvents or material of low molar mass that can act as a plasticiser		Dominates relaxation at low temperature

[a] C = crystalline; A = amorphous.
[b] Reproduced from *Nuclear Magnetic Resonance in Solid Polymers* by V. J. McBrierty and K. J. Packer © Cambridge University Press 1993.

different types of amorphous material, which have less mobile and more mobile regions, respectively. It is probable that ring flips take place only in the less mobile amorphous regions.

5.8 Further reading

(1) *Polymer Single Crystals*, by P. H. Geil, John Wiley & Sons, New York, 1963, reprinted by Robert Krieger Publishing Co. Inc., New York, 1973. A good review up to the time of publication.

(2) *Macromolecular Physics*, by B. Wunderlich, three volumes, Academic Press, New York, 1973–80. These volumes deal with crystal structure, morphology and defects; crystal nucleation, growth and annealing; and crystal morphology.

(3) *Principles of Polymer Morphology*, by D. C. Bassett, Cambridge Solid State Science Series, Cambridge University Press, 1981. This book gives a detailed account of the methods used to study morphology, together with a very wide range of the results obtained.

(4) *Nuclear Magnetic Resonance in Solid Polymers*, by V. J. McBrierty and K. J. Packer; and *NMR of Polymers*, by F. A. Bovey and P. A. Mirau. See section 2.9.

5.9 Problems

5.1. Prove equation (5.1).

5.2. Prove equation (5.2).

5.3. Show that the mass crystallinity χ_m of a polymer sample is given by the expression $A(1 - \rho_a/\rho_s)$, where A depends on the polymer but not on the degree of crystallinity and ρ_a and ρ_s are the densities of the amorphous component and of the sample, respectively. The values of χ_m for two samples of a polymer with densities 1346 and 1392 kg m^{-3} are found by X-ray diffraction to be 10% and 50%, respectively. Calculate the densities ρ_a and ρ_c of the crystalline and amorphous material and the mass crystallinity of a sample of density 1357 kg m^{-3}.

5.4. Starting from the condition that the change in Gibbs free energy per unit mass on melting is zero for large crystals and assuming that the increases in enthalpy and entropy per unit mass on melting are independent of temperature, deduce equation (5.7).

5.5. The table shows the time dependence of the magnetisation of the crystalline phase as a fraction of

t (ms)	5	10	20	40	80	160
M/M_{max}	0.20	0.30	0.41	0.48	0.57	0.69

the maximum magnetisation in a spin-diffusion experiment on a sample of polyethylene known to consist of lamellar stacks. Use these data to obtain the mean value of the thickness of the non-crystalline layers, assuming that the two-phase model is appropriate and that the appropriate spin-diffusion coefficient $D = 8.6 \times 10^{-16}$ m^2 s^{-1}. (Assume that $M = 0$ at the start of the FID.)

5.6. A polymer sample is examined using the arrangement shown in fig. 2.4 and a pattern of the type shown in fig. 5.16 is obtained on the film. The wavelength of laser light used is 633 nm, the distance from the polymer sample to the film is 25 cm and the distance of the maxima from the centre of the pattern is 21 mm. Assuming that the spherulites in the sample can be approximated as isolated anisotropic spheres of radius R embedded in an isotropic matrix, calculate R.

5.7. By plotting a suitable graph, calculate the activation energy for a process for which the jump frequency v at temperature T is given by the following table.

T (°C)	−32	−11	5	21	44	63	85
v (10^4 s^{-1})	1.9	4.0	7.1	15	21	28	57

5.8. Assuming that, at 360 K, each chain in a polyethylene crystal makes 180° helical jumps at the rate of 2×10^4 per second and that the jumps take place in either direction completely at random, calculate the time that it would take, on average, for a particular CH_2 unit to pass completely through a lamellar crystallite of thickness 15 nm. If the activation energy for the process is 105 kJ mol^{-1}, what would the corresponding time at 300 K be? The dimensions of the polyethylene unit cell are given in section 4.4.1. (Note that, in fact, the jumps are not completely random because of constraints in the amorphous regions between the crystal lamellae.)

5.9. Assuming that the value of T_1 is given by the expression

$$R_1 = 1/T_1 = C\left(\frac{\tau}{1 + \omega^2 \tau^2} + \frac{4\tau}{1 + 4\omega^2 \tau^2}\right)$$

show, either by plotting the function $\omega R_1/C$ against $\omega\tau$ or by differentiating R_1 with respect to τ and solving the resulting equation numerically, that the minimum value of T_1 for a fixed value of ω occurs when $\omega\tau \cong 0.616$.

5.10. The correlation function for a particular relaxation process can be well described by the Kohlrausch–Williams–Watts function $C(\tau) = \exp[-(\tau/\tau_K)^\beta]$ with $\tau_K = 20$ ms and $\beta = 0.32$. Show by plotting suitable graphs, using a spreadsheet or otherwise, that, in the region $\tau = 0.25$–100 ms, the correlation function can be well simulated by the following sum of exponential functions:

$$0.24 \exp[-(20\tau/\tau_K)] + 0.35 \exp[-(\tau/\tau_K)] + 0.25 \exp\{-[\tau/(20\tau_K)]\}.$$

Chapter 6
Mechanical properties I – time-independent elasticity

6.1 Introduction to the mechanical properties of polymers

It must be recognised from the beginning that the mechanical properties of polymers are highly dependent on temperature and on the time-scale of any deformation; polymers are *viscoelastic* and exhibit some of the properties of both *viscous liquids* and *elastic solids*. This is a result of various relaxation processes, as described in sections 5.7.2 and 5.7.3, and examples of these processes are given in sections 5.7.4 and 5.7.5.

At low temperatures or high frequencies a polymer may be *glass-like*, with a value of Young's modulus in the region 10^9–10^{10} Pa, and it will break or yield at strains greater than a few per cent. At high temperatures or low frequencies it may be *rubber-like*, with a modulus in the region 10^5–10^6 Pa, and it may withstand large extensions of order 100% or more with no permanent deformation. Figure 6.1 shows schematically how Young's modulus of a polymer varies with temperature in the simplest case. At still higher temperatures the polymer may undergo permanent deformation under load and behave like a highly viscous liquid.

In an intermediate temperature range, called the *glass-transition range*, the polymer is neither glassy nor rubber-like; it has an intermediate modulus and has viscoelastic properties. This means that, under constant load, it undergoes *creep*, i.e. the shape gradually changes with time, whereas at constant strain it undergoes *stress-relaxation*, i.e. the stress required to maintain the strain at a constant value gradually falls.

Various possible load–extension curves for polymers are shown schematically in fig. 6.2. The whole range of behaviour shown in fig. 6.2 can be displayed by a single polymer, depending on the temperature and the *strain-rate*, i.e. how fast the deformation is performed, and whether tensile or compressive stress is used. These curves are discussed further in sections 8.1, 8.2 and 10.2.2.

In order to discuss the mechanical behaviour in a quantitative way, it is necessary to derive expressions that relate stress and strain. An *ideal elastic*

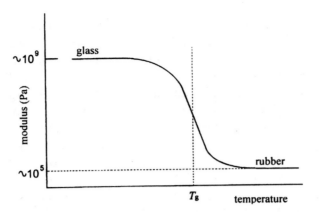

Fig. 6.1 Young's modulus versus temperature for a model polymer. T_g is the glass-transition temperature.

solid obeys *Hooke's law*; $\sigma = Ee$, where the linear strain e is the change in length divided by the original length when a tensile stress σ, the force per unit cross-sectional area, is applied to stretch a piece of the material of uniform cross-section. E is *Young's modulus* of the material. The application of a stress σ leads to an instantaneous strain e and, on removal of the stress, the strain instantaneously reverts to zero. The strain is normally restricted to small values (e less than about 1%) before fracture.

There are five important ways in which the mechanical behaviour of a polymer may deviate from this ideal. The polymer may exhibit:

(i)   *time-dependence* of response;
(ii)   *non-recovery* of strain on removal of stress, i.e. *yield*;
(iii) *non-linearity* of response (e not proportional to σ), which does *not* imply non-recovery;
(iv)   *large strains* without fracture; and
(v)   *anisotropy* of response.

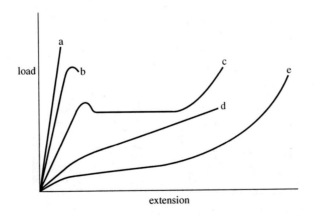

Fig. 6.2 Possible forms of the load–extension curve for a polymer: (a) low extensibility followed by brittle fraction; (b) localised yielding followed by fracture, (c) necking and cold drawing, (d) homogeneous deformation with indistinct yield and (e) rubber-like behaviour.

These are essentially independent effects; a polymer may exhibit all or any of them and they will all be temperature-dependent. Section 6.2 is concerned with the small-strain elasticity of polymers on time-scales short enough for the viscoelastic behaviour to be neglected. Sections 6.3 and 6.4 are concerned with materials that exhibit large strains and non-linearity but (to a good approximation) none of the other departures from the behaviour of the ideal elastic solid. These are *rubber-like* materials or *elastomers*. Chapter 7 deals with materials that exhibit time-dependent effects at small strains but none of the other departures from the behaviour of the ideal elastic solid. These are *linear viscoelastic* materials. Chapter 8 deals with *yield*, i.e. non-recoverable deformation, but this book does not deal with materials that exhibit *non-linear viscoelasticity*. Chapters 10 and 11 consider *anisotropic materials*.

6.2　Elastic properties of isotropic polymers at small strains

6.2.1　The elastic constants of isotropic media at small strains

As stated above, *Young's modulus E* is defined by the equation

$$\sigma = Ee \tag{6.1}$$

Poisson's ratio v is defined by

$$v = -e_{\text{perp}}/e \tag{6.2}$$

where e_{perp} is the linear strain in the direction perpendicular to the tensile stress producing the tensile strain e. The minus sign is introduced in order to make v positive for most materials, for which e_{perp} has the opposite sign to that of e. Figure 6.3 shows a unit cube before and after applying a tensile stress perpendicular to one pair of parallel faces.

The *bulk modulus K* is defined by

$$1/K = -(1/V)(\mathrm{d}V/\mathrm{d}p) \tag{6.3}$$

where V is the volume of the material and p is the applied pressure. The negative sign is introduced once again so that K is positive.

Fig. 6.3 A unit cube before and after applying a tensile stress σ.

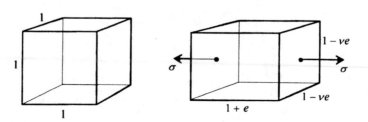

By assuming the linearity of response of the material and considering the effect of applying a tensile stress σ in all three directions simultaneously, which is equivalent to applying a pressure $-\sigma$, it can be shown (see example 6.1) that

$$K = \frac{E}{3(1 - 2v)} \tag{6.4}$$

The *shear* (or *rigidity*) *modulus G* is defined in terms of the *shear strain* θ produced by the *shear stresses* σ, shown in fig. 6.4 (where θ is assumed to be very small), so that

$$G = \sigma/\theta \tag{6.5}$$

Like the bulk modulus, G can also be related to Young's modulus and Poisson's ratio (see problem 6.1). The relationship is

$$G = \frac{E}{2(1 + v)} \tag{6.6}$$

Equations (6.4) and (6.6) show that there are only two independent elastic moduli for an isotropic elastic solid. A further useful relationship can be obtained by assuming that tensile stresses σ_1, σ_2 and σ_3 are applied simultaneously parallel to three mutually perpendicular axes $Ox_1x_2x_3$ chosen in the material and letting the corresponding strains be e_1, e_2 and e_3. The average of the three normal stresses σ_1, σ_2 and σ_3 is equal to minus the hydrostatic pressure (pressure is positive when directed inwards). It can then be shown (see problem 6.2) that

$$e_i = [(1 + v)\sigma_i + 3vp]/E \qquad \text{for } i = 1, 2 \text{ or } 3 \tag{6.7}$$

If $K \gg E$ the material is sometimes said to be 'incompressible', because its shape can be changed very much more easily than its volume and it is thus a good approximation for many purposes to set E/K for such a material equal to zero. It then follows from equations (6.4) and (6.6) that

$$v = \tfrac{1}{2} \qquad \text{and} \qquad G = E/3 \tag{6.8}$$

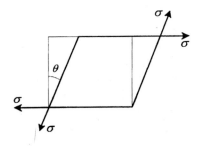

Fig. 6.4 Application of shear stresses σ to the unit cube. The forces σ are applied tangentially to the faces and uniformly spread over them.

so that an 'incompressible' elastic solid has only one independent elastic modulus. It also follows from equation (6.7) that, for an 'incompressible' solid,

$$2e_i = 3(\sigma_i + p)/E \qquad \text{for } i = 1, 2 \text{ or } 3 \tag{6.9}$$

Example 6.1

Prove equation (6.4).

Solution

By virtue of the assumption of linearity, the effects of applying three equal tensile stresses σ parallel to the three axes of a unit cube must be the sum of the effects of applying each of the three stresses σ separately. In each direction the total strain e_{tot} is thus the sum of the tensile strain $e = \sigma/E$ produced by the stress σ parallel to that direction and the two strains equal to $-ve$ produced by the perpendicular stresses (see fig. 6.3). Thus

$$e_{\text{tot}} = e - 2ve = e(1 - 2v)$$

However, for small strains $\mathrm{d}V/V = 3e_{\text{tot}}$ and $1/K = -(1/V)(\mathrm{d}V/\mathrm{d}p) = (\mathrm{d}V/V)/\sigma$. Thus $1/K = 3e(1 - 2v)/\sigma = 3(1 - 2v)/E$, or $K = E/[3(1 - 2v)]$.

6.2.2 The small-strain properties of isotropic polymers

This section is concerned only with the short-term, essentially instantaneous, response of polymeric solids to applied stresses that give rise to very small strains. The response at longer times is considered in the next chapter. The emphasis in this section is on the comparison of various types of polymer with each other and with non-polymeric solids.

There are basically four categories of polymer to consider: amorphous polymers above and below their glass-transition temperatures, T_g, and semicrystalline polymers in which the amorphous regions are above or below T_g.

Amorphous polymers well above T_g behave either as liquids or, if they are cross-linked, as rubbers; the properties of rubbers are discussed in the next section. In the region close to T_g the viscoelastic properties dominate even at small strains and relatively short times and these are considered in the next chapter. This means that the static small-strain properties of amorphous polymers can be discussed meaningfully only when the polymers are well below T_g. Semicrystalline polymers are really composite materials. At temperatures well below the T_g of the amorphous regions the material has small-strain elastic properties that depend on the properties of both components. At temperatures well above the T_g but below the melting temperature of the crystallites the material may be rubbery if the crystallinity is low but will otherwise still have a low modulus dominated by the behaviour of the rubbery amorphous regions. In the region close to T_g the viscoelastic properties again dominate even at small strains and relatively short times.

Table 6.1 shows the crystallinity, the room-temperature elastic modulus E, the density ρ and Poisson's ratio ν for three metals, for glass and for a number of common polymers. The ratio E/ρ for each material and the values of T_g for the polymers are also shown. Lead and tungsten have the lowest and highest values of E for metals and it is seen that glass has a value about one third that of steel. The polymers have even lower values, but, in general, the higher the crystallinity and the higher the value of T_g the higher the modulus. For an amorphous polymer such as poly(methyl methacrylate) that is well below its T_g at room temperature the value of E is not far below that of an inorganic glass.

The comparison of polymers with other materials in terms of their moduli is even more favourable if it is done on the basis of a weight-for-weight comparison, which may be done on the basis of the column showing E/ρ. Polymers that are below the glass transition then have stiffnesses that are comparable to or exceed that for an inorganic glass and are only an order of magnitude less than those for steels, rather than two orders of magnitude less when judged simply on the value of E. It is this factor (and a similar conclusion about their breaking strengths) that gives polymers an important role as structural materials, especially since their properties can be further improved by orientation (see chapters 10 and 11).

In attempting to predict the moduli of glassy polymers it is tempting to compare them with organic compounds consisting of small molecules. Such materials are, however, usually crystalline or liquid at room temperature. The forces involved in the elasticity of these materials are van der Waals-type forces between non-bonded atoms and, at first sight, it might be thought that, in amorphous polymers, these would also be the most important forces governing the elasticity, so that, for instance, the bulk modulus of a polymer might be comparable to that of a small-molecule organic liquid. Typical values for this quantity are in the range $(0.5-1) \times 10^9$ Pa. Assuming such a value for a polymer, noting that the value of Poisson's ratio for a polymer is most unlikely to be below 0.2 and using equation (6.4) then leads to the prediction that E is likely to be less than $(1-2) \times 10^9$ Pa. Table 6.1 shows that this value is a factor of two or more less than observed values for glassy amorphous polymers well below T_g.

Table 6.1. *Small-strain elastic properties of polymers compared with other materials (E, ρ and v at room temperature, all values are approximate)*

Material	Crystallinity	E $(10^9$ Pa$)$	Density, ρ $(10^3$ kg m$^{-3})$	E/ρ $(m^2$ s$^{-2})$	Poisson's ratio, v	T_g (°C)
Lead		16	11.34	1.4	0.43	
Steel	} >0.95	210	7.8	26.9	0.29	
Tungsten		390	19.3	20.2		
Glass	0	72	2.6	27.7	0.22	
Polyethylene						−20
Low density	0.5	0.12	0.91	0.13	0.5	
High density	0.8	0.58	0.95	0.6		
Polypropylene	0.5	1.4	0.9	1.6		−10
Poly(ethylene terephthalate)	0–0.5	2.2	1.3	1.7	0.4	80–120
Nylon-6,6	0.1–0.6	3.3	1.1	3.0	0.33	50
Poly(vinyl chloride)	0.1	3	1.38	2.2	0.38	82
Polystyrene (atactic)		3.3	1.06	3.1	0.33	100
Poly(methyl methacrylate) (atactic)		3.3	1.19	2.8	0.38	105

The reason for this difference is that, in a glassy polymer, there is a much larger proportion of bonded interactions between atoms and the conformations of the polymer chain are frozen in, so that the forces required in order to stretch bonds, change bond angles and rotate segments of polymer chain around bonds are very important in determining the elastic properties. These forces are stronger than the van der Waals forces and cause the higher modulus. The only way that all these interactions can be taken into account for an amorphous polymer is by molecular-modelling techniques.

The problem of calculating the moduli of semicrystalline polymers is even more difficult. It involves in principle four steps: (i) the calculation of the modulus of the amorphous material, (ii) the calculation of the elastic constants of the anisotropic crystalline material, (iii) the averaging of the elastic constants of the crystalline material to give an effective isotropic modulus and (iv) the averaging of the isotropic amorphous and crystalline moduli to give the overall modulus. The second of these steps can now be done fairly accurately, but the other three present serious difficulties.

Example 6.2

Assuming that there is uniform stress throughout a semicrystalline polymer sample and that the average crystal and amorphous moduli, E_c and E_a, are 5×10^9 Pa and 0.25×10^9 Pa, respectively, calculate the modulus of the sample if its volume crystallinity $\chi_v = 0.6$.

Solution

Because the polymer is under uniform stress the modulus will be the same as the effective modulus of a rod of material of uniform unit cross-section consisting of lengths equal to the volume fractions of the crystalline and amorphous materials placed in series. Applying a stress σ to such a rod results in a strain e equal to $\chi_v \sigma / E_c + (1 - \chi_v) \sigma / E_a$ and a modulus E equal to $\sigma / e = 1 / [\chi_v / E_c + (1 - \chi_v) / E_a]$.

Substitution of the given values leads to $E = 0.58 \times 10^9$ N m^{-2}.

The problem with the amorphous material is that, even though it may be in a rubbery state, there are likely to be constraints on the chains due to the crystallisation process, which will give the material different properties from those of a purely amorphous rubbery polymer. The difficulty with the two averaging steps is that the states of stress and strain are not homogeneous in materials made up of components with different elastic properties. The simple assumption of uniform stress often gives results closer to experiment than does the assumption of uniform strain, but neither is physically realistic. For polyethylene, values of the average crystal modulus E_c and the average amorphous modulus E_a are found to be about 5×10^9 Pa and 0.25×10^9 Pa, respectively.

6.3 The phenomenology of rubber elasticity

6.3.1 Introduction

Phenomenological descriptions of the behaviour of a material attempt to describe the behaviour on very general grounds, taking no account of the detailed microstructure of the material. Two such models for the stress–strain relationships for a rubber are now considered. They are suggested by (i) attempts to generalise small-strain elastic behaviour to behaviour at large strains and (ii) attempts to 'guess' possible *strain–energy functions*. The extent to which the descriptions obtained in this way describe correctly the behaviour of real rubbers is then considered. Theories of rubber elasticity more closely related to the microstructure are dealt with in section 6.4.

6.3.2   The transition to large-strain elasticity

In section 6.2 the formalism of the elastic behaviour of an ideal linear elastic solid for small strains was considered. Rubbers may, however, be reversibly extended by hundreds of per cent, implying that a different approach is required. The previous ideas suggest a possible plausible generalisation, as follows.

Start from the following *assumptions*:

(i)   a rubber is *isotropic* in the undeformed state, i.e. it has the same properties in all directions; and

(ii)   changes of volume on deformation are very small and can be neglected, i.e. assume that the rubber is *incompressible* in the sense discussed in section 6.2.1.

For finite strain in isotropic media, only states of *homogeneous pure strain* will be considered, i.e. states of uniform strain in the medium, with all shear components zero. This is not as restrictive as it might first appear to be, because for small strains a shear strain is exactly equivalent to equal compressive and extensional strains applied at 90° to each other and at 45° to the original axes along which the shear was applied (see problem 6.1). Thus a shear is transformed into a state of homogeneous pure strain simply by a rotation of axes by 45°. A similar transformation can be made for finite strains, but the rotation is then not 45°. All states of homogeneous strain can thus be regarded as 'pure' if suitable axes are chosen.

Extension ratios λ in the directions of the three axes can then be used, where λ is defined as (new length)/(original length) $= l/l_o$ (see fig. 6.5).

Let Δ_i be the change in length of unit length in the Ox_i direction, so that Δ_i corresponds to the strain e_i in the small-strain theory. Thus (see fig. 6.5)

$$\lambda_i = 1 + \Delta_i \tag{6.10}$$

and

$$\lambda_i^2 = 1 + 2\Delta_i + \Delta_i^2 \tag{6.11}$$

which, in the limit of small Δ_i, when $\Delta_i = e_i$, gives

$$\lambda_i^2 = 1 + 2e_i \tag{6.12}$$

Fig. 6.5 Definition of λ and Δ. $\lambda = l/l_o$.

Equation (6.9) for the small-strain elasticity of an incompressible solid, viz.

$$2e_i = 3(\sigma_i + p)/E \qquad \text{for } i = 1, 2 \text{ or } 3 \tag{6.13}$$

can be rewritten

$$1 + 2e_i = 3(\sigma_i + p)/E + 1 = (3/E)(\sigma_i + p + E/3) \tag{6.14}$$

Comparison of equations (6.12) and (6.14) suggests a plausible relationship between the extension ratio and stress for a rubber, viz.

$$\lambda_i^2 = (3/E)(\sigma_i^* + p^*) \tag{6.15}$$

where the quantity p^* is analogous to $p + E/3$ and is independent of i and σ_i^* is an appropriate measure of stress. For large strains, the cross-sectional area changes significantly with stress. The *true stress* is defined as the force per unit area of the *deformed* body, whereas the *nominal stress* is defined as the force per unit area of the *undeformed* body. It will be assumed that the appropriate stress for use in equation (6.15) is the true stress and the asterisk on the stress will be dropped. The asterisk will, however, be left on p^*, because it does not correspond exactly to a pressure.

The application of equation (6.15) to simple extension parallel to Ox_3, i.e. to *uniaxial extension*, leads to an important result. For this type of stress $\sigma_1 = \sigma_2 = 0$ and $\lambda_1 = \lambda_2$ by symmetry. Let $\lambda_3 = \lambda$ and $\sigma_3 = \sigma$. Equation (6.15) then leads to

$$\lambda^2 = (3/E)(\sigma + p^*) \tag{6.16}$$

$$\lambda_1^2 = \lambda_2^2 = (3/E)p^* \tag{6.17}$$

Rubbers are 'incompressible' in the sense discussed in section 6.2.1 and, in terms of the extension ratios, the 'incompressibility' is expressed by

$$\lambda_1 \lambda_2 \lambda_3 = 1 \tag{6.18}$$

This equation merely states that the fractional change in volume is essentially zero compared with the fractional changes in the linear dimensions, which cannot, of course, all be greater than unity. It follows that

$$\lambda = \lambda_3 = 1/(\lambda_1 \lambda_2) = 1/\lambda_2^2 = E/(3p^*) \qquad \text{or} \qquad p^* = E/(3\lambda) \tag{6.19}$$

Substitution into equation (6.16) leads to

$$\lambda^2 = (3/E)[\sigma + E/(3\lambda)] \tag{6.20}$$

or

$$\sigma = (E/3)(\lambda^2 - 1/\lambda) \tag{6.21}$$

Consider the *nominal stress* σ_n, the force/unit area of *unstrained* medium. Because $\lambda_1\lambda_2\lambda = 1$, it follows that $\sigma_n = \sigma\lambda_1\lambda_2 = \sigma/\lambda$. Thus

$$\sigma_n = (E/3)(\lambda - 1/\lambda^2) = G(\lambda - 1/\lambda^2) \tag{6.21a}$$

For small strains $\lambda = 1 + e$, so that, *in the limit of small strains,*

$$\sigma_n = \sigma = (E/3)[(1 + e)^2 - (1 + e)^{-1}] \simeq (E/3)[1 + 2e - (1 - e)] = Ee$$

which is Hooke's law.

Thus the relationships (6.21) and (6.21a) are compatible with the isotropy and 'incompressibility' of a rubber and reduce to Hooke's law at small strains. Materials that obey these relationships are sometimes called *neo-Hookeian solids.* Equation (6.21a) is compared with experimental data in fig. 6.6, which shows that, although equation (6.21a) is only a simple generalisation of small-strain elastic behaviour, it describes the behaviour of a real rubber to a first approximation. In particular, it describes qualitatively the initial fall in the ratio of σ_n to λ that occurs once λ rises above a rather low level. It fails, however, to describe either the extent of this fall or the subsequent increase in this ratio for high values of λ.

Fig. 6.6 A comparison of the predictions of the neo-Hookeian equation (6.21a) with experimental results. The value of G for the theoretical curve is chosen to give the experimental slope at the origin. (Reproduced by permission of the Royal Society of Chemistry.)

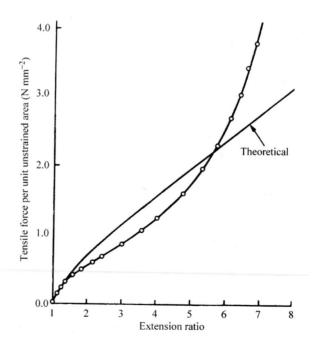

Example 6.3

A particular type of rubber behaves like a neo-Hookeian solid with a value of $G = 4 \times 10^5$ Pa. Calculate (a) the force F required to extend a long piece of this rubber of unstretched cross-section 1 cm^2 to twice its original length and (b) the force required to compress a thin sheet of cross-section 1 cm^2 to half its original thickness if this could be done in such a way that it could expand freely in the lateral directions. What would be the true stresses for (a) and (b)?

Solution

(a) $F = G(\lambda - 1/\lambda^2) \times$ unstrained area $= 4 \times 10^5 (2 - 1/4) \times 10^{-4} = 70$ N.

(b) $F = 4 \times 10^5 (1/2 - 4) \times 10^{-4} = -140$ N.

For (a) and (b) the cross-sectional areas under stress are 0.5 and 2 cm^2, respectively, and the true stresses are $70/(0.5 \times 10^{-4}) = 14 \times 10^5$ Pa and $-140/(2 \times 10^{-4}) = -7 \times 10^5$ Pa, respectively.

6.3.3 Strain–energy functions

A further phenomenological theory, which uses the concept of *strain–energy functions*, deals with more general kinds of stress than uniaxial stress. When a rubber is strained work is done on it. The strain–energy function, U, is defined as the *work done on unit volume* of material. It is unfortunate that the symbol U is conventionally used for the strain–energy function and it will be important in a later section to distinguish it from the *thermodynamic internal-energy function*, for which the same symbol is also conventionally used, but which is *not* the same quantity.

The shear strains are assumed to be zero, so that U depends only on the extension ratios λ_1, λ_2 and λ_3, i.e.

$$U = f(\lambda_1, \lambda_2, \lambda_3) \tag{6.22}$$

The choice of which is the Ox_1 axis, etc. is arbitrary, so that $U = f(\lambda_1, \lambda_2, \lambda_3)$ must be independent of the permutation of the subscripts 1, 2 and 3 on the λs. The simplest functions that satisfy this requirement are

$$\lambda_1 + \lambda_2 + \lambda_3, \qquad \lambda_1^2 + \lambda_2^2 + \lambda_3^2, \qquad \lambda_1\lambda_2 + \lambda_2\lambda_3 + \lambda_3\lambda_1, \qquad \lambda_1\lambda_2\lambda_3 \tag{6.23}$$

but there are clearly infinitely many such functions.

Rivlin first made the assumption, on rather dubious grounds, that $f(\lambda_1, \lambda_2, \lambda_3)$ involves only even powers of the λ_i. The simplest functions from which f can be constructed are then

$$I_1 = \lambda_1^2 + \lambda_2^2 + \lambda_3^2, \qquad I_2 = \lambda_1^2\lambda_2^2 + \lambda_2^2\lambda_3^2 + \lambda_3^2\lambda_1^2, \qquad I_3 = \lambda_1^2\lambda_2^2\lambda_3^2 \tag{6.24}$$

However, $I_3 = \lambda_1^2 \lambda_2^2 \lambda_3^2 = 1$ ('incompressibility'), thus

$$I_2 = 1/\lambda_1^2 + 1/\lambda_2^2 + 1/\lambda_3^2 \tag{6.25}$$

Because $U = 0$ when $\lambda_1 = \lambda_2 = \lambda_3 = 1$, the simplest possible form for U is then

$$U = C(I_1 - 3) = C(\lambda_1^2 + \lambda_2^2 + \lambda_3^2 - 3) \tag{6.26}$$

It is shown below that this equation describes a neo-Hookeian solid as defined earlier if $C = E/6$, where E is the modulus for vanishingly low uniaxial stress. A neo-Hookeian solid can thus more generally be defined as one that obeys equation (6.26).

6.3.4　The neo-Hookeian solid

The predictions of the theory for a solid that obeys equation (6.26) are now considered for various types of deformation.

(i) *Uniaxial stress* parallel to Ox_3.
For this type of stress

$$\sigma_1 = \sigma_2 = 0, \qquad \sigma_3 = \sigma, \qquad \lambda_3 = \lambda, \qquad \lambda_1 = \lambda_2 = 1/\sqrt{\lambda}$$

and equation (6.26) becomes

$$U = C(\lambda^2 + 2/\lambda - 3)$$

so that

$$dU/d\lambda = 2C(\lambda - 1/\lambda^2)$$

For a unit cube, however, $\lambda = l$ and $dU/d\lambda = dU/dl = \sigma_n$, the force per unit of original area. Substitution leads immediately to equation (6.21a) if

$$6C = E = 3G \tag{6.27}$$

i.e. it leads to the neo-Hookeian formula, which has already been compared with experiment in fig. 6.6 for extensional stress. For compressional stress the neo-Hookeian formula is found to describe experimental results very well, with the same value of G that fits the data for low extension.
(ii) *Equibiaxial extension (or inflation)*.
Equal stresses σ_1 and σ_2 are applied parallel to Ox_1 and Ox_2 (fig. 6.7) so that

$$\sigma_1 = \sigma_2 = \sigma, \qquad \sigma_3 = 0, \qquad \lambda_1 = \lambda_2 = \lambda, \qquad \lambda_3 = 1/\lambda_1\lambda_2 = 1/\lambda^2$$

Equation (6.26) with $C = G/2$ then leads to

$$U = (G/2)(2\lambda^2 + 1/\lambda^4 - 3) \tag{6.28}$$

The strain energy U_a per unit of *original unstrained* area is thus

$$U_a = (G/2)(2\lambda^2 + 1/\lambda^4 - 3)t \tag{6.29}$$

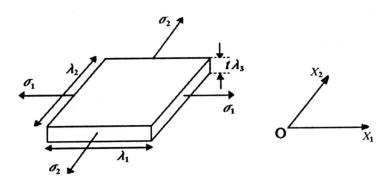

Fig. 6.7 Equibiaxial extension. The stresses σ_1 and σ_2 applied to an original unit square of material of thickness t cause it to deform to the dimensions shown.

where t is the original thickness. If F is the force acting on unit length of *strained* material then $F\lambda$ is the force acting on each of the edges (of length λ) of a square of originally unit edge. When λ changes by $d\lambda$ each pair of opposite forces does work $F\lambda\,d\lambda$, so that $2F\lambda = dU_a/d\lambda$ and

$$2F\lambda = (G/2)(4\lambda - 4/\lambda^5)t$$

so that

$$F = G(1 - 1/\lambda^6)t \tag{6.30}$$

When λ exceeds about 2 this force F becomes essentially constant and rather like a surface tension.

Biaxial stretching can be done by clamping a circular rubber sheet round its circumference and inflating it with a known pressure. Near the centre of the sheet the strain and stress are uniform, so that the sheet becomes part of the surface of a sphere in this region. The force F per unit length can be determined from the radius r of the sphere using the equation $p = 2F/r$ and the strain λ can be determined by measuring the arc length between two marked points.

The results shown in fig. 6.8 for the same rubber as that used for the simple uniaxial extension considered earlier (fig. 6.6) show that the fit between experiment and theory is good for low extension ratios with the same value of G as was used for the simple extension. At higher extension ratios the experimental force is higher than predicted.

(iii) *Simple shear.*

Simple shear is defined to be a constant-volume operation of the type illustrated in fig. 6.4. For large shearing angles θ the shear is usually defined as $\gamma = \tan\theta$. It can then be proved that, for a neo-Hookeian rubber, $\sigma = G\gamma$; experiment showed that this equation applied for the rubber of figs 6.6 and 6.8, with the same value of G as before, up to $\gamma = 1$, or $\theta = 45°$. For higher values of θ the stress was slightly lower than that given by the equation.

Fig. 6.8 A comparison of experimental data for equibiaxial stretching with the prediction of the neo-Hookeian formula for the same rubber as that for fig. 6.6, with the same value of *G* for the theoretical curve. (Reproduced by permission of the Royal Society of Chemistry.)

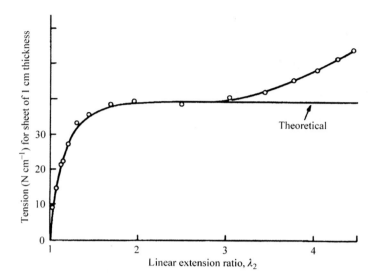

The conclusions to be drawn from the results above are that, although the predictions of equation (6.26) for the simple neo-Hookeian solid do not describe the behaviour of rubbers well at high extensional strains, they describe it well for low extensional strains, for compressional strains and for quite large simple shear strains. Discussion of modifications to the neo-Hookeian equation is deferred until section 6.5, after consideration of a more physical theory of rubber elasticity in the next section.

6.4   The statistical theory of rubber elasticity

6.4.1   Introduction

Section 6.3 deals with purely phenomenological theories. In this section the predictions of a theory based on the microstructure of a rubber are considered. By 1788 at the latest the term *rubber* was being applied to the material obtained from the latex of the tree *Hevea braziliensis* because of its ability to remove pencil marks from paper. The first printed account of this use for 'wiping off from paper the marks of black lead pencil' was given by Joseph Priestley as early as 1770. This material is now called *natural rubber* and its chemical structure is shown in fig. 6.9.

Many synthetic materials have similar physical properties. These are the *synthetic rubbers*, a subgroup of polymers often called *elastomers*. The repeat units of some important natural and synthetic rubbers are shown in fig. 6.10.

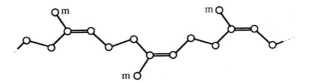

Fig. 6.9 The chemical structure of natural rubber, m = methyl group. (Hydrogen atoms are not shown.)

The double bonds in natural rubber are all in the *cis* configuration and no rotation can take place around them, but a change in molecular conformation corresponding to a rotation around any of the three single bonds per repeat unit requires very little energy. The high extensibility and elasticity of rubber are a result of the great flexibility that this confers on the molecular chains. An important feature of all the rubbers shown in fig. 6.10 is that there are many single bonds in the backbone, leading to the easy formation of non-linear structures. The other very important feature is that the *glass-transition temperatures*, T_g, for all these materials are below room temperature, so they are in the rubbery state at room temperature.

Natural rubber in its raw state will flow under continuously applied stress and is therefore not very useful. *Cross-linking* the molecules prevents flow and this can be done by heating natural rubber with sulphur, in the process called *vulcanisation*, which was discovered by Goodyear in 1839. Other chemical reagents are used more generally to cross-link the mole-

$-CH_2-C=CH-CH_2-$
$\quad\quad |$
$\quad\quad CH_3$

polyisoprene (natural rubber or gutta percha)

$-CH_2-CH=CH-CH_2-$

polybutadiene (*cis* form)

$-CH_2-C=CH-CH_2-$
$\quad\quad |$
$\quad\quad Cl$

polychloroprene (Neoprene)

$\quad\quad CH_3$
$\quad\quad |$
$-CH_2-C-$
$\quad\quad |$
$\quad\quad CH_3$

polyisobutylene (basis of 'butyl' rubber)

$-(CH_2-CH=CH-CH_2)-(CH_2-CH)-$
$\quad\quad\quad\quad\quad\quad\quad\quad\quad\quad\quad\quad |$
$\quad\quad\quad\quad\quad\quad\quad\quad\quad\quad\quad\quad \bigcirc$

*butadiene-styrene (BSR) rubber

$-(CH_2-CH=CH-CH_2)-(CH_2-CH)-$
$\quad\quad\quad\quad\quad\quad\quad\quad\quad\quad\quad\quad\quad |$
$\quad\quad\quad\quad\quad\quad\quad\quad\quad\quad\quad\quad\quad CN$

*butadiene-acrylonitrile ('nitrile') rubber

$\quad\quad CH_3$
$\quad\quad |$
$-O-Si-$
$\quad\quad |$
$\quad\quad CH_3$

poly(dimethyl siloxane) (silicone rubber)

Fig. 6.10 Repeat units of some important natural and synthetic rubbers. In the copolymers marked with an asterisk the respective monomer units occur in a random sequence along the chain (but see also section 12.3.3).

cules of elastomers. Molecular entanglements can act as natural cross-links, but they are not permanent and rubber that has not been vulcanised exhibits *stress-relaxation* or *flow* at long loading times.

6.4.2 The fundamental mechanism of rubber elasticity

The elasticity of rubbers is very different from that of materials such as metals or even glassy or semicrystalline polymers. Young's moduli for metals are typically of the order of 10^5 MPa (see table 6.1) and the maximum elastic extension is usually of order 1%; for higher extensions fracture or permanent deformation occurs. The elastic restoring force in the metal is due to interatomic forces, which fall off extremely rapidly with distance, so that even moderate extension results in fracture or in the slipping of layers of atoms past each other, leading to non-elastic, i.e. non-recoverable, deformation.

As discussed in section 6.2.2, the values of Young's modulus for isotropic glassy and semicrystalline polymers are typically two orders of magnitude lower than those of metals. These materials can be either brittle, leading to fracture at strains of a few per cent, or ductile, leading to large but non-recoverable deformation (see chapter 8). In contrast, for rubbers, Young's moduli are typically of order 1 MPa for small strains (fig. 6.6 shows that the load–extension curve is non-linear) and elastic, i.e. recoverable, extensions up to about 1000% are often possible. This shows that the fundamental mechanism for the elastic behaviour of rubbers must be quite different from that for metals and other types of solids.

Consider a single polymer chain in a liquid solvent. The chain has a high degree of flexibility and buffeting by the thermal motion of the surrounding molecules leads to a *randomly coiled* structure of the kind discussed in sections 3.3.3 and 3.3.4, i.e. to a *state of high entropy*. In this state the end-to-end distance of the chain is very much less the contour length of the chain. Figure 6.11 shows a photograph of a model of such a randomly coiled chain.

In order to move the ends of the chain further apart a force is required, which increases with the separation of the ends, because the chain becomes progressively less randomly coiled as the ends move apart and this is opposed by the randomising effect of the impacts of the surrounding molecules. There is thus an *entropic restoring force*. This force increases with increasing temperature, because this causes harder, more frequent impacts by the surrounding molecules.

Now imagine the chain to be 'dissolved' in a random assembly of other chains of the same kind, i.e. to be one of the chains in a piece of rubber. If $T > T_g$, the atoms of the surrounding chains behave like the molecules of

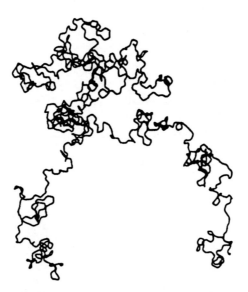

Fig. 6.11 The randomly coiled chain. A photograph of a model of a polymethylene chain containing 1000 links made by setting the links at the correct valence angle and choosing the position of each successive link in the circle of rotation as one of the six equally spaced positions obtained by throwing a die. (Reproduced by permission of Oxford University Press.)

the solvent. Stretching the rubber stretches each of the individual chains from its random-coil conformation and leads to a net entropic restoring force on the material as a whole. This is the most important mechanism for rubber elasticity. For large extensibility a large number of single bonds between cross-link points is required, because this ensures that there is a large ratio between the fully extended length of such a chain segment and the distance between the ends of the segment in the randomly coiled state (see equation (3.5)).

Before developing the entropic, or statistical, theory of rubber elasticity in a quantitative way, it is important to be sure that this really is the most important contribution, i.e. to be sure that any contribution to the elasticity due to changes in the internal energy on stretching is very small compared with the contribution due to changes of entropy. This is shown to be so in the following section.

6.4.3 The thermodynamics of rubber elasticity

As early as 1935, Meyer and Ferri showed experimentally that the stretching force F required to maintain a rubber at constant strain is very nearly proportional to the absolute temperature T, i.e.

$$F = aT \tag{6.31}$$

where a is a constant for a given rubber. It is important to note that the strain must be referred to the unstressed length at the temperature at which the force is measured in order to eliminate the effect of the thermal expansion of the rubber.

The thermodynamic argument that follows shows that, if equation (6.31) holds, the internal energy of the rubber is independent of its length at constant temperature and, to this approximation, the force F thus *depends only on the rate of change of entropy, S, with length.*

The first law of thermodynamics states that

$$dQ = dU + dW \tag{6.32}$$

where dQ is the heat entering the system, dU is the increase in the internal energy of the system, dW is the work done by the system and, for a reversible process, $dQ = T\,dS$. If a force F acting on a filament of rubber produces an increase in length dl the filament does work $-F\,dl$ and equation (6.32) leads to

$$F\,dl = dU - T\,dS \quad \text{or} \quad F = \left(\frac{\partial U}{\partial l}\right)_T - T\left(\frac{\partial S}{\partial l}\right)_T \tag{6.33}$$

Differentiating the Helmholtz function $A = U - TS$ leads to

$$dA = dU - T\,dS - S\,dT = F\,dl - S\,dT \tag{6.34}$$

Thus

$$F = \left(\frac{\partial A}{\partial l}\right)_T \quad \text{and} \quad -S = \left(\frac{\partial A}{\partial T}\right)_l \tag{6.35a,b}$$

For any function f of two variables, say $f(x, y)$, it is always true that

$$\frac{\partial}{\partial x}\left(\frac{\partial f}{\partial y}\right)_x = \frac{\partial}{\partial y}\left(\frac{\partial f}{\partial x}\right)_y \tag{6.36}$$

Applying this equality with $f = A, x = l$ and $y = T$ to equations (6.35) leads to

$$\left(\frac{\partial S}{\partial l}\right)_T = -\left(\frac{\partial F}{\partial T}\right)_l \tag{6.37}$$

Substitution in equation (6.33) then leads to

$$\left(\frac{\partial U}{\partial l}\right)_T = F - T\left(\frac{\partial F}{\partial T}\right)_l \tag{6.38}$$

However, equation (6.31) shows that $F - T(\partial F/\partial T)_l = 0$, which proves that $(\partial U/\partial l)_T = 0$, i.e. that the internal energy is independent of length at constant temperature. In the statistical theory of rubber elasticity it is assumed that this result can be generalised to any kind of deformation of the rubber. The aim of the statistical theory is therefore to calculate the change in entropy when a rubber is deformed and, by relating this to the work done in the deformation, to arrive at the form of the strain–energy function for isothermal stretching.

6.4.4 Development of the statistical theory

When polymer chains are cross-linked to form a non-flowing rubber, a *molecular network* is obtained. It is shown in section 3.3.4 that the freely jointed random-link model of polymer chains is applicable to rubbers provided that the equivalent random link is correctly chosen. In considering the network the following *simplifying assumptions* will therefore be made, leading to the simplest form of the theory.

(i)   Each molecular segment (chain) between two adjacent cross-links is a *freely jointed Gaussian chain* of n links, each of length l, where n may vary from one segment to another but l is the same for all segments.

(ii) Each cross-link remains *fixed* at its mean position in any state, stretched or unstretched.

(iii) When the network is stretched it undergoes an *affine deformation*, i.e. the components of any vector joining cross-link points are multiplied by the corresponding extension ratios λ, as illustrated for two dimensions in fig. 6.12.

The change in entropy of the ideal network caused by the deformation is the sum of the changes for all chains. It is shown in section 3.3.4 that the entropy of a single chain is given by

$$S = c - kb^2 r^2 \tag{6.39}$$

Fig. 6.12 Affine deformation. Projection onto the OX_1X_2 plane of a parallelepiped within the rubber with dimensions a, b and c parallel to the OX_1, OX_2 and OX_3 axes, respectively. (a) Chains before deformation, showing a particular vector r joining two cross-link points, ●, and its components. (b) The same chains and the vector r' joining the same two cross-link points after deformation. The component r_3 is transformed into $\lambda_3 r_3$.

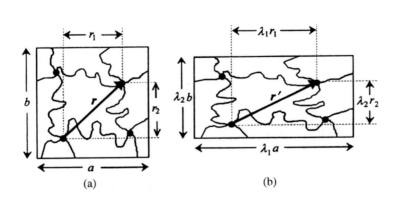

(a)   (b)

where r is the end-to-end distance of the chain, c is independent of r and $b^2 = 3/(2nl^2)$.

Consider first all chains with a particular value of n, and hence of b, and let there be N_b such chains. The entropy s_b contributed by one of these chains is given, from equation (6.39), by

$$s_b = c - kb^2(x^2 + y^2 + z^2) \tag{6.40}$$

provided that the origin of the rectangular co-ordinate system $Oxyz$ is at one end of the chain and the other end of the chain is at (x, y, z).

If, after stretching, the origin is moved to the same end of the chain as before (or the rubber is moved correspondingly) the new contribution of this chain, s'_b, is given by

$$s'_b = c - kb^2(\lambda_1^2 x^2 + \lambda_2^2 y^2 + \lambda_3^2 z^2) \tag{6.41}$$

The increase in entropy due to these N_b chains is thus $\Delta S_b = \sum(s'_b - s_b)$, where the sum is taken only over the N_b chains, so that

$$\begin{aligned}
\Delta S_b &= -kb^2 \sum[(\lambda_1^2 - 1)x^2 + (\lambda_2^2 - 1)y^2 + (\lambda_3^2 - 1)z^2] \\
&= -kN_b b^2[(\lambda_1^2 - 1)\langle x^2 \rangle + (\lambda_2^2 - 1)\langle y^2 \rangle + (\lambda_3^2 - 1)\langle z^2 \rangle]
\end{aligned} \tag{6.42}$$

where the angle brackets $\langle \, \rangle$ denote the average values for the N_b chains.

The end-to-end vectors of the N_b chains are, however, originally directed equally in all directions because of the assumed *isotropy* of the medium before stretching. Thus, if r_b is the length of a chain,

$$\langle x^2 \rangle = \langle y^2 \rangle = \langle z^2 \rangle = \langle r_b^2 \rangle/3 \tag{6.43}$$

It is a reasonable assumption that, in the unstressed state, $\langle r_b^2 \rangle$ is equal to the mean-square length of totally free chains with the given value of b. However, $b^2 = 3/(2nl^2)$ and equation (3.5) shows that $\langle r_b^2 \rangle = l^2 n$, so that

$$\langle x^2 \rangle = \langle y^2 \rangle = \langle z^2 \rangle = \langle r_b^2 \rangle/3 = 1/(2b^2) \tag{6.44}$$

Thus

$$\Delta S_b = -\tfrac{1}{2}N_b k(\lambda_1^2 + \lambda_2^2 + \lambda_3^2 - 3) \tag{6.45}$$

This equation shows that, although ΔS_b is proportional to N_b, as expected, it does not depend explicitly on the value of b, so that the total change of entropy ΔS per unit volume of the network is

$$\Delta S = -\tfrac{1}{2}Nk(\lambda_1^2 + \lambda_2^2 + \lambda_3^2 - 3) \tag{6.46}$$

where N is the *total* number of chains in unit volume of the network.

Because the internal energy of a rubber at constant temperature is independent of the deformation, the first law of thermodynamics, equation (6.32), shows that $dW = dQ$, where dW is the work done *by* the rubber and dQ is the heat entering the rubber during deformation. If the deformation is performed reversibly, $dQ = T\, dS$, so that

$$dW = T\, dS \qquad (6.47)$$

Thus the total work done *on* unit volume of the rubber system during deformation, which was defined in section 6.3.3 to be the strain–energy function, is $-\int T\, dS = -T\, \Delta S$, with ΔS given by equation (6.46).

As already noted in section 6.3.3, the strain–energy function is usually given the symbol U, in spite of the fact that it is *not* equal to the internal energy. Using this terminology, equation (6.46) shows that

$$U = -T\, \Delta S = \tfrac{1}{2}NkT(\lambda_1^2 + \lambda_2^2 + \lambda_3^2 - 3) \qquad (6.48)$$

which is the neo-Hookeian form for U given in equation (6.26) provided that $C = G/2 = E/6 = \tfrac{1}{2}NkT$. In particular, equation (6.21) for the true stress under uniaxial extension becomes

$$\sigma = NkT(\lambda^2 - 1/\lambda) \qquad (6.49)$$

The strain–energy function is actually equal to the change in the *Helmholtz function A* for the rubber, since $\Delta A = -T\, \Delta S$ for an isothermal change for a medium like rubber in which there is no change in the internal energy at constant temperature.

The behaviour predicted for a rubber with U given by an equation of the form (6.48) has been discussed and compared with experimental data in sections 6.3.2 and 6.3.4. The prediction is either good or at least describes the data to a first approximation at low strain, depending on the type of strain. The statistical theory is, however, an improvement on the phenomenological theories discussed there in three ways:

(i) it incorporates the most important features of the molecular structure of a real rubber;

(ii) it predicts that the modulus E should be directly proportional to the absolute temperature T; and

(iii) it predicts that the modulus should be directly proportional to the degree of cross-linking, specified by N. (For the ideal network, N is equal to twice the cross-link density.)

Example 6.4

A cross-linked rubber of undeformed cross-section 30 mm × 2 mm is stretched at 300 K to twice its original length by subjecting it to a load of 15 N. If the density ρ of the rubber is 950 kg m^{-3}, what is the mean molar mass M of the network chains?

Solution

If the length is doubled, the cross-sectional area is halved. The true stress σ applied to the rubber is thus $2F/A$, where F is the applied force and A is the original cross-sectional area. Substituting this value of σ and the values 2 and 300 K for λ and T into equation (6.49) leads to

$\sigma = 2 \times 15/(30 \times 2 \times 10^{-6}) = Nk \times 300(4 - \frac{1}{2})$ and hence to $N = 3.45 \times 10^{25}$ m^{-3}. Thus the mass m per chain is $m = \rho/N = 950/(3.45 \times 10^{25}) = 2.75 \times 10^{-23}$ kg and $M = mN_A = 2.75 \times 10^{-20} \times 6.02 \times 10^{23} = 1.66 \times 10^4$ g mol^{-1}.

6.5   Modifications of the simple molecular and phenomenological theories

More elaborate versions of the molecular theory may account for entanglement, loose ends or loose loops, deviations from Gaussian statistics, the statistics of actual chains, excluded-volume effects (real chains with finite thickness cannot interpenetrate each other), the movement of junction points, the contribution of the internal energy to elasticity, changes in volume, etc.

Developments of the phenomenological theories are also possible and many have been attempted. Ogden, for instance, assumed a more general form for U, viz.

$$U = \sum_i \frac{\mu_i}{\alpha_i} (\lambda_1^{\alpha_i} + \lambda_2^{\alpha_i} + \lambda_3^{\alpha_i} - 3) \tag{6.50}$$

He showed that a satisfactory fit to experimental data for tension, pure shear and equibiaxial tension could then be obtained with a three-term expression. Such expressions are very useful for comparing rubbers, but it is not possible to justify them on fundamental grounds.

6.6   Further reading

(1) *An Introduction to the Mechanical Properties of Solid Polymers*, by I. M. Ward and D. W. Hadley, John Wiley & Sons, Chichester, 1993 and *Mechanical Properties of Solid Polymers*, by I. M. Ward, 2nd Edn, John Wiley & Sons, 1983. A much more detailed account of the mechanical properties than that given in chapters 6–8 and 11,

at a similar or slightly more advanced level (particularly in the second book). The books describe methods of measurement as well as results. A new edition of the first book, by I. M. Ward and J. Sweeney, is in preparation.

(2) *The Physics of Rubber Elasticity*, by L. R. G. Treloar, 3rd Edn, Oxford, 1975. This book gives a much more detailed account of rubber elasticity than that in this chapter, but it is nevertheless very readable.

6.7 Problems

6.1. Prove equation (6.6). (Hint: resolve each of the stresses σ in fig. 6.4 into two components acting at 45° to the corresponding face of the cube and at right angles to each other and then consider the resulting tensile forces parallel to the diagonals of the cube.)

6.2. Prove equation (6.7).

6.3. A thin sheet of material is subjected to a biaxial stress field in the Ox_1x_2 plane, so that the shear strains are zero and $\sigma_1 = 9$ MN m^{-2}, $\sigma_2 = 6$ MN m^{-2} and $\sigma_3 = 0$. If Young's modulus E of the material is 3 GN m^{-2} and its Poisson ratio ν is 0.30, calculate the extensional strains e_1, e_2 and e_3.

6.4. A cube of a material that may be considered incompressible to a good approximation with axes $Ox_1x_2x_3$ along the cube axes is subjected to the following stress field: $\sigma_1 = 8$ MPa, $\sigma_2 = 7$ MPa and $\sigma_3 = 5$ MPa. Given that Young's modulus for small strains is 4 GPa, calculate the strain in the Ox_1 direction. If σ_3 were reduced to zero, what values would be required for σ_1 and σ_2 if the state of strain of the material were to remain unchanged?

6.5. Calculate Young's modulus of a sample with the same average crystalline and amorphous moduli and crystalline volume fraction as in example 6.2 but assuming that there is uniform strain throughout the sample.

6.6. A cylindrical piece of rubber 10 cm long and 2 mm in diameter is extended by a simple tensile force F to a length of 20 cm. If the rubber behaves as a neo-Hookeian solid with a Young's modulus of 1.2 N mm^{-2}, calculate (i) the diameter of the stretched cylinder, (ii) the value of the nominal stress, (iii) the value of the true stress and (iv) the value of F.

6.7. A spherical balloon is made from the same type of rubber as in problem 6. If the balloon has diameter 2 cm and wall thickness 1 mm when it is open to the atmosphere, calculate the pressure inside it when it is blown up to diameters of 2.2, 2.5, 3, 5 and 10 cm.

6.8. Calculate the force required to increase the length of a rubber cord by 50% if the rubber has 2×10^{25} cross-links per m^3 and the cross-section of the cord is a circle of radius 1 mm. Assume that the simple statistical theory applies and that the stretching is performed at 300 K. Consider carefully the assumptions needed when relating the number of cross-links per unit volume to the number of chains per unit volume.

6.9. The increase in force necessary to extend a piece of rubber to which equation (6.49) applies from twice its natural length to three times its natural length at 300 K is 1 N. If the piece of rubber has cross-sectional area 1 mm^2, calculate the number of chains per unit volume in the rubber and the rise in temperature that would require the same increase in force to maintain the rubber at twice its natural length.

6.10. The elastic behaviour of a particular type of rubber is well described by an equation of the form of equation (6.50) with only one term in the summation, with $\alpha = 1.5$ and $\mu/\alpha = 4.2 \times 10^5$ N m^{-2}. A square of this material, of thickness 1 mm and edge length 10 cm, is stretched by applying a total force f to each edge in such a way that it remains square and its area doubles. Calculate the value of f.

Chapter 7
Mechanical properties II – linear viscoelasticity

7.1 Introduction and definitions

7.1.1 Introduction

In section 6.1 it is pointed out that, at low temperatures or high frequencies, a polymer may be glass-like, whereas at high temperatures or low frequencies it may be rubber-like. In an intermediate range of temperature or frequency it will usually have *viscoelastic properties*, so that it undergoes *creep* under constant load and *stress-relaxation* at constant strain. The fundamental mechanisms underlying the viscoelastic properties are various relaxation processes, examples of which are described in section 5.7. The present chapter begins with a macroscopic and phenomenological discussion of linear viscoelasticity before returning to further consideration of the fundamental mechanisms.

Consider first the deformation of a *perfect elastic solid*. The work done on it is stored as energy of deformation and the energy is released completely when the stresses are removed and the original shape is restored. A metal spring approximates to this ideal. In contrast, when a *viscous liquid* flows, the work done on it by shearing stresses is dissipated as heat. When the stresses causing the flow are removed, the flow ceases and there is no tendency for the liquid to return to its original state. Viscoelastic properties lie somewhere between these two extremes.

An *isotropic perfectly elastic solid* obeys equation (6.5) or

$$\sigma = G\theta \tag{7.1}$$

if σ is the shearing force and the shearing angle is θ (see fig. 7.1(a)), whereas a *perfect Newtonian liquid* obeys the equation (see fig. 7.1(b))

$$\sigma = \eta\frac{d\theta}{dt} \tag{7.2}$$

where η is the *viscosity* of the liquid. The simplest assumption to make about the behaviour of a *viscoelastic solid* would be that the shear stress depends linearly both on θ and on $d\theta/dt$, i.e. that

Fig. 7.1 The responses to shear stress of (a) an elastic solid and (b) a Newtonian liquid. The arrows in (b) show the variation of the velocity v across the sample.

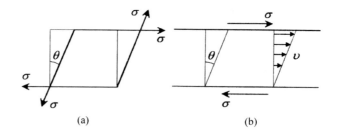

$$\sigma = G\theta + \eta\frac{\mathrm{d}\theta}{\mathrm{d}t} \tag{7.3}$$

This is essentially the result obtained for one of the simple models for viscoelasticity to be considered below.

The models will assume *linearity*, as expressed in equations (7.1) and (7.2). These equations apply only at *small strains* and considerations in this chapter are restricted to such small strains. It should, however, be noted that real polymers are often non-linear even at small strains. The subject of *non-linear viscoelasticity* lies outside the scope of this book but is considered in the books referred to as (1) in section 7.7.

Before the models are described, the two simple aspects of viscoelastic behaviour already referred to – *creep* and *stress-relaxation* – are considered. For the full characterisation of the viscoelastic behaviour of an isotropic solid, measurements of at least two moduli are required, e.g. Young's modulus and the rigidity modulus. A one-dimensional treatment of creep and stress-relaxation that will model the behaviour of measurements of either of these (or of other measurements that might involve combinations of them) is given here. Frequently *compliances*, rather than moduli, are measured. This means that a stress is applied and the strain produced per unit stress is measured, whereas for the determination of a modulus the stress required to produce unit strain is measured. When moduli and compliances are time-dependent they are not simply reciprocals of each other.

7.1.2 Creep

Figure 7.2 shows the effect of applying a stress σ, e.g. a tensile load, to a linear viscoelastic material at time $t = 0$. The resulting strain $e(t)$ can be divided into three parts:

(i) e_1, an essentially instantaneous response, similar to that of an elastic solid;

(ii) $e_2(t)$, which tends to a constant value as t tends to ∞; and

(iii) $e_3(t)$, which is linear in time.

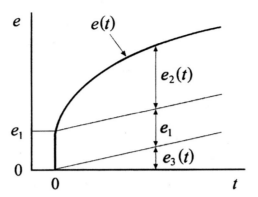

Fig. 7.2 Creep of a viscoelastic solid under a constant stress.

Assuming linearity, i.e. that each part of the strain is proportional to the applied stress, a *time-dependent creep compliance J(t)* can be defined as

$$J(t) = \frac{e(t)}{\sigma} = \frac{e_1}{\sigma} + \frac{e_2(t)}{\sigma} + \frac{e_3(t)}{\sigma} = J_1 + J_2(t) + J_3(t) \qquad (7.4)$$

The term $J_3(t)$ corresponds to *flow* and will be assumed to be zero. (For cross-linked polymers it is zero and for crystalline polymers it is approximately zero). The J_1 term really corresponds to a response that is faster than can be observed experimentally, rather than an instantaneous one; this is the response discussed in section 6.2.2 of the previous chapter. The strain e_1 is sometimes called the *unrelaxed response*, in contrast to the *relaxed response* observed at long times, $e(\infty)$. J_1 and $J_2(t)$ are not usually considered separately in what follows, so that $J(t)$ will mean $J_1 + J_2(t)$.

Figure 7.3 shows log $J(t)$ plotted against log t at constant temperature T over a very wide range of time t for an idealised amorphous polymer with

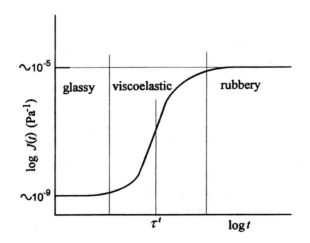

Fig. 7.3 Compliance as a function of time for an idealised amorphous polymer.

only one relaxation transition. There are three regimes, depending on the time-scale Δt or frequency ν used in any experiment:

(a) $\Delta t \ll \tau'$ (high ν) – the polymer is *glassy*, with a constant low compliance;

(b) $\Delta t \gg \tau'$ (low ν) – the polymer is *rubbery*, with a constant high compliance; and

(c) $\Delta t \sim \tau'$ – the polymer is *viscoelastic* with an intermediate time-dependent compliance

The time τ' is called the *retardation time* and its value depends on the nature of the polymer and on the temperature, T. If T increases, the frequency of molecular rearrangements rises and τ' falls. This *time–temperature equivalence* is considered in detail later.

It is important to realise that the very different shapes of the curves in figs 7.2 and 7.3, which are really expressing the same thing, arise because in fig. 7.3 it is assumed that $J_3(t)$ is zero (no flow) and because the scales are logarithmic in fig. 7.3 but linear in fig. 7.2.

Example 7.1: creep

A straight rod of polymer is 10 mm in diameter ($2r$) and 1 m long. The polymer behaves in a linear viscoelastic manner with a tensile creep compliance that can be well approximated by $J(t) = (2 - e^{-0.1t})$ GPa^{-1}, where t is in hours. The rod is suspended vertically and a mass M of 10 kg is hung from it. Find the change in length after (i) 1 h, (ii) 10 h and (iii) 100 h.

Solution

$e(t) = J(t)\sigma \qquad \sigma = Mg/(\pi r^2) = 10 \times 9.81/[\pi \times (5 \times 10^{-3})^2] = 1.249 \times 10^6$ Pa

$e(1) = J(1)\sigma = 1.249 \times 10^6(2 - e^{-0.1 \times 1}) \times 10^{-9} = 1.37 \times 10^{-3} \equiv 1.37$ mm
change in 1 m

$e(10) = J(10)\sigma = 1.249 \times 10^6(2 - e^{-0.1 \times 10}) \times 10^{-9} = 2.04 \times 10^{-3} \equiv 2.04$ mm
change in 1 m

$e(100) = J(100)\sigma = 1.249 \times 10^6(2 - e^{-0.1 \times 100}) \times 10^{-9} = 2.50 \times 10^{-3} \equiv 2.50$ mm
change in 1 m

7.1.3 Stress-relaxation

If a sample is subjected to a *fixed strain* the stress rises 'immediately' and then falls with time to a final constant value, assuming that there is no flow, as shown in fig. 7.4.

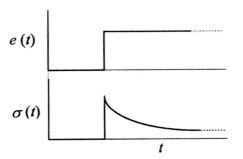

Fig. 7.4 Stress-relaxation. The upper graph shows the applied strain as a function of time and the lower graph the resulting stress.

The stress-relaxation modulus $G(t)$ is defined by

$$G(t) = \sigma(t)/e \tag{7.5}$$

A plot of $\log G(t)$ against $\log t$ is similar to an inverted form of $\log J(t)$ against $\log t$. A characteristic *relaxation time* τ analogous to τ' can be defined.

7.1.4 The Boltzmann superposition principle (BSP)

Boltzmann extended the idea of linearity in viscoelastic behaviour to take account of the time dependence He assumed that, in a creep experiment;

(i)   the strain observed at any time depends on the entire stress history up to that time and

(ii) each step change in stress makes an independent contribution to the strain at any time and these contributions add to give the total observed strain.

This leads to the following interpretation of the creep compliance $J(t)$: any incremental stress $\Delta\sigma$ applied at time t' results in an incremental strain $\Delta e(t)$ at a later time t given by $\Delta e(t) = \Delta\sigma J(t - t')$, where $t - t'$ is the time that has elapsed since the application of $\Delta\sigma$.

As an example, fig. 7.5 illustrates a two-step loading programme. The dashed curve shows the strain $\Delta\sigma_1 J(t - t_1)$ that would have been present at time $t > t_2$ if the second step in stress $\Delta\sigma_2$ had not been applied at t_2. The actual strain is this strain plus the strain $\Delta\sigma_2 J(t - t_2)$ produced by the second step in stress applied at t_2.

It is easy to generalise to a *multi-step loading programme*:

$$e(t) = \Delta\sigma_1 J(t - t_1) + \Delta\sigma_2 J(t - t_2) + \Delta\sigma_3 J(t - t_3) + \cdots \tag{7.6}$$

or to a *continuously changing stress*:

$$e(t) = \int_{-\infty}^{t} J(t - t') \, d\sigma(t') = \int_{-\infty}^{t} J(t - t') \frac{d\sigma(t')}{dt'} \, dt'$$

where $\sigma(t)$ is the stress at time t.

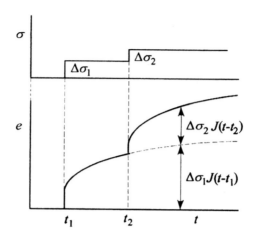

Fig. 7.5 A two-step creep experiment. See the text for explanation and symbols.

An interesting and important application of the BSP is creep followed by recovery. Consider the loading programme shown in fig. 7.6: apply stress σ at time t_1; hold the stress constant until time t_2 and then reduce the stress to zero. This last step is equivalent to applying an *additional stress* $-\sigma$ at t_2. Thus, according to the BSP, at $t > t_2$, $e(t) = \sigma J(t - t_1)$ $-\sigma J(t - t_2)$. Note that recovery e_r is not defined as one might expect; it is defined as the difference between the existing strain at t and the strain that would have been observed at t if the stress had not been removed

The BSP applies in an exactly analogous way to stress-relaxation, so that the stress-relaxation modulus $G(t - t')$ can also be interpreted in an analogous way:

$$\sigma(t) = \int_{-\infty}^{t} G(t - t')\, \mathrm{d}e(t') = \int_{-\infty}^{t} G(t - t') \frac{\mathrm{d}e(t')}{\mathrm{d}t'}\, \mathrm{d}t' \tag{7.8}$$

Creep and stress-relaxation are different manifestations of the viscoelastic nature of a polymer and must be related to each other. It is possible to show that

$$\int_{0}^{t} G(t')J(t - t')\, \mathrm{d}t' = t \tag{7.9}$$

Fig. 7.6 Creep and recovery. The upper graph shows the applied stress as a function of time and the lower graph the resulting strain.

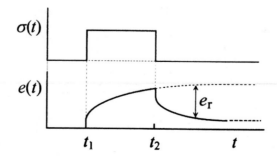

In practice the relationship between creep and stress-relaxation data is usually considered in terms of *relaxation- or retardation-time spectra* (see section 7.2.5) or by the use of various approximate methods.

7.2   Mechanical models

7.2.1   Introduction

So far the ideas of relaxation and retardation times have been defined rather loosely in terms of plots of $J(t)$ or $G(t)$ against $\log t$, each being specified as the value of t at the centre of the region of rapid change in the value of $J(t)$ or $G(t)$. In order to specify the meaning of τ or τ' more closely, it is necessary to consider the simplest form of creep or stress-relaxation that might be expected. It is supposed that, when a constant stress σ is applied to a polymer, it creeps to the equilibrium strain in such a way that the rate of change of strain at any time t is proportional to the difference between the equilibrium strain σJ and the existing strain e, so that

$$\frac{de}{dt} = \frac{\sigma J - e}{\tau'} \tag{7.10}$$

where J is the creep compliance at infinite time. This equation then defines the retardation time τ' and a similar definition of the relaxation time τ for the simplest kind of stress-relaxation can be given. Integration of either equation leads to an exponential relaxation of strain or stress to the equilibrium value, which is consistent with the simplest forms of relaxation discussed in sections 5.7.2 and 5.7.3.

Viscoelastic behaviour is often represented by mechanical models consisting of *elastic springs* that obey Hooke's law and *dashpots* containing viscous liquids that obey Newton's law, equation (7.2). The two simplest models each use one spring and one dashpot and lead to simple exponential relaxation:

(a) the *Maxwell model* – spring and dashpot in series; and
(b) the *Kelvin* or *Voigt model* – spring and dashpot in parallel.

These models have relaxation and retardation behaviour of the simplest kind described by equation (7.10) and its equivalent for stress-relaxation and it can be shown (see problem 7.2) that both models obey the BSP. In discussing the models, which are really one-dimensional, no distinction is made between load and stress, which is equivalent to considering unit cross-section.

7.2.2　The Maxwell model

This model consists of a spring and dashpot in series, as shown schematically in fig. 7.7. If a fixed strain is suddenly applied, the spring responds immediately by extending and a stress is produced in it, which is therefore also applied to the dashpot. The dashpot cannot be displaced instantaneously but begins to be displaced at a rate proportional to the stress. The strain and stress in the spring thus decay to zero as the dashpot is displaced at a decreasing rate and is eventually displaced by the same amount as the spring was originally displaced. This is therefore a model for stress-relaxation, but the stress relaxes to zero, which is not always the case for real polymers. Under constant stress, the spring remains at constant length, but the dashpot is displaced at a constant rate, so that the model cannot describe creep.

For the spring

$$\sigma_s = Ee_s \tag{7.11}$$

and for the dashpot

$$\sigma_d = \eta \frac{de_d}{dt} \tag{7.12}$$

For a series system

$$\sigma = \sigma_s = \sigma_d \quad \text{and} \quad e = e_s + e_d \tag{7.13}$$

Thus

$$\frac{de}{dt} = \frac{de_s}{dt} + \frac{de_d}{dt} = \frac{1}{E}\frac{d\sigma}{dt} + \frac{\sigma}{\eta} \tag{7.14}$$

Fig. 7.7 The Maxwell model: spring and dashpot in series.

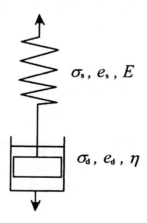

σ_s, e_s, E

σ_d, e_d, η

For stress-relaxation at constant strain e is independent of t, so that

$$\frac{de}{dt} = 0 \quad \text{and} \quad \frac{1}{E}\frac{d\sigma}{dt} + \frac{\sigma}{\eta} = 0 \qquad (7.15)$$

Integration leads to

$$\sigma = \sigma_o \exp(-Et/\eta) = \sigma_o \exp(-t/\tau) \qquad (7.16)$$

with $\sigma_o = eE$. The stress decays exponentially to zero with a *relaxation time* $\tau = \eta/E$. It follows from equation (7.14) that, if σ is constant, $de/dt = \sigma/\eta$ and is constant. This is the formal expression of the fact that this model cannot describe creep, as has already been indicated.

Example 7.2: the Maxwell model

A Maxwell element consists of an elastic spring of modulus $E = 10^9$ Pa and a dashpot of viscosity 10^{11} Pa s. Calculate the stress at time $t = 100$ s in the following loading programme: (i) at time $t = 0$ an instantaneous strain of 1% is applied and (ii) at time $t = 30$ s the strain is increased instantaneously from 1% to 2%.

Solution

$\tau = \eta/E = 10^{11}/10^9 = 100$ s.
According to the BSP, for $t > t_2 > t_1$
$\sigma = \sigma_1 \exp[-(t - t_1)/\tau] + \sigma_2 \exp[-(t - t_2)/\tau]$.
However, $\sigma_1 = \sigma_2 = 0.01E = 10^7$ Pa and
$t_1 = 0$ and $t_2 = 30$ s.
Thus the stress at $t = 100$ s is
$\sigma(100) = 10^7(e^{-100/100} + e^{-70/100}) = 10^7(e^{-1} + e^{-0.7}) = 8.6 \times 10^6$ Pa

---

7.2.3 The Kelvin or Voigt model

This model consists of a spring and dashpot in parallel, as shown schematically in fig. 7.8. If a fixed stress is suddenly applied, the dashpot cannot be displaced instantaneously, so that the spring does not change in length and carries none of the stress. The dashpot is then displaced at a decreasing rate as the spring strains and takes up some of the stress. Eventually the dashpot and spring have both been displaced far enough for the spring to take the whole load. This is thus a model for creep. For constant strain, there is no relaxation of stress and there is also no way to apply an immediate finite

Fig. 7.8 The Kelvin or Voigt
model: spring and dashpot
in parallel.

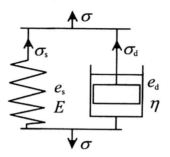

strain, because this would require an infinite stress to be applied to the
dashpot. The model cannot therefore describe stress-relaxation.

Equations (7.11) and (7.12) still apply to the stresses, but for a parallel
system

$$e = e_s = e_d \quad \text{and} \quad \sigma = \sigma_s + \sigma_d \tag{7.17}$$

Thus

$$\sigma = Ee + \eta \frac{\mathrm{d}e}{\mathrm{d}t} \tag{7.18}$$

For creep under constant load, $\sigma = \sigma_0$ and it then follows, as can easily be
checked by differentiation, that

$$e = \frac{\sigma_o}{E}\left[1 - \exp\left(\frac{-E}{\eta}t\right)\right] \tag{7.19}$$

The strain rises exponentially to σ_0/E, with *retardation time* $\tau' = \eta/E$.

The fact that this model cannot describe stress-relaxation is formally
expressed by equation (7.18), which shows that, if $\mathrm{d}e/\mathrm{d}t = 0$, $\sigma = Ee$ and is
constant. Notice that equation (7.18) reduces to equation (7.10) if η/E is
replaced by τ', E is replaced by $1/J$ and a slight rearrangement is made.

7.2.4   The standard linear solid

The *standard linear solid* (SLS) is a more complicated model than the two
previously considered. It combines series and parallel elements, as shown
in fig. 7.9, and can describe both stress-relaxation and creep. For *stress-
relaxation* the spring 'a' remains at the original strain and only E_b and η are
involved in the relaxation. Hence $\tau = \eta/E_b$, but the stress relaxes to eE_a,
not to zero. For creep it can be shown that $\tau' = (1/E_a + 1/E_b)\eta$. Unlike the
Voigt model, the SLS exhibits an *immediate response*, $e = \sigma/(E_a + E_b)$,
because the two springs in parallel can extend immediately. Thus the
SLS is a much better model than either of the simpler models.

Example 7.3: the Kelvin (Voigt) model

The behaviour of a polymer is described by a Voigt element whose creep strain e under constant load σ_0 applied for a time t is given by
$e = (\sigma_0/E)[1 - \exp(-Et/\eta)]$, where E is the elastic modulus of the spring and η is the viscosity of the dashpot. If $E = 10^{10}$ Pa and $\eta = 10^{12}$ Pa s, calculate the strain at 100 s for the following loading programme: a stress of 3×10^8 Pa is applied at time $t = 0$ and is reduced to a stress of 2×10^8 Pa at $t = 50$ s.

Solution
$\tau' = \eta/E = 10^{12}/10^{10} = 100$ s.
By the BSP, for $t > t_2 > t_1$ the strain e is
$e = (\sigma_1/E)\{1 - \exp[-(t - t_1)/\tau']\} +$
$(\sigma_2/E)\{1 - \exp[-(t - t_2)/\tau']\}$
with $t_1 = 0$, $t_2 = 50$ s, $\sigma_1 = 3 \times 10^8$ Pa and
$\sigma_2 = (2 \times 10^8 - 3 \times 10^8)$ Pa $= -10^8$ Pa. Thus
$e = (3 \times 10^8/10^{10})[1 - \exp(-100/100)] +$
$(-10^8/10^{10})[1 - \exp(-50/100)]$
$= 10^{-2}\{3[1 - \exp(-1)] - [1 - \exp(-\tfrac{1}{2})]\} = 1.5 \times 10^{-2} = 1.5\%$

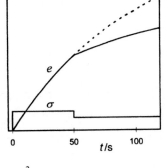

7.2.5 Real materials – relaxation-time and retardation-time spectra

All the simple models discussed exhibit exponential behaviour for relaxation or creep, but real solids do not. As described in section 5.7, there can be several different relaxation processes with different relaxation times and the glass transition itself cannot be represented by a single exponential relaxation. This more complicated behaviour can be modelled by a number of Maxwell elements in parallel. The total stress is the sum of the stresses in the individual elements, each of which relaxes with its own characteristic

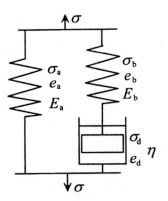

Fig. 7.9 The standard linear solid model.

relaxation time. This leads to the idea of a *relaxation-time spectrum*. A similar treatment with a number of Voigt elements in series leads to a *retardation-time spectrum*.

7.3   Experimental methods for studying viscoelastic behaviour

In order to characterise the viscoelastic behaviour of a material as fully as possible it is important to work over a very wide time-scale or, for oscillatory methods, to use as wide a range of frequencies as possible. The methods that have been used fall into five main classes covering the approximate frequency ranges shown in Table 7.1. Some of them are now discussed.

7.3.1   Transient measurements

As formalised in the BSP, the viscoelastic response of a sample of polymer depends on the *mechanical history* of the sample up to the time when the response is measured. It is therefore necessary to *condition* the sample before making measurements by subjecting it to successive creep and recovery cycles in the following way:

  (i)   apply the maximum load to be used in the measurement for the maximum period of loading to be used;
 (ii)   remove the load for about ten times the loading period; and
(iii)   repeat (i) and (ii) several times.

The simplest method for measuring *extensional creep* is the *dead-loading* method. The simplest form of this uses a cathetometer to measure, at various times, the separation of two marks on a long sample hanging under a fixed load. A cathetometer is simply a device consisting of a vertical stand along which a horizontally mounted microscope (or telescope) can be moved. A more sophisticated version of this experiment

Table 7.1. *Methods for characterising viscoelastic behaviour*

Method	Approximate frequency range (Hz)
Creep and stress-relaxation measurements	$<10^{-6}$–1
Low-frequency free-oscillation methods	0.1–10
Forced vibration methods	10^{-2}–10^{2}
High-frequency resonance methods	10^{2}–10^{4}
Wave-propagation methods	$>10^{4}$

makes use of a linear displacement transducer, as shown in fig. 7.10. The transducer, which may consist of the 'slug' of a linear differential transformer, is in series with the sample. The displacement is recorded as a function of time, e.g. on a chart recorder or computer. The sample may be surrounded by a fluid to change its temperature.

The simplest method for measuring stress-relaxation is to place the polymer sample in series with a spring or springs of stiffness very much greater than that of the sample. The strain in the spring(s), which is very much smaller than that of the sample, is then a measure of the *stress* in the sample and can be measured by, for instance, a linear differential transformer. The polymer is not strictly at constant strain, because of the small change in strain of the spring(s) in series with it, but this can be allowed for or the sample can be maintained at constant length by means of a servomechanism.

7.3.2 Dynamic measurements – the complex modulus and compliance

Dynamic methods involve oscillatory deformation and it is therefore necessary to consider the strain generated by a sinusoidal applied stress.

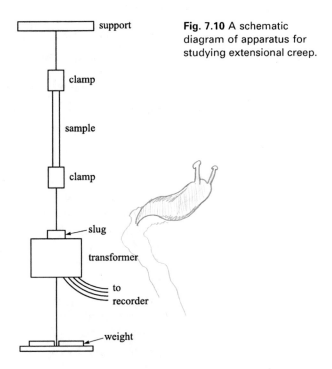

Fig. 7.10 A schematic diagram of apparatus for studying extensional creep.

Linearity implies that the strain is sinusoidal, with an amplitude proportional to that of the stress, but there will be a *phase lag* δ. The strain and stress at angular frequency ω can thus be represented by

$$e = e_0 \exp(i\omega t) \qquad \text{and} \qquad \sigma = \sigma_0 \exp[i(\omega t + \delta)] \qquad (7.20)$$

A *complex modulus* **G** is then defined by

$$\sigma/e = \mathbf{G} = (\sigma_0/e_0)\exp(i\delta) = (\sigma_0/e_0)(\cos\delta + i\sin\delta) = G_1 + iG_2 \qquad (7.21)$$

It can be shown (see problem 7.6) that the energy $\Delta\mathcal{E}$ dissipated per cycle per unit volume of material is given by

$$\Delta\mathcal{E} = \pi G_2 e_0^2 \qquad (7.22)$$

Thus the imaginary part G_2 is often called the *loss modulus* and the real part G_1 is called the *storage modulus*. The ratio $G_2/G_1 = \tan\delta$.

Except near the glass transition, G_2 is usually very much smaller than G_1, so that $G = |\mathbf{G}| \approx G_1$ and $\delta \approx \tan\delta = G_2/G_1$. Typical values are $G_1 = 10^9$ Pa, $G_2 = 10^7$ Pa and $\tan\delta = 0.01$.

A complex compliance is defined by

$$\mathbf{J} = J_1 - iJ_2 = 1/\mathbf{G} \qquad (7.23)$$

Hence

$$J_1 = \frac{G_1}{G_1^2 + G_2^2} \qquad \text{and} \qquad J_2 = \frac{G_2}{G_1^2 + G_2^2} \qquad (7.24a,b)$$

All the quantities G_1, G_2, J_1 and J_2 are *frequency-dependent*. Insertion of

$$\sigma = \sigma_0 \exp(i\omega t) = (G_1 + iG_2)e \qquad (7.25)$$

into the basic differential equation (7.14) for the Maxwell model shows that G_1 and G_2 are given by

$$G_1 = \frac{E\omega^2\tau^2}{1 + \omega^2\tau^2} \qquad \text{and} \qquad G_2 = \frac{E\omega\tau}{1 + \omega^2\tau^2} \qquad (7.26a,b)$$

so that G_1, G_2 and $\tan\delta$ for the Maxwell model have the frequency dependences shown in fig. 7.11.

As expected, these frequency dependences are not consistent with those for real materials. The most important differences for even the simplest real polymers are that (i) G_1 does not fall to zero at low frequency so that $\tan\delta$ remains finite and has a maximum to the low-frequency side of the peak in G_2; (ii) the viscoelastic region, i.e. the region in which the peaks in G_2 and $\tan\delta$ and the transition from a high to a low value of G_1 occur, is usually much wider than the three or four decades of ω suggested by the Maxwell model; and (iii) the maximum value of G_2

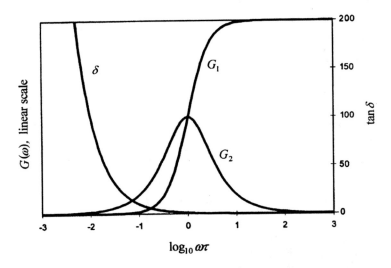

Fig. 7.11 G_1, G_2 and δ as functions of ω for the Maxwell model.

is usually much less than half the high-frequency value of G_1, the value given by the model. The SLS model can correct (i) qualitatively, but makes little difference to (ii) and (iii) and, as indicated in section 7.2.5, it is necessary to use a relaxation-time spectrum or some other method to model a real system well.

Equations (7.26) are called *dispersion relations* and analogous equations can be derived for J_1 and J_2 (see problem 7.7) and for more general models. They can also be derived for the real and imaginary parts of the dielectric constant (see section 9.2.4). The limiting values of G_1 and J_1 at low frequencies are called the *relaxed modulus* and *compliance*, G_r and J_r, and the limiting values at high frequencies are called the *unrelaxed modulus* and *compliance*, G_u and J_u.

There are formal mathematical relationships between the complex and stress-relaxation moduli, but a practical approach has often been to use approximate relationships to derive the relaxation-time spectrum from either type of measurement. Similar approximations give retardation-time spectra from complex or creep compliances. Further details can be found in the books referred to as (1) in section 7.7.

7.3.3 Dynamic measurements; examples

(i) *The torsion pendulum*

The simplest type of dynamic measurement uses a torsional-pendulum apparatus of the type shown in fig. 7.12. The system is allowed to undergo free torsional oscillations, which usually occur at frequencies of the order

Fig. 7.12 A torsion
pendulum for the
determination of shear
modulus and damping at
frequencies near 1 Hz.
The support wire has
negligible torsional
rigidity. (Reproduced by
permission of L. C. E.
Struik.)

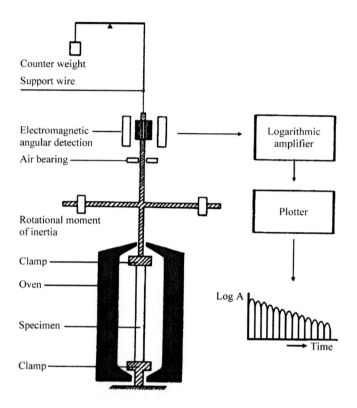

of 1 Hz, and the amplitude decays exponentially with time. It can be shown
that, for small losses,

$$G_1 = 2lI\omega^2/(\pi r^4) \tag{7.27}$$

and

$$\tan \delta = \Lambda/\pi \tag{7.28}$$

where r is the radius, l the length of the rod sample, G is the shear modulus
and I is the moment of inertia of the system. The quantity Λ is the *logarithmic decrement*, defined by $\Lambda = \ln(A_n/A_{n+1})$, where A_n is the amplitude
of the nth oscillation.

(ii) *Modern commercial equipment for dynamic studies*

Modern commercial equipment for dynamic studies allows automatic measurements to be made in tension, compression or bending or in shear or
torsion, although these features are not necessarily all available for every
piece of equipment. Figure 7.13 shows the principles of operation of one

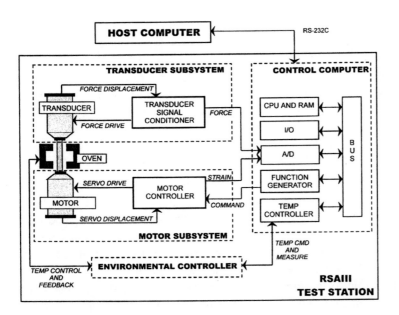

(a)

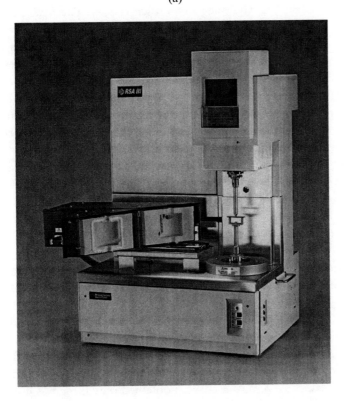

(b)

Fig. 7.13 The Rheometrics Solids Analyser RSA III. (a) A schematic diagram of the control system. (b) The oven doors are shown open with the oven drawn aside and a sample is shown ready for a bending test. (Courtesy of Rheometric Scientific GmbH.)

such system. An actuator applies the (alternating) strain and a transducer measures the stress. Measurements of stress-relaxation or of dynamic (complex) moduli can be made and a wide range of temperature can be used, from -150 to $600\,°C$, and a wide range of frequencies, from 0.0001 to 500 rad s^{-1}. Some data obtained on equipment of this kind are shown in the next section following a discussion of *time–temperature equivalence*.

7.4   Time–temperature equivalence and superposition

The relaxations that lead to viscoelastic behaviour are the result of various types of molecular motions, some of which are described in section 5.7, and they occur more rapidly at higher temperature. The compliance $J(t)$, for instance, is therefore a function of temperature, T, which means that it should really be written as $J(t, T)$. Suppose that the effect of a rise in temperature from some chosen standard temperature T_o is to speed up every stage in a relaxation process by a constant factor that depends on the new temperature T. This is equivalent to saying that the interval of time required for any small change in strain to take place is divided by a factor a_T that depends on T and has the value 1 when $T = T_0$. This means that, if measured values of $J(t, T)$ are plotted against ta_T, curves for all temperatures should be superposed. Similarly, if values of $J(\omega, T)$ are plotted against ω/a_T, curves for all temperatures should be superposed on the curve for T_0.

Because the time or frequency scales over which the compliance changes are very large, it is usual to use logarithmic scales for t or ω. When this is done it is necessary to shift the curves for different temperatures by a constant amount $\log a_T$ or $-\log a_T$ along the t or ω axes, respectively, in order to get superposition with the curve for T_0. The quantity a_T is therefore called the *shift factor*. Figure 7.14 shows data for a set of measurements before and after the shifts have been applied. The idea that the same effect can be produced on the compliances or moduli either by a change of temperature or by a change of time-scale is called *time–temperature equivalence*.

(Ferry suggested on the basis of a molecular theory of viscoelasticity that there should be a small vertical shift factor $T_o\rho_o/(T\rho)$, where ρ is the density at T and ρ_o is the density at the reference temperature T_o. These shifts have been made in fig. 7.14, but they are usually small and are often neglected.)

The curve obtained after the shifts have been applied is called a *master curve*. The important thing about this curve is that it effectively gives the response of the polymer over a very much wider range of frequencies than any one piece of apparatus is able to provide. It is important to realise, however, that a master curve is obtained only if there is just one important relaxation process in the effective range of frequencies studied. This is because there is no reason why the shift factors for two different relaxation processes should be the same at any temperature. Time–temperature super-position is, however, observed over a wide range of frequencies for many non-crystalline polymers in which the glass transition is the only important relaxation process.

It is clear from fig. 7.14 that measurements made at fixed ω over a wide enough temperature range can show the complete viscoelastic behaviour. For instance, measurements made at 100 Hz over the temperature range -14 to $130\,^\circ$C for this polymer show almost the full range of values of $J(\omega, T)$. It is much easier to vary the temperature of a sample than it is to obtain measurements over the equivalent wide range of frequencies, so measurements as a function of T at fixed ω are often used to study relaxation mechanisms in polymers. Figure 7.15 shows some data of this type obtained for PVC on a commercial apparatus. It is important to note, however, that such data cannot be used to construct curves showing the

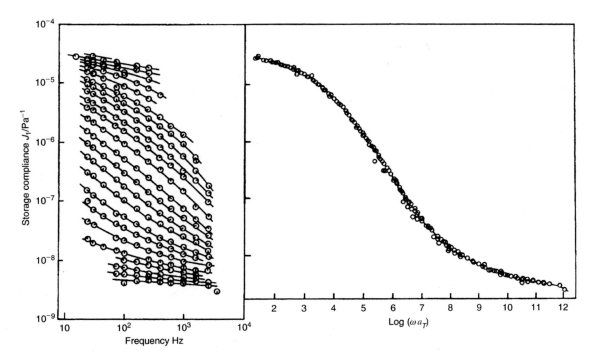

Fig. 7.14 The storage compliance of poly(n-octyl methacrylate) in the glass-transition region plotted against frequency on a logarithmic scale. Left-hand side: values obtained at 24 temperatures, ranging from $-14.3\,^\circ$C (bottom curve) to $129.5\,^\circ$C (top curve); right-hand side: the master curve obtained by using appropriate shifts log a_T. Note that the horizontal scales differ by a factor of two for the two parts of the figure. (Adapted by permission of Academic Press.)

modulus or compliance as a function of ω for any specific temperature unless the shift factors are known and that, when there is more than one relaxation process, there is no single shift factor for any temperature.

7.5 The glass transition in amorphous polymers

7.5.1 The determination of the glass-transition temperature

A curve such as that for G_1 shown in fig. 7.15 illustrates well the most obvious feature of the glass transition, the very large change in modulus that takes place over a fairly restricted temperature range. In this particular example the modulus increases by a factor of about 200 as the temperature falls from 120 to 60 °C, with a fastest rate of change centred near 85 °C, where the modulus changes by a factor of ten over a range of about 10 °C.

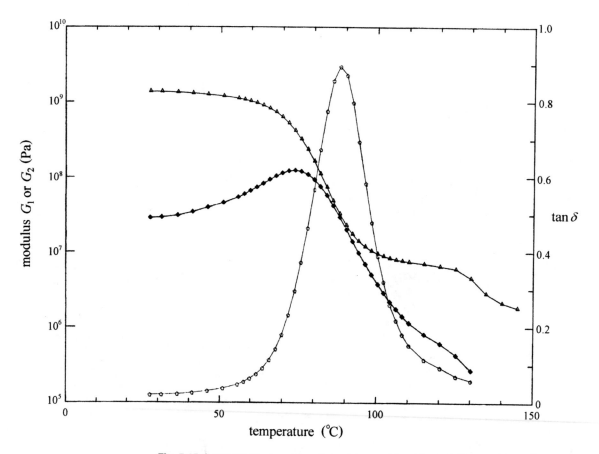

Fig. 7.15 A temperature sweep of modulus and tan δ for a PVC sample: △, G_1; ◆, G_2; and ○, tan δ. (Courtesy of Dr I. Karaçan.)

Tan δ peaks at a temperature slightly greater than this. If the data had been obtained at a different frequency or the temperature had been changed at a different rate, slightly different curves would have been obtained, with slightly different values for these characteristic temperatures.

Several other kinds of data reveal evidence of the glass transition. For instance, as noted in the discussion of the use of differential scanning calorimetry in section 2.2, the specific heat of a polymer undergoes a step-like change at T_g. Figure 2.3 shows that here also the transition is not sharp and takes place over a range of about 10 °C. If measurements are made at different rates of temperature scanning the transition is again observed to move slightly. Is it therefore important to consider whether it is possible to specify a definite value for the glass-transition temperature T_g, i.e., whether there is an underlying sharp transition that would take place at a well-defined temperature T_g under suitable conditions. This question is considered in the following sections.

Another property that changes as the temperature is varied through the glass transition is the thermal expansion. The value of T_g is usually determined using this effect. Provided that the temperature is changed slowly enough, particularly in the transition region, it is found that the specific volumes V_s (reciprocals of the densities) above and below T_g change approximately linearly with temperature, but with different slopes. The transition from one slope to another usually takes place over a range of only about 5 °C and T_g is then taken to be the temperature at which the two straight lines extrapolated from above and below the transition meet (see fig. 7.16). It is important to emphasise that, in order to obtain an accurate value of T_g, the temperature may have to be held constant for several days after each change, particularly in the transition region. In practical determinations the temperature is

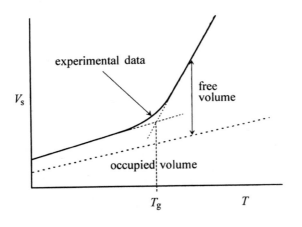

Fig. 7.16 A schematic diagram showing the determination of T_g from a plot of specific volume against temperature and the definition of the free volume.

often changed at a rate of 1 K min^{-1} and the value obtained may then be a few degrees too high.

7.5.2 The temperature dependence of the shift factor: the VFT and WLF equations

In section 5.7.3 it was shown that the relaxation time τ for a process that can be described by the site model should vary with temperature according to the equation

$$\tau(T) = A' \exp[\Delta H/(RT)] \tag{7.29}$$

where A' is a constant and ΔH is an enthalpy difference, both of which depend on the nature of the sites involved. If this equation is assumed to apply to the glass transition and T_g is taken as a standard temperature, the shift factor for any other temperature should be given by

$$a_T = \frac{\tau(T_g)}{\tau(T)} = \frac{\exp[\Delta H/(RT_g)]}{\exp[\Delta H/(RT)]} = \exp\left[\frac{\Delta H}{R}\left(\frac{1}{T_g} - \frac{1}{T}\right)\right];$$

$$\ln a_T = C\left(\frac{T - T_g}{T}\right) \tag{7.30}$$

where C is equal to $\Delta H/(RT_g)$. This equation is *not* found to apply to the observed variation with temperature of the shift factor for the glass transition.

An equation that does fit the data originated from attempts to fit the temperature dependence of the viscosity. The *Vogel–Fulcher*, or *Vogel–Fulcher–Tamman* (VFT), *equation* for $\tau(T)$, viz.

$$\tau(T) = \tau_0 \exp\left(\frac{C_0}{T - T_{\text{VFT}}}\right) \tag{7.31}$$

where τ_0 and C_0 are constants and T_{VFT} is a standard temperature, leads immediately to

$$a_T = \frac{\tau_0}{\tau(T)} = \exp\left(\frac{-C_0}{T - T_{\text{VFT}}}\right) \tag{7.32}$$

This equation is found to fit many sets of data over many orders of magnitude with approximately the same values of $C_0 = 2.1 \times 10^3 \text{ K}$ and $T_{\text{VFT}} = T_g - 52 \text{ K}$ for a wide range of amorphous polymers.

Quite independently, Williams, Landel and Ferry found an empirical equation, now called the *WLF equation*, which fits the dependence of the shift factor on temperature for a large number of amorphous polymers. The equation is usually written in the form

$$\log_{10} a_T = \frac{C_1(T - T_g)}{C_2 + (T - T_g)} \tag{7.33}$$

where C_1 and C_2 are approximately the same for all polymers and the shifts are now referred to T_g as the standard temperature.

Although this equation looks at first sight very different from the VFT equation, they are in fact equivalent. It is merely necessary to substitute $T_{VFT} = T_g - C_2$, $C_o = C_1 C_2 / \log_{10} e$ and $\tau_o = \tau(T_g) \exp(-C_1 / \log_{10} e)$ into equation (7.32) and take the logarithm to base 10 to obtain equation (7.33) (see problem 7.9). The mean values of C_1 and C_2 are then 17.4 and 52 K, but the values for individual polymers may deviate significantly from these. The curve in fig. 5.27 is an example of a fit to the WLF equation with $C_1 = 18.2$, $C_2 = 54.2$ K and $\tau(T_g) = 1000\,s^{-1}$. Although the VFT and WLF equations are equivalent, the slightly simpler form of the VFT equation is sometimes preferred.

7.5.3 Theories of the glass transition

At temperatures well above T_g the molecules have a great deal of freedom to move: the molecular theory that explains the high elasticity of rubbers is based on the idea that the chains are free to take up all the conformations allowed by rotations around single bonds. At temperatures well below T_g the small-strain elastic properties can be understood only, as described in section 6.2.2, by assuming that all these conformational changes of the backbone of the molecule are 'frozen out', so that the elastic properties are determined mainly by the larger forces required to stretch bonds or change bond angles. How does this 'freezing out' of motion come about?

Free-volume theories of the glass transition assume that, if conformational changes of the backbone are to take place, there must be space available for molecular segments to move into. The total amount of free space per unit volume of the polymer is called the *fractional free volume* V_f. As the temperature is lowered from a temperature well above T_g, the volume of the polymer falls because the molecules are able to rearrange locally to reduce the free volume. When the temperature approaches T_g the molecular motions become so slow (see e.g. fig. 5.27) that the molecules cannot rearrange within the time-scale of the experiment and the volume of the material then contracts like that of a solid, with a coefficient of expansion that is generally about half that observed above T_g.

If V_g is the fractional free volume at T_g, then ideally, above T_g,

$$V_f = V_g + \alpha(T - T_g) \tag{7.34}$$

where α is the temperature coefficient of expansion of free volume. If it is assumed that the relaxation time for the process occurring at the glass transition rises exponentially as the free volume falls and is given by

$$\tau(T) = \tau_o \exp(b/V_f) \tag{7.35}$$

where τ_{o} and b are constants, the VFT equation, equation (7.31), follows immediately, with $C_{\mathrm{o}} = b/\alpha$ and $T_{\mathrm{VFT}} = T_{\mathrm{g}} - V_{\mathrm{g}}/\alpha$. This leads to $C_1 = b/[V_{\mathrm{g}}\ln(10)]$ and $C_2 = V_{\mathrm{g}}/\alpha$ in the WLF equation, equation (7.33). As already indicated in section 5.7.5, the use of a single relaxation time τ is not really appropriate for the process that gives rise to the glass transition because it cannot be described by a single exponential function. Equation (7.35) should therefore be interpreted as stating that the times for corresponding parts of the relaxation to take place at different temperatures scale as $\exp(b/V_{\mathrm{f}})$.

By assuming that α is equal to the difference between the coefficients of expansion above and below T_{g}, the fractional free volume V_{g} at the glass transition can be found as αC_2. Values thus found are of order 0.025, which is much smaller than would be estimated for the amorphous regions of a semicrystalline polymer from the densities of the crystalline and amorphous regions and the assumption that there is no free volume in the crystallites. The free volume is thus a somewhat ill-defined concept and the theory just described is really a phenomenological theory, which links the observed temperature dependence of the thermal expansion to the observed temperature dependence of the shift factor in time–temperature equivalence, via a simple assumption about the way the relaxation time depends on the hypothetical free volume, i.e. equation (7.35). The success of this theory in linking two rather disparate phenomena suggests that it may be possible to obtain a deeper understanding of the transition by looking more closely at the implications of the assumptions.

Equation (7.34) suggests that $V_{\mathrm{f}} = 0$ when $T = T_{\mathrm{g}} - V_{\mathrm{g}}/\alpha = T_{\mathrm{VFT}}$ and equation (7.35) shows that $\tau(T)$ would then become infinite, as expected; there could be no motion if there were no unoccupied volume. Gibbs and DiMarzio put forward a statistical thermodynamic theory for the glass transition and suggested that there was a temperature, which they called T_2, at which the conformational entropy is equal to zero. Following on from this, Adams and Gibbs derived an equation equivalent to the VFT equation (7.31) with $T_{\mathrm{VFT}} = T_2$. In this theory the number of molecular segments that have to move co-operatively to make a conformational rearrangement is inversely proportional to the entropy and becomes infinite at $T = T_2 = T_{\mathrm{VFT}}$. According to these theories there is therefore a true equilibrium second-order transition at $T = T_2 = T_{\mathrm{VFT}}$ and it is the entropy, rather than a hypothetical free volume, which is zero at this temperature. This equilibrium can never be reached because an infinite time would be required in order to reach it. The observed value of T_{g} is about 52 °C higher than this because the time constant is high at this temperature and increases extremely rapidly below it (see example 7.4).

Example 7.4

Assuming that the constant C_o in the VFT equation (7.31) is 2.1×10^3 K and that this equation correctly describes the behaviour of a polymer with a glass-transition temperature $T_g = T_{VFT} + 52\,°C$ and a value of $\tau = 1$ h at T_g, calculate the values of τ for the following temperatures (i) $T_g \pm 1\,°C$ and (ii) $T_g \pm 10°C$.

Solution

$\tau = \tau_o \exp[C_o/(T - T_{VFT})]$. Thus

$$\tau = \tau_g \exp\left[C_o \left(\frac{1}{T - T_{VFT}} - \frac{1}{T_g - T_{VFT}} \right) \right]$$

(i) Setting $T_g = T_{VFT} + 52\,°C$ and $T = T_g \pm 1\,°C$ leads to

$$\tau = \tau_g \exp\left[2.1 \times 10^3 \left(\frac{1}{52 \pm 1} - \frac{1}{52} \right) \right]$$

which results in $\tau = 0.47$ h for $T = T_g + 1\,°C$ and $\tau = 2.2$ h for $T = T_g - 1\,°C$.
(ii) Similarly, for $T = T_g + 10\,°C$ and $T = T_g - 10\,°C$, τ takes the values 1.5×10^{-3} h (≈ 5.4 s) and 1.5×10^4 h (≈ 1.7 years), respectively.

7.5.4 Factors that affect the value of T_g

In sections 5.7.5 and 7.5.3 the origin of the glass transition is explained as the 'freezing out' of various complicated motions of whole chain segments that can take place at higher temperatures. It is therefore not surprising that the glass-transition temperature of a polymer is influenced by

(i)   main-chain flexibility – inflexible groups increase T_g,
(ii) the nature of any side groups – bulky side groups generally increase T_g and
(iii) a number of other factors, including the presence of plasticisers.

The influence of factors (i) and (ii) can be seen in some of the values given in table 6.1. For instance, the fact that T_g is lower for nylon, in which all the bonds in the backbone are single, than it is for poly(ethylene terephthalate), which contains a large fraction of inflexible phenylene groups in the backbone, is an example of (i). An example of (ii) is the fact that polypropylene has a lower T_g than those of polystyrene and poly(methyl methacrylate), which have larger side groups.

A plasticiser is a small-molecule additive deliberately introduced into a polymer in order to act as a sort of internal lubricant to reduce T_g. A very familiar example is the use of plasticisers to make PVC flexible at room temperature for use in such applications as simulated leather. The glass-transition temperature of unplasticised PVC is about 80 °C.

The molar mass and degree of cross-linking are other influences on T_g. This is most easily understood by reference to the free-volume model of the glass transition. Shorter chains have more ends per unit volume of material, contributing extra free volume and thus lowering T_g. Cross-linking, on the other hand, pulls chains together and generally increases T_g.

7.6   Relaxations for amorphous and crystalline polymers

7.6.1   Introduction

Real polymers do not behave entirely according to the way suggested by the VFT or WLF equations. In general they undergo more than one transition (see e.g. fig. 7.17) and amorphous and crystalline polymers behave differently. In section 5.7 the use of NMR spectroscopy for studying the relaxation mechanisms that underlie the transitions is illustrated by a number of specific examples and a general classification of the types of motion expected to be observable in polymers is given in table 5.2. Table 5.2 is, however, compiled essentially from a theoretical view of the possibilities. In practice, when mechanical observations are made, the problem is that of assigning the observed relaxations to various mechanisms.

Polymers can be divided into two broad classes, amorphous and semi-crystalline. If observations are made at a fixed frequency, or *isochronally*, crystalline polymers often exhibit three major transitions as the temperature is varied, usually labelled α, β and γ in decreasing order of temperature, whereas amorphous polymers generally exhibit two major transitions, labelled α and β in decreasing order of temperature. If other relaxations are seen at lower temperatures, they are labelled γ or δ, respectively.

These labels are arbitrary and do not indicate any fundamental relationship between the relaxations given the same label in different polymers. For an amorphous polymer the transition at the highest temperature, labelled α, always corresponds to the glass transition, whereas for polymers of low crystallinity it is often the transition labelled β that corresponds to the glass transition. For this reason the glass transition in amorphous polymers is sometimes labelled α_a to distinguish it from the α transition in crystalline polymers, which can sometimes be assigned to the crystalline material. Amorphous and crystalline polymers are discussed separately in the following sections.

At the outset it must be understood that a relaxation can manifest itself in mechanical measurements only if the jump or more complex rearrangement to which it corresponds contributes to a change in shape or dimensions of the polymer sample as a whole. For instance, a jump involving changes from *gauche* to *trans* conformations in a chain backbone would tend to straighten the chain and hence increase its end-to-end length and might also tend to displace it laterally. A series of such jumps could thus lead to extension or distortion of the whole sample and such a relaxation process would be observable mechanically. On the other hand, random helical jumps within a crystal, of the type discussed in sections 5.7.4 and 5.7.5, do not distort the crystal and cannot therefore be directly active mechanically. They can, however, sometimes be observed through their effects on the amorphous material, as discussed in section 7.6.3.

In trying to identify relaxations it is often useful to consider their relative strengths. The standard definition of the *relaxation strength* for mechanical relaxations is that it is the difference between the unrelaxed and relaxed modulus or compliance, $G_u - G_r$ or $J_r - J_u$. Determination of these quantities requires measurement over a wide range of frequencies, which is not easy, as well as the separation of overlapping relaxations.

7.6.2 Amorphous polymers

A real amorphous polymer usually exhibits more than one transition. As indicated above, there is a high-temperature transition, usually labelled α or α_a, which is the glass transition and corresponds to the onset of main-chain segmental motion, as discussed in sections 5.7.5 and 7.5.3, and *secondary transitions* at lower temperatures. These are assigned to various types of motion, such as motions of side groups, restricted motion of the main chain or motions of end groups, some of which are discussed in detail in sections 5.7.4 and 5.7.5. The secondary relaxations often show up more clearly in the loss modulus or $\tan \delta$.

As an example, fig. 7.17 shows the behaviour of G_1 and $\tan \delta$ for atactic polystyrene. Four transitions, labelled α, β, γ and δ, can be distinguished, with the strongest, the α transition, being the glass transition. Results from NMR studies suggest strongly that the β transition is due to co-operative restricted oscillations involving both the main chain and the phenyl ring and that the γ transition is associated with 180° ring flips.

7.6.3 Crystalline polymers

These materials exhibit complicated behaviour, which depends on the degree of crystallinity and the detailed morphology. In general, relaxations can occur within the amorphous phase, within the crystalline phase, within both phases or be associated with specific details of the morphology, e.g. with the movement of chain folds. For these reasons the detailed interpre-

Fig. 7.17 Variations of shear
modulus G_1 and $\tan\delta$ with
temperature for
polystyrene. (Reproduced
by permission of Oxford
University Press.)

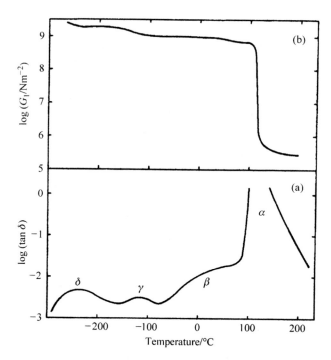

tation of mechanical relaxations in crystalline polymers is considerably
more complicated than that for amorphous materials and can be done
with reasonable certainty only when other information, such as that
obtainable from dielectric studies or, more importantly, from NMR
studies, is also available.

The first stage in assigning transitions in crystalline polymers to
mechanisms is that of distinguishing between those due to the amorphous
regions and those due to the crystalline regions, or possibly to both. A
simple method for doing this appears to be to vary the crystallinity of the
sample and see how the strength of the relaxation changes. This involves
first arriving at a way of measuring the relaxation strength and secondly
the assumption that the strength of a crystalline relaxation increases when
the crystallinity of the sample increases. This second assumption seems at
first obvious, but in fact the behaviour observed depends on the morphology of the sample, the moduli of the crystalline and amorphous regions
and the particular measure of strength used.

It is not easy in mechanical measurements to vary the frequency over a
wide range, as is possible for measurements of dielectric relaxation, so
that direct determination of relaxation strengths as defined in section
7.6.1 is not usually possible. A further complication is that the amorphous and crystalline regions are coupled together mechanically, so that
they do not contribute independently to the spectrum of relaxations

observed. Measurements are usually performed isochronally, with temperature as the variable. It can be shown that, taking the possible effects of mechanical coupling into account, the maximum value of $\tan \delta$ is then a suitable measure of the relaxation strength because this quantity should vary monotonically with the degree of crystallinity, though not necessarily linearly.

The data shown in fig. 7.18 for polytetrafluoroethylene (PTFE) provide a good example of how the variation of shear modulus and $\tan \delta$ with temperature can depend strongly on the degree of crystallinity of the sample. Because $\tan \delta$ for the γ relaxation increases as the crystallinity decreases and $\tan \delta$ for the β relaxation decreases, the former is associated with the amorphous material and the latter with the crystalline material. The behaviour of $\tan \delta$ for the α relaxation strongly suggests that, in this polymer, it is associated with the amorphous regions.

Samples of poly(ethylene terephthalate) (PET) can be prepared with crystallinities varying from a few per cent to about 50%. All undergo two relaxations, an α process in the region of 80–100 °C, which decreases in strength, broadens and moves to higher temperature with increasing crystallinity and represents the glass transition, and a broad β process at about −50 °C. NMR evidence shows that there is a motion of the methylene groups, possibly *trans–gauche* isomerism, which is correlated to the α process, and that flips of the aromatic rings are probably involved in the β process. Involvement of a localised motion is suggested by the insensitivity of this relaxation to the degree of crystallinity. The significant change in the

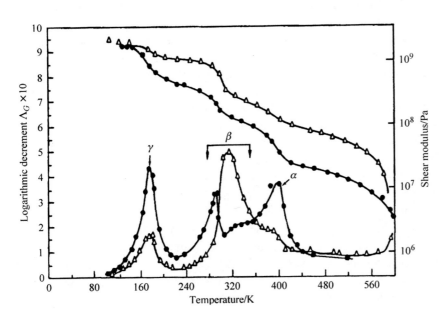

Fig. 7.18 The temperature dependence of the shear modulus (upper curves) and the logarithmic decrement Λ_G (lower curves) at about 1 Hz for PTFE samples of 76% ($\triangle$) and 48% ($\bullet$) crystallinity. (Adapted by permission of John Wiley & Sons, Inc.)

position of the glass transition with crystallinity shows the effect of increasing crystallinity restricting the possible motions in the amorphous regions. In more highly crystalline polymers, such as some forms of polyethylene, such restrictions would be expected to be very important.

The interpretation of the mechanical relaxations in polyethylene is complicated by the fact that there are different types of material, as described in section 1.3.3, some of which may be very highly, but not completely, crystalline. It is now established that all types exhibit three mechanical relaxations, labelled α (in the region of 70 °C), β (in the region of 0 °C) and γ (in the region of −120 °C). The α relaxation is associated with the helical jumps within the crystalline regions discussed in section 5.7.5 in connection with NMR experiments. This is shown by the agreement of the activation energy determined by NMR, 105 kJ mol^{-1}, with that found from mechanical studies. The β relaxation is believed to be the equivalent of the glass transition. It is hard to detect in the more crystalline high-density polyethylenes and is rather broader than the glass-transition region in completely amorphous polymers, probably because of the restraining effect of the crystallites on the motions of the amorphous regions. The γ relaxation is weak or absent in very highly crystalline material and, because it takes place well below the glass transition, it is believed to be due to restricted motions in the non-crystalline material, probably involving conformational changes that do not require large-scale disturbance to surrounding chains, and possibly of a type similar to that described for the butylene group of poly(butylene terephthalate) in section 5.7.4.

As explained in section 7.6.1, helical jumps in the crystalline material are at first sight not expected to produce mechanically observable relaxation. Such jumps do, however, necessarily produce changes in the lengths of the chain segments in the interlamellar amorphous regions. Lengthening of such chains can, under suitable morphology and stressing conditions, lead to the possibility of easier shearing parallel to the lamellar planes, called *interlamellar shear*, and thus provide a mechanically observable relaxation. Various pieces of evidence show that the helical 'jump' is not a rigid rotation and translation of the whole chain stem within the crystal but rather takes place by the propagation through the crystal of a chain distortion. This distortion involves a section of chain about 12 CH$_2$ units long that has the correct crystal alignment at each end but is twisted and compressed uniformly between these alignments. One of the main pieces of evidence for this model is that the activation energy for the process is found to be approximately independent of crystal thickness for very thick crystals, whereas the assumption of rigid-chain rotation/translation would require an activation energy proportional to the crystal thickness.

7.6.4 Final remarks

In chapter 5 the use of NMR spectroscopy in studying the various kinds of motion that can take place in solid polymers is described and in the present chapter the way that these motions contribute to the time-dependent mechanical properties of the polymers is considered. In chapter 9 the effects of motion on the dielectric properties and the use of measurements of dielectric relaxation in the study of relaxation mechanisms are considered.

7.7 Further reading

(1) *An Introduction to the Mechanical Properties of Solid Polymers*, by I. M. Ward and D. W. Hadley; and *Mechanical Properties of Solid Polymers*, by I. M. Ward. See section 6.6.
(2) *Viscoelastic Properties of Polymers*, by J. D. Ferry, 3rd Edn, John Wiley, New York, 1980. This is one of the classic texts on the subject and covers a wide variety of theory, experimental techniques and results.
(3) *Anelastic and Dielectric Effects in Polymer Solids*, by N. G. McCrum, B. E. Read and G. Williams, John Wiley & Sons, London, 1967. Another classic of the subject. After chapters on theory and experimental methods, individual groups of polymers are considered.

7.8 Problems

7.1. The behaviour of a polymer is described by a Maxwell model consisting of a spring element of modulus 10^{10} Pa in series with a dashpot of viscosity 10^{12} Pa s. Calculate the stress in the solid 50 s after the sudden application of a fixed strain of 1%.

7.2. A strain e_1 is applied to a Maxwell element at time t_1 and a second strain increment e_2 is applied at time t_2. Without assuming the applicability of the BSP, calculate the stress in the element immediately after the application of e_2 and hence calculate the stress at any time $t > t_2$. Show that this stress is the same as would have been calculated by assuming applicability of the BSP.

7.3. A grade of polypropylene is found to have the following tensile creep compliance at 35 °C: $J(t) = 1.2t^{0.1}$ GPa^{-1}, where t is expressed in seconds. A sample of this polymer is subjected to the following time sequence of tensile stress at 35 °C: $\sigma = 0$ for $t < 0$; $\sigma = 1$ MPa for $0 \leq t < 1000$ s; $\sigma = 1.5$ MPa for 1000 s $\leq t < 2000$ s; and $\sigma = 0$ for $t \geq 2000$ s. Find the tensile strain at the following times t: (i) 1500 s and (ii) 2500 s. Assume that, under these conditions, polypropylene is linearly viscoelastic and obeys the BSP.

7.4. A nylon bolt of diameter 8 mm is used to join two rigid plates. The nylon can be assumed to be linearly viscoelastic with a tensile-stress-relaxation modulus approximated by $E = 5\exp(-t^{1/3})$ GPa, where t is in hours. The bolt is tightened quickly so that the initial force in the bolt at $t = 0$ is 1 kN. Find (a) the strain in the bolt and (b) the force remaining after 24 h.

7.5. Show that, if a polymer responds to a sudden applied strain e by relaxing to the equilibrium stress corresponding to e in such a way that the rate of change of stress at any time t is proportional to the difference between the existing stress σ and the equilibrium stress, then $d\sigma/dt = -(\sigma - eG)/\tau$, where G is the stress-relaxation modulus at infinite time and τ is a constant with the dimensions of time (the relaxation time). Show that this equation is equivalent to equation (7.15) derived for the Maxwell model only when $G \to 0$.

7.6. By considering the work done when a tensile stress σ applied perpendicular to a pair of opposite faces of a unit cube of material increases the tensile strain by δe, prove equation (7.22). Remember that only the real parts of the right-hand sides of equations (7.20) are physically meaningful.

7.7. Show by substitution into equation (7.10) that, if a stress $\sigma = \sigma_0 \cos(\omega t)$ is applied to the polymer and a strain $e = e_1 \cos(\omega t) + e_2 \sin(\omega t)$ results, the compliance components J_1 and J_2 are given by

$$J_1 = \frac{e_1}{\sigma_0} = \frac{1/E}{1 + \omega^2 \tau^2} \quad \text{and} \quad J_2 = \frac{e_2}{\sigma_0} = \frac{\omega\tau/E}{1 + \omega^2 \tau^2}$$

Either evaluate $J_1 E$ and $J_2 E$ for $\omega\tau = 0.01, 0.1, 0.316, 1, 3.16, 10$ and 100 and hence sketch on the same graph the dependences of $J_1 E$ and $J_2 E$ on $\log(\omega t)$, or plot more accurate graphs using a spreadsheet or programmable calculator.

7.8. The table gives the values of the logarithm (to base 10) of the dynamic shear compliance J_2 obtained for poly(vinyl acetate) at various frequencies f and temperatures T, where J_2 is expressed in Pa^{-1}.

f (Hz)	90 °C	80 °C	70 °C	60 °C	55 °C	50 °C
43	−5.83	−6.19	−6.81			
95	−6.00	−6.41	−7.20			
206	−6.16	−6.64	−7.51	−8.36	−8.60	
406	−6.35	−6.89	−7.77	−8.52	−8.71	−8.87
816			−8.13	−8.61	−8.77	−8.91
1030	−6.64	−7.31	−8.16			
2166		−7.58	−8.33	−8.72	−8.84	−8.95
3215				−8.76	−8.85	−8.94
4485			−8.50	−8.80		

Produce a master curve for 70 °C from these data and show that, to a good approximation, the required shift factors fit the WLF equation (7.33) with C_1 and C_2 equal to 17.4 and 52 K and $T_g = 30$ °C.

7.9. Deduce the WLF equation (7.33), from equation (7.32). See the text below equation (7.33) and remember that a_T is defined slightly differently in the two equations.

7.10. For a particular polymer the real part of the compliance for $T = 100$ °C can be approximated by $\log_{10} J_1(100, \omega) = 5 + 4/[\exp(L - 6) + 1]$, where $J_1(T, \omega)$ is expressed in pascals and $L = \log_{10} \omega$ with ω in s^{-1}. (a) Assuming that this expression holds for the whole range, plot a graph showing $\log_{10} J_1(100, \omega)$ against L in the range $0 < L < 12$. (b) If the glass-transition temperature T_g of the polymer is 50 °C and the polymer obeys the WLF equation (7.33) with $C_1 = 17.4$ and $C_2 = 52$ K, calculate the shift factor $\log_{10} a_{100}$ which is appropriate to the temperature 100 °C and hence write down the expression for $\log_{10} J_1(T_g, \omega)$. (c) Now write down the expression for $\log_{10} J_1(T, \omega)$ for any general values of T and ω and hence plot a graph of $\log_{10} J_1(T, \omega)$ for $\omega = 1$ s^{-1} and 40 °C $< T < 80$ °C, assuming that the WLF equation applies throughout this range. It is suggested that a spreadsheet should be used for plotting the graphs.

Chapter 8
Yield and fracture of polymers

8.1 Introduction

When polymers are applied for any practical purpose it is necessary to know, among other things, what maximum loads they can sustain without failing. Failure under load is the subject of the present chapter.

In chapter 7 the phenomenon of creep in a viscoelastic solid is considered. For an ideal linear viscoelastic medium the deformation under a constant stress eventually becomes constant provided that $e_3(t)$ in equation (7.4) is zero. If the load is removed at any time, the ideal material recovers fully. For many polymers these conditions are approximately satisfied for low stresses, but the curves (b) and (c) in fig. 6.2 indicate a very different type of behaviour that may be observed for some polymers under suitable conditions. For stresses above a certain level, the polymer *yields*. After yielding the polymer either *fractures* or retains a *permanent deformation* on removal of the stress.

Just as linear viscoelastic behaviour with full recovery of strain is an idealisation of the behaviour of some real polymers under suitable conditions, so ideal yield behaviour may be imagined to conform to the following: for stresses and strains below the yield point the material has time-independent linear elastic behaviour with a very low compliance and with full recovery of strain on removal of stress; at a certain stress level, called the *yield stress*, the strain increases without further increase in the stress; if the material has been strained beyond the yield stress there is no recovery of strain. This ideal behaviour is illustrated in fig. 8.1 and the differences between ideal viscoelastic creep and ideal yield behaviour are shown in table 8.1.

Ideal yielding behaviour is approached by many glassy polymers well below their glass-transition temperatures, but even for these polymers the stress–strain curve is not completely linear even below the yield stress and the compliance is relatively high, so that the deformation before yielding is not negligible. Further departures from ideality involve a strain-rate and temperature dependence of the yield stress. These two features of behaviour are, of course, characteristic of viscoelastic behaviour.

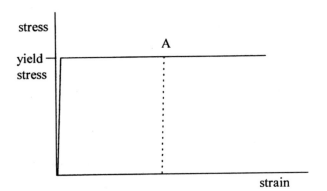

stress

yield
stress

A

strain

Fig. 8.1 Ideal yield behaviour. The full lines represent the stress–strain relationship on loading and the dotted line represents the relationship when unloading takes place starting at the point A.

Just as real materials may have behaviour close to ideal yielding behaviour but with some features similar to those of viscoelastic materials, materials that exhibit behaviour close to the ideal linear viscoelastic may exhibit features similar to those of yield, particularly for high stresses and long times of application. If the term $e_3(t)$ in equation (7.4) is not zero, a linear viscoelastic material will not reach a limiting strain on application of a fixed stress: at long times the strain will simply increase linearly with time. The material may also depart from linearity at high or even moderate stresses, so that higher stresses produce disproportionately more strain. These are features characteristic of yield.

These observations suggest that, for real materials, there need not be a very clear-cut distinction between yield and creep. In fact polymers present a complete spectrum of behaviours between the two ideal types. Fortunately, however, many polymers under conditions of temperature and strain-rate that are within the ranges important for applications do have behaviours close to one of these ideals. It is therefore useful both from a practical and from a theoretical point of view to try to understand these approximately ideal behaviours before attempting to study the more complicated behaviours exhibited by other materials. In chapter 7 this approach is used in discussing creep and linear viscoelasticity and in the present chapter it is used in discussing yield.

Table 8.1. *Idealised creep and yield behaviours*

Behaviour	Creep	Yield
Is a threshold stress required?	No	Yes
Does stress increase with strain?	Yes	No
Is strain recoverable on removal of stress?	Yes	No

Polymers do not always fail mechanically by yielding, i.e. by becoming *ductile*. This is illustrated by curve (a) of fig. 6.2, which represents the behaviour of a polymer that fails by *brittle fracture*. As was pointed out in section 6.1, under suitable conditions any particular polymer can usually exhibit all of the behaviours illustrated in fig. 6.2. Not only the types of behaviour but also the loads, or stresses, at which they happen vary with the conditions. Experiment shows that the ways in which the stresses for yielding and for brittle fracture vary with temperature and *strain-rate* are different for a given polymer, as illustrated schematically in fig. 8.2. (Strain-rate $= de/dt$, where e is strain and t is time.) The two processes can be thought of as being in competition; which of them occurs as the stress is increased under particular conditions depends on which occurs at the lower stress under those conditions. At the temperature T_1 in fig. 8.2 the polymer will fail by brittle fracture at each of the strain-rates illustrated and the fracture stress will be higher for the higher strain-rate. At the temperature T_2 failure will take place by yielding for the lower strain-rate and by brittle fracture for the higher strain-rate, whereas at the temperature T_3 failure will take place by yielding for both strain-rates.

There is, however, no generally accepted theory for predicting the *brittle–ductile transition* or relating it to other properties of the polymer, although for some polymers it is closely related to the glass transition. The type of failure is also affected by geometrical factors and the precise nature of the stresses applied. *Plane-strain* conditions, under which one of the principal strains is zero, which are often found with thick samples, favour brittle fracture. *Plane-stress* conditions, under which one of the principal stresses is zero, which are often found with thin samples, favour ductile fracture. The type of starting crack or notch often deliberately introduced when fracture behaviour is examined can also have an important effect;

Fig. 8.2 'Competition' between yield and brittle fracture. The curves show schematically the dependences of brittle-fracture stress and yield stress on temperature. The dashed lines correspond to higher strain-rates than do the full lines. See the text for discussion. (Adapted by permission of I. M. Ward.)

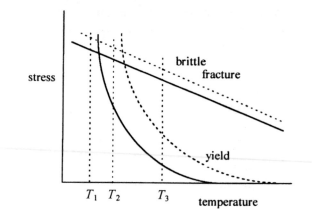

sharp cracks favour brittle fracture, blunt notches favour ductile fracture. Brittle fracture is related not only to the properties of the polymer but also to the presence of flaws. Polymers that fail by brittle fracture under tension tend to yield if they are tested in compression or pure shear. Because of these complications, yield and fracture are treated separately in the following sections.

8.2   Yield

8.2.1   Introduction

As already mentioned in the previous section, the extension that takes place before a polymer yields can be significant, which implies that there is a change in area perpendicular to the stress. It is therefore necessary in defining the yield stress to take account of the difference between the true stress and the nominal stress (see section 6.3.2). For polymers that behave in the manner illustrated by curve (c) in fig. 6.2 the yield stress can then be defined as the true stress at the maximum observed load. For polymers that behave in the manner illustrated by curve (d) in fig. 6.2, such that there is merely a change in slope rather than a drop in load, the yield point is usually defined as the point of intersection of the tangents to the initial and final parts of the curve.

   The simplest type of stress that can be applied to a polymer is a tensile stress, so the behaviour of polymers under such a stress is described before more general forms of stress are considered. This leads directly to a discussion of *necking* and *cold drawing*.

8.2.2   The mechanism of yielding – cold drawing and the Considère construction

The yield behaviour corresponding to the curve (c) in fig. 6.2 involves *necking* and is often called *cold drawing* because it can often take place at or near room temperature. The appearance of a sample undergoing this process is shown schematically in fig. 8.3.

   Yield involves an irreversible deformation and takes place by a shearing mechanism in which molecules slide past one another. If molecules are to slide past each other, energy barriers have to be overcome. Raising the temperature of the polymer will make it easier for these barriers to be overcome, as discussed already in relation to the site-model theory of mechanical relaxation in section 5.7.3 and as discussed in section 8.2.5 of the present chapter in relation to yielding.

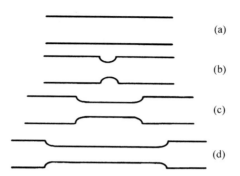

Fig. 8.3 Necking and cold drawing (schematic): (a) represents a length of undrawn sample; (b) shows the start of neck formation at a point of slight weakness; (c) and (d) show progressive elongation of the neck, which eventually spreads through the whole sample.

It was therefore suggested in early attempts to explain necking and cold drawing that the work done on the sample during drawing caused a rise in temperature that softened the material and thus led to yielding and the formation of a neck. Subsequently all the work done in drawing (and the consequent heating) would be concentrated at the 'shoulder' region of the neck (see fig. 8.3), where all subsequent yielding takes place. It has more recently been shown, however, that, although a considerable rise in temperature can take place in cold drawing under conditions in the usual range of strain-rates of order 10^{-2} s^{-1} or higher, cold drawing can still take place for a significant number of polymers at very low extension rates at which there is no significant rise in temperature. This suggests that yielding takes place because the shape of the isothermal *true-stress*–strain curve leads to a maximum in the *nominal-stress*–strain curve.

Let the true tensile stress on a sample of initial cross-sectional area A_o and initial length l_o be σ. If the length and area are l and A, respectively, when the strain is e, then for constant volume

$$A = \frac{A_o l_o}{l} = \frac{A_o l_o}{l_o + l_o e} = \frac{A_o}{1 + e} \tag{8.1}$$

The nominal stress σ_n, which is given by

$$\sigma_n = \frac{\sigma A}{A_o} = \frac{\sigma}{1 + e} \tag{8.2}$$

has a maximum or minimum value when

$$\frac{d\sigma_n}{de} = 0, \quad \text{i.e.} \quad \frac{1}{1 + e}\frac{d\sigma}{de} - \frac{\sigma}{(1 + e)^2} = 0 \tag{8.3}$$

or

$$\frac{d\sigma}{de} = \frac{\sigma}{1 + e} \tag{8.4}$$

Equation (8.4) shows that maxima or minima will occur in the *nominal* stress at points where lines drawn from the point $e = -1$ on the strain axis are tangential to the *true-stress*–strain curve. This construction is called the *Considère construction*.

Figure 8.4 shows in a highly schematic way the various possible forms of true-stress–strain curves, the possible tangents that can be drawn to them from the point $e = -1$ and the corresponding nominal-stress–strain curves. Figures 8.4(b), (c) and (d) show regions where the true-stress–strain curve becomes less steep with increasing stress, which is called *strain softening*. Figs 8.4(c) and (d) also show regions where the true-stress–strain curve becomes steeper with increasing stress, which is called *strain hardening*. The 'curve' shown in fig. 8.1 is an idealised form of the type of curve shown in fig. 8.4(b).

No tangent can be drawn in fig. 8.4(a), a single tangent in fig. 8.4(b) and two tangents in each of figs 8.4(c) and (d). If a sample of polymer corresponding to fig. 8.4(b) is drawn, it will deform homogeneously at first. Eventually, at some point in the sample that is slightly weaker than the rest, either because it is slightly thinner or because its properties are slightly

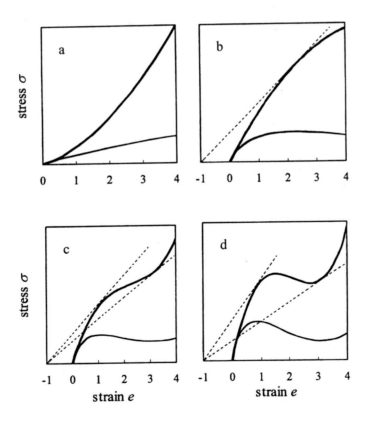

Fig. 8.4 The Considère construction. Possible forms of true-stress–strain curves (upper curves), the tangents drawn to them from the point $e = -1$ on the strain axis and the corresponding nominal-stress–strain curves. The curves are schematic for clarity in showing the construction; for real polymers the initial part of the curve would generally be much steeper and the maximum, when it occurs, is generally at a much lower value of strain than that shown. See example 8.1 for a more realistic version of the curves in (c).

different, the maximum in the nominal stress will be reached. Further elongation of the sample will then take place near this point with rapidly decreasing load and the sample will fail. This corresponds to curve (b) in fig. 6.2.

A sample of polymer corresponding to fig. 8.4(c) or (d) will also deform homogeneously until eventually, at some region in the sample that is slightly weaker than the rest, the maximum in the nominal stress is reached. The stress then falls and this region alone continues to extend, forming a *neck*, until the minimum in the nominal stress is reached. Further elongation causes the nominal stress to rise and this is sufficient to cause adjacent regions of the polymer to yield and the neck propagates along the sample at constant load. Finally, when the neck has extended throughout the whole sample, the stress rises again as the whole sample deforms homogeneously.

There remains the question of whether the drop in load observed at yielding arises from the purely geometrical strain softening associated with a true-stress–strain curve of the form shown in fig. 8.4(c), where there is no drop in the true stress but merely a reduction in slope of the stress–strain curve, or whether there is actually a maximum in the true-stress–strain curve as shown in fig. 8.4(d). Experiments on polystyrene and PMMA in compression, under which the geometrical effect cannot take place, show that a drop in load is still observed. Results from extensive studies of PET under a variety of loading conditions also support the idea that a maximum in the true-stress–strain curve may occur in a number of polymers.

The polymer chains in the necked region are highly oriented towards the stretching direction and the properties of the material are consequently highly anisotropic. The subject of oriented polymers is dealt with in chapters 10 and 11. This chapter is concerned with what happens below (and more specifically at) the yield point of the initially isotropic polymer. Yield in oriented polymers is rather more complicated and is not considered in the present book. Cold drawing is associated with a *natural draw ratio*, the ratio of any length within the necked region to its original length, which is discussed in section 10.2.4. The remainder of section 8.2 is concerned with yielding under more general conditions than the application of a simple tensile stress.

8.2.3   Yield criteria

Yielding under a tensile load is the simplest form of yielding to consider, but polymers are often subjected to more complicated stresses and it is important to ask whether any general expressions can be developed for predicting when a polymer will yield under any type of stress field. This leads to the idea of *yield criteria*.

Example 8.1

The nominal-stress–strain curve for a particular polymer in tension can be approximated by the expression $\sigma_n(e) = C(ae - be^2)$ in the region $0 < e < 0.184$ and by the expression $\sigma_n(e) = C(c - de + 2e^2)$ in the region $0.184 < e < 6$, where C is a constant depending on the cross-section of the sample, $a = 356.6$, $b = 998$, $c = 34.03$ and $d = 12.34$. Calculate the strain at which the minimum occurs in $\sigma_n(e)$ and show by calculation that the true-stress–strain curve does not exhibit a minimum. Confirm the results by plotting both curves.

Solution

Differentiation of the first expression for $\sigma_n(e)$ shows that the stress increases monotonically until $e = 356.6/(2 \times 998) = 0.179$ and then begins to fall monotonically. The minimum must therefore lie in the region described by the second expression. Differentiation shows that it lies at $e = d/4 = 3.09$. The true-stress–strain curve is obtained by multiplying $\sigma_n(e)$ by $1 + e$ and any minimum in it must lie in the region described by the second expression for $\sigma_n(e)$. Differentiation of $(1 + e)\sigma_n(e)$ leads to the condition $(a - b) + (4 - 2b)e + 6e^2 = 0$ for the value of e at which any minimum occurs. Substitution of the values of a and b shows that this quadratic expression has no real roots, so that there is no minimum. The curves are as shown.

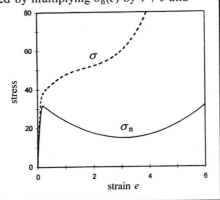

In general, stress must be represented by a second-rank tensor with six independent components (see the appendix). If, however, only isotropic materials are considered, it is always possible to choose reference axes so that only the principal components σ_1, σ_2 and σ_3 of the stress tensor are non-zero (see the appendix and section 6.3.2). A yield criterion then takes the form of the following equation:

$$f(\sigma_1, \sigma_2, \sigma_3) = C \tag{8.5}$$

with C a constant, i.e. it is assumed that it is possible to find a function f of the three principal stresses such that, when f reaches the value C as the stresses are increased, the polymer will yield. Equation (8.5) represents a surface in a three-dimensional space in which the axes represent σ_1, σ_2 and σ_3. It is useful to consider possible simple forms that the yield surface might take before comparing these with experimental data.

The simplest assumption to make is that f is independent of the pressure, p. This means that, if a particular point $(\sigma_1, \sigma_2, \sigma_3)$ lies on the surface and each of σ_1, σ_2 and σ_3 is increased by the same amount, which corresponds to a decrease in pressure, the new point must still lie on the surface. The surface must thus take the form of a cylinder with its axis making equal angles with the three principal axes of stress, i.e. the axis must be parallel to the 111 direction. Since the choice of subscripts 1, 2 and 3 is arbitrary, it follows that the cylinder must have at least threefold symmetry around the 111 direction and its axis must pass through the origin. If it is further assumed that the effects of stresses σ_i and $-\sigma_i$ are equivalent when $\sigma_j = \sigma_k = 0$, it follows that the cylinder must have at least sixfold symmetry about the 111 direction.

The equation $\sigma_1 + \sigma_2 + \sigma_3 = 0$ represents the plane through the origin normal to the 111 direction and it also represents the condition that the pressure $p = -(\sigma_1 + \sigma_2 + \sigma_3)/3$ is zero. If $(\sigma_1, \sigma_2, \sigma_3)$ represents a particular point on the yield surface and $p' = -p$, then $(\sigma_1 - p', \sigma_2 - p', \sigma_3 - p')$ represents a point on the surface that also lies on the plane through the origin normal to the 111 direction. The distance of this point from the origin is $\sum_{i=1}^{3}(\sigma_i - p')^2$. If it is assumed that all such points must lie on a circle of radius $\sqrt{C}$, i.e. that the yield-criterion cylinder is a circular cylinder of radius $\sqrt{C}$ rather than a hexagonal cylinder, the yield criterion becomes

$$(\sigma_1 - p')^2 + (\sigma_2 - p')^2 + (\sigma_3 - p')^2 = C \tag{8.6}$$

This is the so-called *von Mises yield criterion*. It can also be written (see problem 8.1) in the form $\tau_{\text{oct}} = \sqrt{C/3}$, where τ_{oct} is the so-called *octahedral shear stress* given by

$$9\tau_{\text{oct}}^2 = (\sigma_1 - \sigma_2)^2 + (\sigma_2 - \sigma_3)^2 + (\sigma_3 - \sigma_1)^2 \tag{8.7}$$

In the second form the von Mises criterion expresses directly the fact that the yield depends equally on the three shear stresses $(\sigma_i - \sigma_j)/2$. A somewhat simpler criterion, the *Tresca yield criterion*, makes the slightly different assumption that yield takes place when the largest of these three shear stresses reaches a critical value. The surface in σ-space that represents the criterion is therefore defined by the six equations

$$\begin{aligned}
\sigma_1 - \sigma_2 &= C', & \sigma_2 - \sigma_1 &= C' \\
\sigma_2 - \sigma_3 &= C', & \sigma_3 - \sigma_2 &= C' \\
\sigma_3 - \sigma_1 &= C', & \sigma_1 - \sigma_3 &= C'
\end{aligned} \tag{8.8}$$

where C' is another constant and it is assumed in each case that the value of the principal stress not involved in an equation lies between those of the two that are involved. Each pair of equations represents two planes in σ-space that are parallel to each other and to the 111 direction and these planes intersect the plane $\sigma_1 + \sigma_2 + \sigma_3 = 0$ in a regular hexagon. Only the parts of the planes that enclose the hexagonal cylinder represent parts of the yield surface.

These yield criteria, which assume that the yield is independent of the pressure and depends only on shear stresses, are expected to be useful for materials of low compressibility, such as metals. Polymers, on the other hand, have high compressibilities, so that the application of hydrostatic pressure to a polymer significantly affects the structure and it is necessary to take into account the pressure dependence of yield. There are in fact several differences between polymers and materials that exhibit *ideal plastic behaviour*. A material is said to exhibit ideal plastic behaviour if the material does not deform in any way under the application of any set of stresses below those required to produce yield. For an ideal material the time taken to reach any stress level is unimportant, because the material does not respond to any set of stresses below those that cause instantaneous yield. Polymers do not behave in this way; they undergo very significant changes of dimensions at stresses below the yield stress and the stress developed at a given strain depends on the strain-rate. The behaviour of polymers is also very temperature-dependent. It is not surprising, therefore, that the Tresca and von Mises yield criteria have not been found to describe the yield behaviour of polymers well, even for relatively incompressible glassy polymers such as poly(methyl methacrylate) and polystyrene.

One way of attempting to allow for the non-ideal properties of polymers in the yield criteria would be to assume that C or C' in equation (8.6) or (8.8) was not a constant but some function $C(\dot{e}, T, p)$ or $C'(\dot{e}, T, p)$, where $\dot{e}$ represents the strain-rate, T the temperature and p the pressure. A somewhat simpler approach, based on the so-called *Coulomb yield criterion*, has, however, been found useful and is described below.

If yielding is to occur by sliding parallel to any plane, it seems reasonable to suppose that there must be a critical shear stress τ parallel to that plane. It is also physically reasonable to assume that this critical stress τ will be increased if the compressive stress $-\sigma_n$ normal to this plane is increased, because this will force the molecules closer together and make it more difficult for them to slide past each other or to deform parallel to the plane. (In the remainder of section 8.2 σ_n will mean normal stress, as just defined, not nominal stress. Remember also that the usual convention for stress makes $+\sigma_n$ the *tensile* stress.) The simplest assumption is that τ depends linearly on σ_n and has the value τ_c when $\sigma_n = 0$. This leads to the Coulomb yield criterion:

$$\tau = \tau_c - \mu\sigma_n \qquad (8.9)$$

where μ is a constant.

The simplest kind of experiment that might be used for testing the applicability of the Coulomb criterion is the application of a tensile or compressive stress along the axis of a rod-like specimen. It can then be shown (see example 8.2 and problem 8.3) that if the Coulomb criterion applies, yielding will take place by shearing along a plane whose normal makes the angle $\theta = \pi/4 \pm \phi/2$ with the axis of the rod, where $\phi = \tan^{-1} \mu$ and the plus sign applies for compressive stress and the minus sign for tensile stress.

The next simplest type of stress field consists of two stresses applied at right angles. It can then be shown (see problem 8.4) that, if one of the stresses, σ_2, is held constant for this type of experiment and the other, σ_1, is varied until yield occurs, the value of σ_1 at yield is given by

$$\sigma_1 = a + b\sigma_2 \tag{8.10}$$

Example 8.2

The compressive stress σ applied parallel to the axis of a rod of material that obeys the Coulomb yield criterion with $\tau_c = 10^7$ Pa and $\mu = 0.4$ is gradually increased. Calculate the stress σ at which the rod will yield and the angle θ between the yield plane and the axis of the rod.

Solution

Writing $\tan \phi$ for μ in equation (8.9) gives $\tau = \tau_c - \sigma_n \tan \phi$. The area of the yield plane is $1/\cos \theta$ times that of the rod, so that the shear stress and the normal stress on this plane are $\sigma \sin \theta \cos \theta$ and $-\sigma \cos^2 \theta$, respectively (the *tensile* stress applied is $-\sigma$). According to the yield criterion, yield occurs when

$$\sigma \sin \theta \cos \theta = \tau_c + \sigma \tan \phi \cos^2 \theta$$

i.e. when

$$\sigma(\sin \theta \cos \theta - \tan \phi \cos^2 \theta) = \tau_c \tag{a}$$

Yield will take place at the angle θ for which σ is a minimum, so that $(\sin \theta \cos \theta - \tan \phi \cos^2 \theta)$ must be a maximum. Equating the differential coefficient of this expression with respect to θ to zero gives

$$\tan \phi \tan(2\theta) = -1, \text{ or } \theta = \pi/4 + \phi/2 \tag{b}$$

Inserting the value of $\tan \phi = \mu = 0.4$ into equation (b) gives $\theta = 55.9°$. Equation (a) gives the value of σ for yield as

$$\tau_c/(\sin \theta \cos \theta - \tan \phi \cos^2 \theta)$$

and inserting the values of τ_c and θ gives $\sigma = 3.0 \times 10^7$ Pa.

where a and b are constants that depend on the values of τ_c and μ. This expression was found to apply to the results of a plane-strain compression experiment on poly(methyl methacrylate) carried out by Bowden and Jukes, in which σ_2 was a tensile stress applied to a strip of the polymer and σ_1 was a compressive stress applied perpendicular to it (see fig. 8.5). The values of a and b were found to be -11.1 and 1.365 when the (true) stresses were expressed in units of 10^7 Pa. Although there were some problems in reconciling the predicted and experimental values of the direction of the yield plane, these results suggest that the Coulomb criterion is a useful one for at least some polymers.

If the pressure on a sample is increased the normal force on every plane is increased and the Coulomb criterion thus suggests that there is a dependence of yield on pressure. The pressure dependence of yield is therefore considered in the following section.

8.2.4 The pressure dependence of yield

Equation (8.9) can be rewritten

$$\tau + \mu(\sigma_n + p) = \tau_c + \mu p \tag{8.11}$$

where $p = -\frac{1}{3}(\sigma_1 + \sigma_2 + \sigma_3)$ is the hydrostatic-pressure component of the system of applied stresses. For any plane in the material the LHS of this equation is a function of the applied stresses that is independent of the hydrostatic pressure (because an increase in p by Δp reduces σ_n by Δp) and the RHS increases linearly with the hydrostatic pressure. Thus, according to the Coulomb criterion, for yield to occur the stresses must be such that, for some plane in the material, the value of a function of the stresses that is independent of the pressure attains a value that depends linearly on the

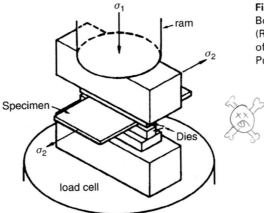

Fig. 8.5 The apparatus of Bowden and Jukes. (Reprinted by permission of Kluwer Academic Publishers.)

pressure. There is also experimental evidence that yield does depend linearly on pressure for a variety of polymers under a variety of loading conditions.

The application of the yield criterion represented by equation (8.11) is not straightforward, however, because it is necessary to find the plane for which the LHS is a maximum under the particular system of applied stresses. It has been found that a simpler criterion, based on the von Mises criterion, is useful for describing yield in a variety of multi-axial stressing experiments. This criterion is expressed by the equation

$$\tau_{oct} = \tau_{oct}^o + \alpha p \qquad (8.12)$$

where τ_{oct} is the octahedral yield stress, τ_{oct}^o is the value at zero pressure and α is a constant. It must be emphasised that, although the LHS of this equation is also independent of pressure, the equation does *not* follow from equation (8.11) and the corresponding yield criteria are in fact different.

8.2.5   Temperature and strain-rate dependences of yield

The temperature dependence of yield can be modelled by using ideas very similar to those involved in the site model introduced in section 5.7.3 in order to explain the temperature dependence of relaxation processes associated with constant activation energies. It is assumed in that model that (reversible) creep takes place because the application of stress causes a change in the numbers of molecules jumping in the two directions over an asymmetric energy barrier. This leads to a change in the length of the sample because jumps in the direction that causes elongation are favoured over those that cause shortening. The application of stress thus causes a change in the equilibrium concentrations in the two sites. Removal of the stress leads to a relaxation back to the original distribution.

Suppose, however, that, instead of two sites, there is a row of sites all with equal energies and separated by equal energy barriers ΔG. Application of a stress now causes a flow, because the stress lowers the energy of any one site with respect to the adjacent one in the direction opposite to the flow, so that jumps in the direction of flow are favoured over those in the reverse direction. Because there is a series of energy minima, there is in this case no net change in the number of occupants of any site, so that, on release of the stress, there is no tendency for the flow to continue or reverse; jumps take place in either direction equally, at a very low rate.

In adapting the model of chapter 5 to the description of the flow that takes place on yielding the following changes need to be made: (i) the applied stress now produces a large change in the free energies of the sites, so that the approximation made in equations (5.22) cannot be made here; and (ii) the equilibrium number of occupants of any site is now constant. Setting $v_{12}^{0} = v_{21}^{0} = v$ leads to the replacement of equation (5.22) by

$$v = A \exp[-\Delta G/(RT)] \tag{8.13}$$

for jumps from either side of the barrier to the other when no stress is applied. Assuming that the application of the stress σ changes the effective barrier heights on the two sides of any barrier to $\Delta G \pm v\sigma$, where v has the dimensions of volume, and denoting the number of occupied sites by n, the difference between the transition rates in the two directions (i.e. the net transition rate) becomes

$$nA\left[\exp\left(\frac{-\Delta G + v\sigma}{RT}\right) - \exp\left(\frac{-\Delta G - v\sigma}{RT}\right)\right] \tag{8.14}$$

Assuming that

$$v\sigma/(RT) \gg 1 \tag{8.15}$$

the second exponential in expression (8.14) is negligible compared with the first. Inserting $\Delta G = \Delta H - T\Delta S$ and assuming that the strain-rate $\dot{e}$ is proportional to the net transition rate then leads to

$$\dot{e} = C \exp\left(\frac{-\Delta H - v\sigma}{RT}\right) \tag{8.16}$$

where C is a constant proportional to $\exp(\Delta S/R)$. Rearranging equation (8.16) leads finally to

$$\frac{\sigma}{T} = \frac{R}{v}\left[\frac{\Delta H}{RT} + \ln\left(\frac{\dot{e}}{C}\right)\right] \tag{8.17}$$

Figure 8.6 shows a set of experimental data conforming to the prediction of equation (8.17). From the fit of the data to the equation the value of v, called the *activation volume*, can be obtained. In the equations above, the value of v is the activation volume per mole of sites and must be divided by Avogadro's number, N_A, to obtain the activation volume per site. The value found depends on the type of stress applied, i.e. whether it is shear, tensile or compressive, and is generally 2–10 times the volume of the effective random link deduced from data on dilute solutions. This suggests that yield involves the co-operative movement of a relatively large number of chain segments.

The yield stresses of several polymers at low temperatures and high strain-rates have been found to increase more rapidly with increasing strain-rate and decreasing temperature than predicted by equation (8.17). This behaviour can often be described by the addition of terms representing two activated processes with different activation volumes.

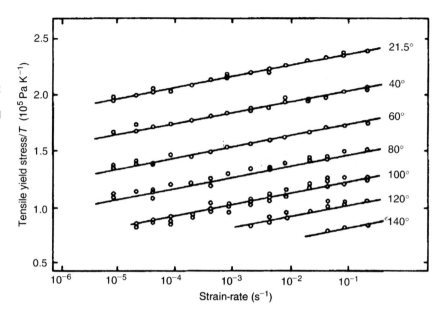

Fig. 8.6 The measured ratio of yield stress to temperature as a function of the logarithm of strain-rate for polycarbonate. The set of parallel straight lines is calculated from equation (8.17). (Reprinted by permission of John Wiley & Sons, Inc.)

The model just discussed predicts that, after yield has taken place there is no tendency for recovery on removal of the stress. Many glassy polymers will, however, recover their original shape if they are heated above T_g. This can be accommodated in the model by assuming that the successive barrier heights are not identical, but very slightly different in a direction favouring the original shape. As long as the difference is very much less than $v\sigma$ the argument above is not significantly affected, but raising the temperature sufficiently can now lead to recovery by the same arguments as those used in the discussion of creep in chapter 5. Physically this corresponds to the fact that the molecules slide past each other in yield but entanglements ensure that there is a memory of the original state.

8.3 Fracture

8.3.1 Introduction

As indicated in section 8.1, fracture can take place in essentially two ways, either following macroscopic yield, when the fracture is said to be *ductile fracture*, or without macroscopic yield, when it is called *brittle fracture*. The present section is concerned with the brittle fracture of polymers. Experimental studies of brittle fracture can be divided broadly into two types, those that are undertaken with a view to understanding the details of what happens during fracture and those that are aimed at providing engineering data about a polymer. Experiments of the former type are often designed to test the predictions of a theoretical model, whereas experiments of the latter type are performed in a way that accords with some recognised international standard. The *Izod, Charpy and falling-weight-impact tests* fall into the second category and are discussed after the theories of fracture and the experiments to test them have been discussed.

The process of brittle fracture involves two stages, *crack initiation* and *crack propagation*. Traditionally it has been assumed that, in practice, minor cracks or flaws always exist in real polymer samples, so that the propagation stage is the one of practical consequence and is usually treated first, forming the subject of *fracture mechanics*. This order is used here. Another reason why it is useful to consider fracture mechanics first is that it leads to a consideration of what happens very close to the tip of a crack, which is very closely related to the initiation stage.

8.3.2 Theories of fracture; toughness parameters

There are two principal theories, or models, that attempt to describe what happens during brittle fracture, the *Griffith fracture theory* and the *Irwin model*. Both assume that fracture takes place through the presence of pre-existing cracks or flaws in the polymer and are concerned with what happens near such a crack when a load is applied. Each leads to the definition of a *fracture-toughness parameter* and the two parameters are closely related to each other. The Griffith theory is concerned with the elastically stored energy near the crack, whereas the Irwin model is concerned with the distribution of stresses near the crack. Both theories apply strictly only for materials that are perfectly elastic for small strains and are therefore said to describe *linear fracture mechanics*.

As a crack propagates from a pre-existing flaw it produces new surface area and it is assumed in the Griffith theory that the energy needed to produce this area comes from elastically stored energy, which is not uniformly distributed throughout the polymer but is concentrated near the flaw. Imagine a single small crack propagating under a (constant) stress applied to the sample. Assume that the crack length l increases by dl, producing new surface area dA, during a time in which the work done by the applied stress is dW and the elastically stored energy increases by dU. It follows that

$$\frac{dW}{dl} - \frac{dU}{dl} \geq \gamma \frac{dA}{dl} \tag{8.18}$$

where γ is the free energy per unit area of surface. Using a solution for a plate pierced by a small elliptical crack stressed at right angles to its major axis, Griffith derived an expression that interrelates the stored energy, the applied stress and the crack length l. From this relationship he showed that, when the term dW/dl is zero because the forces applying the stress do no work (constant macroscopic strain), the breaking stress σ_B, corresponding to the condition for equality in expression (8.18), is given by

$$\sigma_B = \sqrt{4\gamma E^*/(\pi l)} \tag{8.19}$$

provided that the plate is very wide compared with the length of the crack. The quantity E^* is the *reduced modulus*, equal to Young's modulus E for a thin sheet in plane stress and to $E/(1 - v^2)$ for a thick sheet in plane strain, where v is Poisson's ratio.

The general form of equation (8.19) can be explained by a very simple argument, as follows. A sheet of material without a crack under a uniform tensile stress σ has an elastic stored energy density $\sigma^2/(2E)$. When a crack is present the energy stored in any small volume near the crack is changed. For a crack of length l in a sheet of thickness B the total volume affected can be written kBl^2, where k is expected to be of the order of unity because the stress will differ from the applied stress only at distances from the crack comparable to its length. Within this volume the reduction in average energy density must be of the same order as the pre-existing energy density, say $k'\sigma^2/(2E)$, where k' is also of the order of unity. The reduction ΔU in stored energy U due to a crack of length l must thus be given by $\Delta U = kk'[\sigma^2/(2E)]Bl^2$. Thus $dU/dl = -kk'B\sigma^2 l/E$ and, if the external

---

Example 8.3

A long strip of polymer 1 cm wide and 1 mm thick, which contains no cracks, is subjected to a tensile force of 100 N along its length. This produces a strain of 0.3%. Another strip of the same polymer is identical except that it has a crack of length 1 mm perpendicular to its long axis at its centre. If the value of γ for the polymer is 1500 J m^{-2}, estimate the tensile load necessary to break the second strip.

Solution

Young's modulus E is given by the equation $\sigma = Ee$, where σ is the applied stress and e the corresponding strain. Thus $E = F/(Ae)$, where F is the applied force and A is the cross-sectional area over which it is applied. Substituting the data given leads to

$$E = 100/(10^{-5} \times 3 \times 10^{-3}) = 3.33 \times 10^9 \text{ Pa}$$

Assuming that plane-stress conditions apply, equation (8.19) shows that $F_B = A\sigma_B = A\sqrt{4\gamma E/(\pi l)}$, where F_B is the breaking force and l the length of the crack. Substitution leads to

$$F_B = 10^{-5}\sqrt{4 \times 1500 \times 3.33 \times 10^9/\pi \times 10^{-3}} = 797 \text{ N}$$

forces do no work as the crack extends, this must be equal, from relation (8.18), to $2B\gamma$. This leads immediately to $\sigma_B = \sqrt{2\gamma E/(kk'l)}$, which is equivalent to equation (8.19) if $kk' = \pi/2$.

Figure 8.7 shows how the stress increases near the tips of a crack in a sheet to which a uniform tensile stress is applied normal to the crack. Near the tip of the crack the tensile component of stress must be higher than the applied stress because the stress within the crack is zero and the stress a long way from it must be equal to the applied stress. Near the crack the other components of stress are not all zero and, in the model due to Irwin, it is shown that the whole system of stresses near the tip of a crack loaded in this way can be written in terms of the applied stress, geometrical factors and a single further quantity K called the *stress-intensity factor*. For an infinite sheet with a central crack of length l, it can further be shown that

$$K = \sigma\sqrt{\pi l/2} \tag{8.20}$$

where σ is the uniform applied stress. It is then assumed that the fracture stress is reached when K reaches a critical value K_c, the *critical stress-intensity factor*, which is a measure of the *fracture toughness*, so that

$$\sigma_B = K_c\sqrt{2/(\pi l)} \tag{8.21}$$

Comparison of equations (8.19) and (8.21) shows that

$$\gamma = K_c^2/(2E^*) \tag{8.22}$$

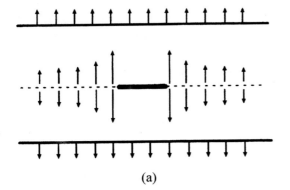

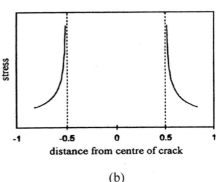

| | (a) | | (b) |

Fig. 8.7 The tensile stress near a crack in a sheet of polymer in tension. (a) A schematic representation of the stress. The arrows at the top and bottom of the diagram represent a uniform tensile stress applied to the polymer sheet in a direction normal to the crack, represented by the short thick line at the centre. The arrows pointing upwards from the dotted centre line represent the forces exerted by the upper half of the sheet on the lower half, whereas the arrows pointing downwards represent the forces exerted by the lower half of the sheet on the upper half. (b) The form of the tensile stress given by Irwin's equations for a crack of length one unit. See also section 8.3.4.

Equation (8.22) is often written in a slightly different way, making use of the *strain-energy release rate*, G, which is the energy available per unit increase in crack area. Using the notation introduced for relation (8.18) it follows that

$$G = \frac{1}{B}\left(\frac{dW}{dl} - \frac{dU}{dl}\right) = 2\gamma \tag{8.23}$$

where B is the thickness of the sample and the factor 2 appears because here the area concerned, $B\,dl$, is the area of new crack, whereas in equation (8.18) it is the area of new surface, $2B\,dl$. It is then assumed that fracture takes place when G is greater than some critical value G_c, *the critical strain-energy-release rate*, that takes account of all the energy required to produce new crack area. G_c includes the energy required to produce any accompanying plastic deformation of the surrounding material, whereas γ was intended to represent only the surface energy. This is equivalent to replacing γ by $G_c/2$ in equation (8.19), so that equation (8.22) becomes

$$G_c = K_c^2/E^* \tag{8.24}$$

The value of G_c can be related to experimentally measurable quantities as follows. Consider the application of tensile forces F to a sample of thickness B in such a way that the forces *just* cause a crack of length l to propagate, so that there is no kinetic energy involved. Suppose that, when the crack extends by dl, the forces F cause the sample to extend by dx in the line of action of the forces. The total work dW done by the forces is then $F\,dx$. Let $S = x/F$, where x is the total extension of the sample, in the line of action of the forces, due to the opening of the crack. S is then a measure of the compliance of the sample when the crack length is l. By virtue of the assumption of linearity, the elastically stored energy is given by $U = \frac{1}{2}Fx$ and the change in the elastically stored energy when the crack opens by dl is thus

$$dU = \tfrac{1}{2}(F\,dx + x\,dF) \tag{8.25}$$

The energy $G_c B\,dl$ expended in forming the new length dl of crack is thus

$$G_c B\,dl = dW - dU = F\,dx - \tfrac{1}{2}(F\,dx + x\,dF) = \tfrac{1}{2}(F\,dx - x\,dF) \tag{8.26}$$

However, $x = SF$, so that $dx = S\,dF + F\,dS$ and equation (8.26) leads to

$$G_c = \frac{F^2}{2B}\frac{dS}{dl} \tag{8.27}$$

Equation (8.27) is known as the *Irwin–Kies relationship* and it allows G_c to be determined experimentally provided that dS/dl can be determined experimentally or theoretically. Equation (8.27) then leads to an expression relating G_c to the crack length l and the applied force F, or the distance x.

8.3.3 Experimental determination of fracture toughness

Figure 8.8 shows the principles of one method that can be used for determining the value of G_c. The sample takes the form of a double cantilever beam. Forcing the ends of the beams apart by applying forces F causes a crack to propagate in the region where the beams are joined, so cleaving the sample. It is straightforward to show that, for such an experiment,

$$F = \frac{EBb^3}{64l^3}x \quad \text{so that} \quad S = \frac{x}{F} = \frac{64l^3}{EBb^3} \tag{8.28}$$

where x is the separation of the two beams along the line joining the points of application of the forces F. Substituting into equation (8.27) then leads to

$$G_c = \frac{3x^2b^3E}{128l^4} \tag{8.29}$$

Measurement of x and the distance l of the tip of the crack from this line allows G_c to be calculated.

Values of $\gamma = G_c/2$ can also be obtained from equation (8.19) by direct measurements of the tensile strength, σ_B, of samples containing cracks of known magnitude. This equation is strictly valid only for specimens of infinite width and corrections for finite width must be made.

Table 8.2 gives approximate ranges of values of γ for poly(methyl methacrylate) and polystyrene deduced from various experiments and compares them with the value estimated from the energy required to break a typical bond and the number of polymer chains crossing unit area.

Table 8.2 shows that the observed energies are at least two orders of magnitude greater than the bond-breaking energy, so that some other process or processes must be involved in fracture. These processes are in fact various localised yielding processes, the most important of which is the formation of a *craze* at the crack tip. The nature of crazes is described in the next section. Another yielding phenomenon that can take place during fracture for some polymers, such as polycarbonate, is the production of

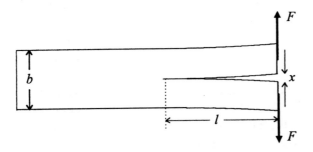

Fig. 8.8 The principle of the double-cantilever-beam method. The thickness of the beam is assumed to be B. The forces F are the least forces that will just cause the length of the crack to increase.

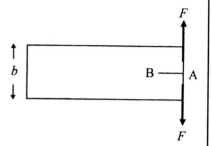

Example 8.4

Show that, if the geometry is as shown, the forces F required to make a crack AB in a sheet of polymer just propagate are inversely proportional to the length l of the crack. If $b = 2$ cm, the thickness of the sheet is 1 mm, the value of the critical energy-release rate G_c is 2500 J m^{-2} and Young's modulus of the polymer is 3×10^9 Pa, calculate the value of F when $l = 1$ cm.

Solution

Equation (8.29) shows that, when the crack just propagates, the separation x of its faces at the line of action of the forces F is given by $x = \sqrt{128l^4 G_c/(3b^3 E)}$. Equation (8.28) shows that the value of F is given by

$$F = EBb^3 x/(64l^3) = [B/(8l)]\sqrt{2Eb^3 G_c/3}$$

where B is the thickness of the sheet. Substituting the given data leads to $F = (10^{-3}/8 \times 10^{-2})\sqrt{(2 \times 3 \times 10^9 \times 8 \times 10^{-6} \times 2500/3)} = 79$ N.

thin lines of plastically deformed material, called *shear lips*, at the edges of the fracture surfaces. Because of the various features ascribable to localised yielding that can accompany fracture, each of which requires work to be done on the polymer in addition to that necessary for the formation of new surface area, it is preferable to use the term *surface-work parameter* $\mathcal{S}$ to describe the total measured energy required to produce unit area of surface during fracture.

8.3.4 Crazing

The equations derived by Irwin for the stress distribution near the tip of a crack show that the stress becomes infinite at the tip itself and is extremely high just beyond this. This cannot correspond to reality; the region just beyond the crack tip must yield and thus reduce the stress. For thick specimens under conditions of plane strain, a region called a *craze* often forms ahead of the crack. By viewing the interference fringes formed in the crack and in the craze in reflected light it can be deduced that the craze is a region of lower density than that of the bulk polymer and that it extends typically a few tens of micrometres from the tip of the crack. The profile of the craze, i.e. its variation in thickness with distance from the crack tip, can

Table 8.2. *Fracture surface energies for polymers*

Polymer	$\gamma(10^2 \text{ J m}^{-2})$
Poly(methyl methacrylate)	1.4–4.9
Polystyrene	7–26
Theory	0.015

also be deduced from the fringes and the thickness is found to decrease steadily to zero.

These measurements show that the shape of the craze is well described by the Dugdale plastic zone model originally proposed for metals. The tip of the craze is approximated by a straight line perpendicular to the direction of advance of the crack and the thickness profile can be deduced from the assumption that there is a constant *craze stress* σ_{cr} normal to the plane of the craze. Provided that $\sigma \ll \sigma_{\text{cr}}$, where σ is the applied tensile stress causing the crack, the length R of the craze ahead of the crack tip is given by

$$R = \frac{\pi}{8}\left(\frac{K_{\text{c}}}{\sigma_{\text{cr}}}\right)^2 \tag{8.30}$$

The thickness profile is given by a rather complicated expression involving R, σ_{cr} and E and is shown in fig. 8.9.

Measurements of the critical angle for reflection at a craze yield a value for the refractive index of the craze and show that it must consist of approximately 50% polymer and 50% void. Investigations by electron microscopy, electron diffraction and small-angle X-ray scattering show

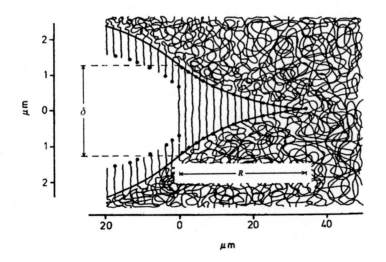

Fig. 8.9 The shape of a craze profile. The thickness of the craze at the crack tip is $\delta = 8\sigma_{\text{cr}}R/(\pi E^*)$. The deformation behaviour of the molecules is indicated schematically, but see also the text and fig. 8.10. The dimensions are those of a particular craze observed in poly(methyl methacrylate). Note the different vertical and horizontal scales. (Adapted with permission of IOP Publishing Limited.)

that a craze consists of fibrils of polymer running normal to the plane of the craze, surrounded by void, as shown in fig. 8.10. Figure 8.10 does not, however, show the gradual reduction of the thickness of the craze towards its tip.

The investigations that show the presence of fibrils also show that they consist of highly oriented material, i.e. that the polymer chains run preferentially parallel to the length of the fibrils. Studies on various polymers have shown that the surface energy rises and the craze dimensions fall with increasing molar mass of the polymer. This immediately suggests that the length of the polymer chains is an important factor in determining the properties of crazes. Attempts to explain the properties in terms of the overall length of the molecule required to pass from one side of the craze to the other have not been satisfactory and it is generally accepted that the material in the fibrils is very similar to that in a polymer that has been cold-drawn. The natural draw ratio associated with cold drawing is discussed in section 10.2.4 and the conclusion is that it depends on the presence of a network of chains. The extensibility of a network is defined by the separation between the network junction points, not the overall length of the polymer chains, and this is believed to apply also in the formation of the fibrils within crazes.

Crazes can also be produced in certain polymers, e.g. glassy poly(methyl methacrylate) and polystyrene, by applying a sufficiently large tensile stress, without there being any associated cracks. (In these cases it is generally believed that the crazes initiate on microscopic flaws or defects.) The crazes form with their planes normal to the direction of the stress. Attempts to find a stress criterion for crazing analogous to that for yielding

Fig. 8.10 A schematic diagram of the cylindrical fibrils oriented normal to a craze plane. For clarity the craze region is shown at (a) and the fibrils within it at (b). (Adapted by permission of Marcel Dekker Inc.)

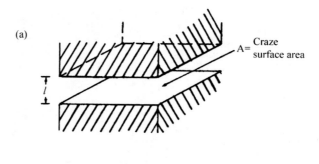

have been made. One suggestion was that crazing occurs when the extensional strain in any direction reaches a critical value, which depends on the hydrostatic component of stress. Some, but not all, experimental evidence supports this idea. The only other theory of crazing generally considered is that due to Argon, which involves detailed micro-mechanical considerations too complex to be considered here. Crazing is strongly affected by the presence of liquids and gases; the critical strain for crazing is decreased as the solubility of the surrounding fluid increases. The phenomenon is called *environmental crazing* and is of great practical importance.

Cracks that have a craze at the tip advance by the craze growing at its tip and the fibrils fracturing at its rear end where it joins the crack. The mechanism of craze growth has been explained as one of *meniscus instability* under the action of a negative pressure gradient. The interface between the polymer at the craze tip and the vacuum just outside it within the craze does not remain straight but develops a waviness with a characteristic wavelength λ and finally breaks up to give columns of polymer just behind the new craze tip. As the tip advances the thickness of the region behind the tip increases so that the columns are stretched to give the highly oriented fibrils in the mature part of the craze. This is illustrated in fig. 8.11.

8.3.5 Impact testing of polymers

The ability of a component to withstand a sudden impact is obviously of great importance for any practical application of the material. The impact testing of polymers is therefore equally important. By using standardised techniques it is fairly straightforward to obtain results that allow the comparison of one type of polymer with another, but it is less easy to obtain fundamental data.

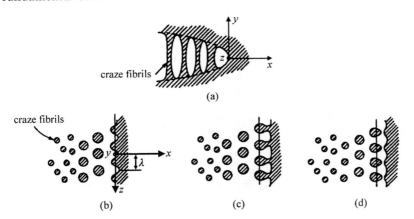

Fig. 8.11 Craze growth by meniscus instability. The x direction is the direction of advance of the craze tip and the y direction is normal to the craze plane. (a) A side view of the craze tip; (b)–(d) sections through the mid-plane of the craze, the xz plane, illustrating the advance of the craze tip and fibril formation by the meniscus-instability mechanism. (Adapted by permission of Taylor & Francis Ltd.)

Fig. 8.12 The various kinds of impact tests.

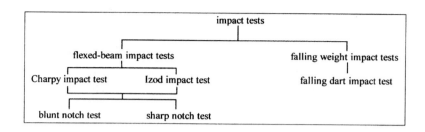

Figure 8.12 shows the principal types of test. In the Charpy test a beam of the polymer is held at each end and is struck at its centre by a hammer with one or two knife edges, giving a three- or four-point impulsive bending stress, respectively. Figure 8.13 shows a diagram of a standard Charpy impact tester. In the Izod test the specimen is held at one end and struck at the other. For either test a notch is cut in the sample at the point where it is to break and the sample is placed in the tester with the notch pointing away

Fig. 8.13 The Charpy impact test: (a) the pendulum tester, ASTM 256; and (b) the sample and striking hammer tip. ((a) Reprinted and (b) adapted, with permission from the American Society for Testing and Materials.)

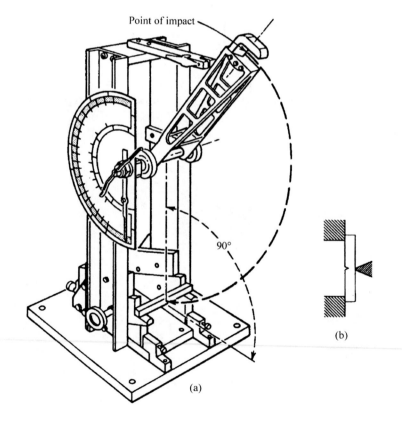

from the hammer, so that the blow tends to widen the notch. If the blow is strong enough the sample cracks at the foot of the notch and fractures. The standard type of specimen has a blunt notch, which is a V-shaped notch with a tip radius of 0.25 mm, but sometimes a sharper notch is made, into the foot of which a razor blade is driven to provide a starting place for a crack.

In the falling-dart test a circular sample of polymer is supported horizontally and a pointed steel dart is allowed to fall so that it strikes the centre of the disc. The standard dimensions for this test are radius of test disc 6 cm, thickness of disc 2 mm and radius of dart tip 1 mm. The disc is supported horizontally by an annulus of 4 cm aperture to which it is clamped by a second annulus. The behaviour of the sample can be studied under a variety of different degrees of severity of impact simply by varying the height from which the dart is allowed to fall.

Of the tests described above, the one most amenable to theoretical interpretation is the Charpy test, particularly when sharp notches are used, and this is the only one considered in detail here. Descriptions of the other tests can be found in the books referred to as (1) in section 8.4.

It is assumed that, when the hammer strikes the sample, it begins to impart elastic stored energy to the sample and thereby loses kinetic energy. It is further assumed that the sample is perfectly elastic, so that the assumptions of linear fracture mechanics apply. The elastic stored energy U at any time is thus given by $U = Fx/2 = F^2 S/2$, where S is the appropriate compliance defined in the manner described in section 8.3.2 and F is the instantaneous applied force. As the sample continues to bend under the impact the bending force and the stored energy increase until F reaches a value F_o that satisfies the Irwin–Kies relationship, equation (8.27), and thus represents the force that just causes the crack to propagate, so that

$$G_c = \frac{F_o^2}{2B}\frac{dS}{dl} \tag{8.31}$$

Once the crack starts to propagate dS/dl increases, so that the force required to satisfy the Irwin–Kies relationship falls. The crack therefore propagates extremely rapidly across the sample and the reaction force of the sample on the hammer falls rapidly to zero. If the stored energy corresponding to F_o is U_o then

$$G_c = \frac{U_o}{B}\frac{1}{S}\frac{dS}{dl} \tag{8.32}$$

The hammer continues its swing and the energy U_o is easily obtained from the difference between the potential energies of the pendulum at the beginning and end of the swing. The value of $(1/S)\,\mathrm{d}S/\mathrm{d}l$ can be calculated from the geometry of the sample and so the value of G_c can be deduced.

In practice several corrections to the difference between the potential energies of the pendulum at the beginning and end of the swing have to be made in order to obtain the correct value for U_o. The most obvious of these is that the sample is given kinetic energy as it flies off after fracture, but it is also important to take account of any energy transferred to the rest of the equipment. Once appropriate corrections have been made, values deduced in this way for glassy polymers are usually found to be independent of sample geometry and similar results have been obtained for polyethylene.

Somewhat more sophisticated instruments are frequently used for so-called *instrumented impact tests*. These instruments carry devices for measuring quantities such as the force applied to the sample and its displacement as a function of time and they allow the impact energy to be obtained more directly. Bending tests are, however, frequently carried out in a less sophisticated way using specimens without notches. The energy of fracture divided by the area of fracture is then called the *impact strength*. This quantity cannot be simply related to the quantities such as the critical strain-energy-release rate defined in fracture mechanics.

8.4 Further reading

(1) *An Introduction to the Mechanical Properties of Solid Polymers*, by I. M. Ward and D. W. Hadley; and *Mechanical Properties of Solid Polymers*, by I. M. Ward. See section 6.6.

(2) *Fracture Behaviour of Polymers*, by A. J. Kinloch and R. J. Young, Applied Science Publishers, London and New York, 1983. This book has a general introduction to solid polymers, followed by two major sections on I, Mechanics and Mechanisms, and II, Materials, where different categories of polymers are considered separately.

8.5 Problems

8.1. Derive the von Mises yield criterion in the form $\tau_{\mathrm{oct}} = \sqrt{C/3}$ from equation (8.6) and the definition of the octahedral shear stress τ_{oct}.

8.2. Gradually increasing stresses σ_1 and $\sigma_2 = -\sigma_1$ are applied to a material with the third principal stress σ_3 constant and equal to zero and the magnitudes of σ_1 and σ_2 are found to be σ when the material yields. Assuming that the material obeys the von Mises yield criterion, calculate the magnitudes of σ_1 and σ_2 that would produce yield if $\sigma_2 = \sigma_1$ and $\sigma_3 = 0$.

8.3. Show that, if a tensile stress σ is applied parallel to the axis of a rod of uniform cross-section, the shear stress τ acting on any plane whose normal makes the angle θ with the axis of the rod is $\sigma \sin\theta\cos\theta$ and that the normal stress σ_n on the plane is $\sigma \cos^2 \theta$. Hence, by finding the angle θ for which $\tau + \mu \sigma_n$ is a maximum for a given value of σ, deduce an expression for the angle θ at which yielding takes place if the Coulomb criterion is obeyed by the material of the rod.

8.4. Deduce the values of a and b in equation (8.10) for an experiment of the type carried out by Bowden and Jukes for a polymer for which $\tau_c = 5 \times 10^7$ Pa and $\mu = 0.25$. (This problem should be attempted only after the solution to problem 8.3 has been understood.)

8.5. For a particular polymer rod, yield in compression parallel to the axis of the rod occurs on the plane whose normal makes an angle of $47°$ with the axis of the rod. Calculate the value of μ for the polymer and the corresponding angle at which yield would take place for a tensile stress parallel to the axis of the rod.

8.6. In a tensile-yield experiment carried out at $60\,°C$ on a polymer the stresses shown in the table were required in order to produce yield at the indicated strain-rates. Use these data to calculate the best value of the activation volume per site, expressing your answer in nm^3.

Strain-rate (s^{-1})	0.00001	0.0001	0.001	0.01	0.1
Stress (10^7 Pa)	4.53	4.91	5.23	5.39	5.76

8.7. A rod of polymer containing no cracks and of length 10 cm and diameter 5 mm is subjected to a tensile force of 200 N. This produces an extension of 0.25 mm. A piece of the same polymer 15 mm wide and 1.5 mm thick that has a crack 1 mm long perpendicular to its long axis at its centre requires a force of 1600 N to break it. Assuming that linear elastic behaviour applies, calculate the value of Young's modulus, E, and the critical strain-energy-release rate G for the polymer.

8.8. Show that, in a double-cantilever-beam experiment, the value of the critical strain-release rate G can be deduced, without any need to know the length of the crack, from simultaneous measurements of the applied force F required to make the crack just propagate and the separation x of the faces of the crack at the line of action of the forces, provided that the dimensions of the beam and Young's modulus E of the material are known. Assuming that $E = 2.5 \times 10^9$ Pa and the width b and thickness B of the sample are 2 cm and 4 mm, respectively, calculate G when $x = 2.5$ mm and $F = 10$ N.

8.9. The tensile fracture stress σ_B for a particular polymer is found to be 6.0×10^7 Pa when it contains a crack of length 1 mm perpendicular to the axis of stress. If a craze extends to a distance $R = 30\ \mu m$ from the tip of each end of the crack, calculate the craze stress σ_{cr}.

Chapter 9
Electrical and optical properties

9.1 Introduction

As indicated in chapter 1, the first electrical property of polymers to be valued was their *high electrical resistance*, which made them useful as *insulators* for electrical cables and as the *dielectric media* for capacitors. They are, of course, still used extensively for these purposes. It was realised later, however, that, if *electrical conduction* could be added to the other useful properties of polymers, such as their low densities, flexibility and often high resistance to chemical attack, very useful materials would be produced. Nevertheless, with few exceptions, conducting polymers have not in fact displaced conventional conducting materials, but novel applications have been found for them, including plastic batteries, electroluminescent devices and various kinds of sensors. There is now much emphasis on semiconducting polymers. Figure 9.1 shows the range of conductivities that can be achieved with polymers and compares them with other materials.

Conduction and dielectric properties are not the only electrical properties that polymers can exhibit. Some polymers, in common with certain other types of materials, can exhibit *ferroelectric properties*, i.e. they can acquire a permanent electric dipole, or *photoconductive properties*, i.e. exposure to light can cause them to become conductors. Ferroelectric materials also have *piezoelectric properties*, i.e. there is an interaction between their states of stress or strain and the electric field across them. All of these properties have potential applications but they are not considered further in this book.

The first part of this chapter describes the electrical properties and, when possible, explains their origins. Unfortunately, some of the properties associated with conduction are not well understood yet; even when they are, the theory is often rather difficult, so that only an introduction to the conducting properties can be given. The second part of the chapter deals with the *optical properties*. The link between these and the electrical properties is that the optical properties depend primarily on the interaction of the electric field of the light wave with the polymer molecules. The following

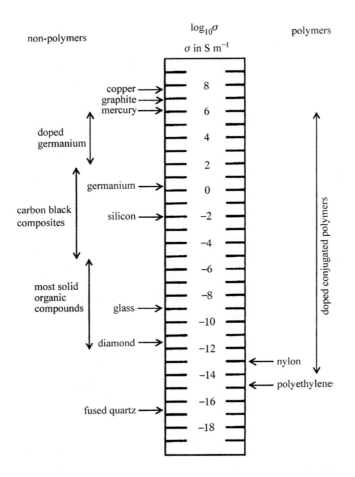

Fig. 9.1 Electrical conductivities of polymers compared with those of other materials. All values are approximate.

section deals with the *electrical polarisation* of polymers, which underlies both their dielectric and their optical properties.

9.2 Electrical polarisation

9.2.1 The dielectric constant and the refractive index

In an ideal insulating material there are no free electric charges that can move continuously in the presence of an electric field, so no current can flow. In the absence of any electric field, the bound positive and negative electric charges within any small volume element dv will, in the simplest case, be distributed in such a way that the volume element does not have any electric dipole moment. If, however, an electric field is applied to this volume element of the material, it will cause a change in the distribution of charges, so that the volume element acquires an electric dipole moment proportional to the field. It will be assumed for simplicity that the material is isotropic; if this is so, the electric dipole will be parallel to the electric field. The dipole moment $d\mu$ of the volume element will then be given by

$$d\mu = P\,dv = pE_L\,dv \tag{9.1}$$

where p is the *polarisability* of the material and P is the *polarisation* produced by the field E_L acting on dv. On the molecular level a *molecular polarisability* α can be defined; and, if N_o is the number of molecules per unit volume, then

$$P = N_o\alpha E_L \tag{9.2}$$

The mechanism for the molecular polarisability is considered in section 9.2.2.

So far no consideration has been given to the origin of the electric field experienced by the small volume element under consideration, but it has been denoted by E_L to draw attention to the fact that it is the *local field* at the volume element that is important. An expression for this field is derived below.

Equation (9.1) shows that P is the dipole moment per unit volume, a vector quantity that has its direction parallel to E_L for an isotropic medium. Imagine a small cylindrical volume of length dl parallel to the polarisation and of cross-sectional area dA. Let the apparent surface charges at the two ends of the cylinder be $\pm dq$. It then follows that $P = dq\,dl/dv$, where dv is the volume of the small cylinder. However, $dv = dl\,dA$, so that $P = dq/dA = \sigma$, where σ is the apparent surface charge per unit area normal to the polarisation. (Note: σ stands for conductivity in section 9.3.)

The simplest way to apply an electric field to a sample of polymer is to place it between two parallel conducting plates and to apply a potential difference V between the plates, which gives rise to an applied electric field normal to the plates of magnitude $E = V/d$, where d is the separation of the plates. This field may be imagined to be produced by layers of charge at the conducting plates. The net charges in these layers are the differences between the free charges on the plates and the apparent surface charges on the polymer adjacent to the plates due to its polarisation. Imagine now that a small spherical hole is cut in the polymer and that this can be done without changing the polarisation of the remaining polymer surrounding it.

New apparent surface charges will be produced at the surface of the spherical hole. The normal to any small surface element of the sphere makes an angle θ with the polarisation vector, so that the apparent charge per unit area on this element is $\sigma\cos\theta = P\cos\theta$. The electric field produced at the centre of the sphere by all these surface elements can be shown by integration (see problem 9.1) to be $P/(3\varepsilon_0)$, where ε_0 is the permittivity of free space, so that the total field at the centre of the sphere is equal to $E + P/(3\varepsilon_0)$. The only way, however, that the polarisation could remain undisturbed outside the imaginary spherical cavity would be for the cavity to be refilled with polymer! The total electric field at the centre of the sphere would then be increased by the field produced there by all the dipoles in this sphere of polymer. Fortunately it can be shown that the latter field is zero if the individual molecular dipoles have random positions within the sphere. In this case the field at the centre of the sphere remains $E + P/(3\varepsilon_0)$, which is therefore the field that would be experienced by a molecule at the centre, i.e. it is the local field E_L. Thus

$$E_L = E + P/(3\varepsilon_0) \tag{9.3}$$

The *dielectric constant*, or *relative permittivity*, ε of the polymer is defined by $\varepsilon = V_0/V$, where V_0 is the potential difference that would exist between the plates if they carried a fixed free charge Q_0 per unit area in the absence of the dielectric and V is the actual potential difference between them when they carry the same free charge in the presence of the dielectric. With the polymer present, the total effective charge per unit area at the plates is $Q = Q_0 - P$ when the effective surface charge on the polymer is taken into account, so that

$$\varepsilon = \frac{V_0}{V} = \frac{E_0}{E} = \frac{Q_0}{Q} = \frac{Q+P}{Q} \tag{9.4}$$

where E_0 is the field that would exist between the plates in the absence of the polymer. However, $E = Q/\varepsilon_0$ (which can be proved by applying a similar method to that of problem 9.1 to two infinite parallel plates with charges $+Q$ and $-Q$ per unit area) so that

$$\varepsilon = \frac{\varepsilon_0 E + P}{\varepsilon_0 E} \quad \text{or} \quad P = \varepsilon_0 E(\varepsilon - 1) \tag{9.5}$$

and finally, from equation (9.3),

$$E_L = \frac{\varepsilon + 2}{3} E \tag{9.6}$$

A derivation of the local field in essentially this way was originally given by H. A. Lorentz.

Since, according to equation (9.2), $P = N_0 \alpha E_L$, equations (9.5) and (9.6) lead to

$$N_0 \alpha \frac{\varepsilon + 2}{3} E = \varepsilon_0 E(\varepsilon - 1) \quad \text{or} \quad \frac{\varepsilon - 1}{\varepsilon + 2} = \frac{N_0 \alpha}{3\varepsilon_0} \tag{9.7}$$

This relationship between the dielectric constant and the molecular polarisability is known as the *Clausius–Mosotti relation*. It can usefully be written in terms of the molar mass M and density ρ of the polymer in the form

$$\frac{\varepsilon - 1}{\varepsilon + 2} \frac{M}{\rho} = \frac{N_A \alpha}{3\varepsilon_0} \tag{9.8}$$

where N_A is the Avogadro constant and the quantity on the RHS is called the *molar polarisation* and has the dimensions of volume.

It follows from Maxwell's theory of electromagnetic radiation that $\varepsilon = n^2$, where ε is the dielectric constant measured at the frequency for which the refractive index is n. Equation (9.8) thus leads immediately to the *Lorentz–Lorenz equation*

$$\frac{n^2 - 1}{n^2 + 2} \frac{M}{\rho} = \frac{N_A \alpha}{3\varepsilon_0} \tag{9.9}$$

relating the refractive index and the molecular polarisability. Some applications of this equation are considered in sections 9.4.3 and 10.4.1. The expression on the LHS is usually called the *molar refraction* of the material, which is thus equal to the molar polarisation at optical frequencies.

9.2.2 Molecular polarisability and the low-frequency dielectric constant

There are two important mechanisms that give rise to the molecular polarisability. The first is that the application of an electric field to a molecule can cause the electric charge distribution within it to change and so induce an electric dipole, leading to a contribution called the *distortional polarisability*. The second is that the molecules of some materials have permanent electric dipoles even in the absence of any electric field. If an electric field is applied to such a material the molecules tend to rotate so that the dipoles become aligned with the field direction, which gives rise to an *orientational polarisation*. Thermal agitation will, however, prevent the molecules from aligning fully with the field, so the resulting polarisation will depend both on the field strength and on the temperature of the material. This contribution is considered first.

It is relatively easy to calculate the observed polarisation of an assembly of non-interacting molecules for a given field E_L at each molecule at a temperature T. Assume that each molecule has a permanent dipole μ and that there is no change in its magnitude on application of the field. Consider a molecule for which the dipole makes the angle θ with E_L. The energy u of the dipole is then given by

$$u = -\boldsymbol{\mu} \cdot \boldsymbol{E}_L = -\mu E_L \cos\theta \qquad (9.10)$$

If there are dn dipoles within any small solid angle $d\omega$ at angle θ to the applied field, Boltzmann's law shows that

$$dn = Ae^{-u/(kT)}\,d\omega \qquad (9.11)$$

where A depends on the total number of molecules. The total solid angle $d\omega$ lying between θ and $\theta + d\theta$ when all angles around the field direction are taken into account is $2\pi \sin\theta\,d\theta$ and the corresponding number of dipoles is thus

$$dn = 2\pi Ae^{-u/(kT)} \sin\theta\,d\theta \qquad (9.12)$$

Each of these dipoles has a component $\mu \cos\theta$ in the direction of the field and their components perpendicular to the field cancel out. They thus contribute a total of $\mu\,dn\cos\theta$ to the total dipole moment parallel to the field. Using the fact that $d(\cos\theta) = -\sin\theta\,d\theta$ and writing $\cos\theta = \xi$ and $\mu E_L/(kT) = x$, the average dipole moment $\langle\mu\rangle$ contributed by all the molecules at all angles to the field is given by

$$\frac{\langle\mu\rangle}{\mu} = \frac{\int_0^\pi 2\pi Ae^{\mu E_L \cos\theta/(kT)} \cos\theta\,d(\cos\theta)}{\int_0^\pi 2\pi Ae^{\mu E_L \cos\theta/(kT)}d(\cos\theta)} = \frac{\int_{-1}^1 \xi e^{x\xi}\,d\xi}{\int_{-1}^1 e^{x\xi}\,d\xi} \qquad (9.13)$$

Integration by parts then leads to

$$\frac{\langle\mu\rangle}{\mu} = \frac{\left[(\xi/x)e^{x\xi} - (1/x^2)e^{x\xi}\right]_{-1}^1}{\left[(1/x)e^{x\xi}\right]_{-1}^1} = \frac{e^x + e^{-x}}{e^x - e^{-x}} - \frac{1}{x} = \coth x - \frac{1}{x} = L(x) \qquad (9.14)$$

where $L(x)$ is the *Langevin function*.

For attainable electric fields, $\mu E_L/(kT) = x$ is very small and the exponentials can be expanded as $e^{\pm x} \cong 1 \pm x + x^2/2 \pm x^3/6 + O(x^4)$. Substitution then shows that the Langevin function reduces approximately to $L(x) = x/3$, so that finally

$$\langle\mu\rangle \cong \mu^2 E_L/(3kT) \qquad (9.15)$$

and the contribution to the molecular polarisability from rotation is to a good approximation $\mu^2/(3kT)$. The total molecular polarisability is therefore $\alpha = \alpha_d + \mu^2/(3kT)$, where α_d is the molecular deformational polarisability. Use of the Clausius–Mosotti relation, equation (9.8), then gives

$$\frac{\varepsilon - 1}{\varepsilon + 2}\frac{M}{\rho} = \frac{N_A\alpha_d}{3\varepsilon_0} + \frac{N_A\mu^2}{9\varepsilon_0 kT} \qquad (9.16)$$

In developing this equation, no account has been taken of any possible interactions between the molecular dipoles. It is therefore expected to be most useful for polar gases and for solutions of polar molecules in non-polar solvents, where the polar molecules are well separated.

At optical frequencies $\alpha = \alpha_d$, because the molecules cannot reorient sufficiently quickly in the oscillatory electric field of the light wave for the orientational polarisability to contribute, as discussed further in section 9.2.4. Setting $\alpha = \alpha_d$ in equation (9.9) gives α_d in terms of the optical refractive index and, if this expression for α_d is inserted, equation (9.16) can then be rearranged as follows:

$$\frac{3(\varepsilon - n^2)}{(\varepsilon + 2)(n^2 + 2)}\frac{M}{\rho} = \frac{N_A\mu^2}{9\varepsilon_0 kT} \tag{9.17}$$

Fröhlich showed that a more exact equation for condensed matter, such as polymers, is

$$\frac{(\varepsilon - n^2)(2\varepsilon + n^2)}{\varepsilon(n^2 + 2)^2}\frac{M}{\rho} = \frac{N_A g\mu^2}{9\varepsilon_0 kT} \tag{9.18}$$

The LHS of this equation differs from that of equation (9.17) because a more exact expression has been used for the internal field and the RHS now includes a factor g, called the *correlation factor*, which allows for the fact that the dipoles do not react independently to the local field. If μ, g, n and ρ are known for a polymer, it should therefore be possible to predict the value of ε. Of these, g is the most difficult to calculate (see section 9.2.5).

9.2.3 Bond polarisabilities and group dipole moments

The idea that the mass of a molecule can be calculated by adding together the atomic masses of its constituent atoms is a very familiar one. Strictly speaking, this is an approximation, although it is an excellent one. It is an approximation because the equivalent mass of the bonding energy is neglected; the mass of a molecule is in fact lower than the sum of the masses of its atoms by only about one part in 10^{14}. In a similar way, but usually to a much lower degree of accuracy, various other properties of molecules can be calculated approximately by neglecting the interactions of the various parts of the molecule and adding together values assigned to those individual parts. The word 'adding' must, however, be considered carefully. For instance, a dipole moment is a vector quantity; if the dipole moments of the various parts of a molecule are known, the dipole moment of the whole molecule can be calculated only if the constituent dipole moments are added *vectorially*.

In section 9.2.2 above it is shown that the refractive index of a medium is related to the high-frequency deformational polarisability of its molecules. Polarisability is treated there as a scalar quantity, but it is in fact a *second-rank tensor* quantity (see the appendix) and the correct way of adding such quantities is somewhat more complicated than the addition of vectors. When the refractive indices of stressed or oriented polymers are considered, as in chapters 10 and 11, this complication must be taken into account, as described later in the present chapter (section 9.4.3). For the moment attention is restricted to media in which the individual molecules are oriented randomly.

The high-frequency polarisation of a molecule consists almost entirely of the slight rearrangement of the electron clouds forming the bonds between the atoms. It is the polarisabilities of individual bonds that are assumed to be additive, to a good approximation. If the molecules are randomly oriented, then so are all the bonds of a particular type. The sum of the polarisabilities of all the bonds of this type must therefore be isotropic, so the polarisability for an individual bond can be replaced by a scalar average value called the *mean value* or *spherical part* of the tensor. These scalar values can be added together arithmetically for the various bonds within the molecule to obtain the spherical part of the polarisability for the molecule. This is what has so far been called α_d.

It is possible to find values of the mean polarisabilities α_i for a set of different types of bond i by determining the refractive indices for a large number of compounds containing only these types of bond. The value of α_d can be found for each compound by means of the Lorentz–Lorenz equation (9.9) with $\alpha = \alpha_d$ and the values of α_i are then chosen so that they add for each type of molecule to give the correct value of α_d. In order to obtain *molar refractions* or *molar bond refractions*, α_d or α_i must be multiplied by $N_A/(3\varepsilon_0)$. The values of a number of molar bond refractions obtained in this way are given in table 9.1. It should be noted that the simple additivity fails for molecules that contain a large number of double bonds, which permit the electrons to be delocalised over a large region of the molecule (see section 9.3.4).

A similar argument to the above can be used to deduce the molecular dipole for a molecule if the dipoles of its constituent parts are known. It has been found that the additivity scheme works best for dipoles if *group dipoles* are used, i.e. each type of molecular group in a molecule, such as a —COOH group, tends to have approximately the same dipole moment independently of the molecule of which it is a constituent part. Equation (9.16), which applies to dilute solutions of polar molecules in a non-polar solvent, can be used to find μ for a range of different molecules. This is done by plotting against $1/T$ the values calculated for the left-hand side of the equation from values of ε measured at various temperatures T. The gradient of the straight line plot is $N_A\mu^2/(9\varepsilon_0 k)$, from which μ is easily

Table 9.1. *Molar bond refractions for the D line of Na*[a]

Bond	Refraction (10^{-6} m^3)
C—H	1.676
C—C	1.296
C=C	4.17
C≡C (terminal)	5.87
C≡C (non-terminal)	6.24
C—C (aromatic)	2.688
C—F	1.44
C—Cl	6.51
C—Br	9.39
C—I	14.61
C—O (ether)	1.54
C=O	3.32
C—S	4.61
C—N	1.54
C=N	3.76
C≡N	4.82
O—H (alcohol)	1.66
O—H (acid)	1.80
N—H	1.76

[a]Reproduced from *Electrical Properties of Polymers* by A. R. Blythe. © Cambridge University Press 1979.

obtained. When this has been done for a range of molecules one can find a set of group dipoles μ_i that by *vector addition* predicts approximately correctly the values of μ found for all the compounds. Since the quantities μ_i are vectors, not only their magnitudes but also their directions within the molecule must be specified. This is usually done by giving the angle between μ_i and the bond joining the group i to the rest of the molecule. Table 9.2 gives dipoles for some important groups in polymers.

9.2.4   Dielectric relaxation

The orientation of molecular dipoles cannot take place instantaneously when an electric field is applied. This is the exact analogy of a fact discussed in chapter 7, namely that the strain in a polymer takes time to develop after the application of a stress. In fact the two phenomena are not simply analogous; the relaxation of strain and the rotation of dipoles are due to the same types of molecular rearrangement. Both viscoelastic

Table 9.2. *Group dipole moments*[b]

Group	Aliphatic compounds		Aromatic compounds	
	Moment $(10^{-30}$ C m)	Angle[a] (degrees)	Moment $(10^{-30}$ C m)	Angle[a] (degrees)
—CH_3	0	0	1.3	0
—F	6.3		5.3	
—Cl	7.0		5.7	
—Br	6.7		5.7	
—I	6.3		5.7	
—OH	5.7	60	4.7	60
—NH_2	4.0	100	5.0	142
—COOH	5.7	74	5.3	74
—NO_2	12.3	0	14.0	0
—CN	13.4	0	14.7	0
—$COOCH_3$	6.0	70	6.0	70
—OCH_3	4.0	55	4.3	55

[a]The angle denotes the direction of the dipole moment with respect to the bond joining the group to the rest of the molecule.
[b]Reproduced from *Electrical Properties of Polymers* by A. R. Blythe. © Cambridge University Press 1979.

Example 9.1

Calculate the refractive index of PVC using the bond refractions given in table 9.1. Assume that the density of PVC is 1.39 Mg m^{-3}.

Solution

For a polymer the Lorentz–Lorenz equation (9.9) can be rewritten

$$\frac{n^2 - 1}{n^2 + 2} = \left(\frac{N_A}{3\varepsilon_o}\frac{\alpha_d}{M}\right)\rho$$

where the quantity in brackets is the molar refraction per repeat unit divided by the molar mass of the repeat unit. The PVC repeat unit $-(CH_2-CHCl)-$, has relative molecular mass $(2 \times 12) + (3 \times 1) + 35.5 = 62.5$ and molar mass 6.25×10^{-2} kg. There are three C—H bonds, one C—Cl bond and two C—C bonds per repeat unit, giving a refraction per repeat unit of $[(3 \times 1.676) + 6.51 + (2 \times 1.296)] \times 10^{-6} = 1.413 \times 10^{-5}$ m^3. Thus $(n^2 - 1) = 1.413 \times 10^{-5} \times 1.39 \times 10^3(n^2 + 2)/(6.25 \times 10^{-2}) = 0.3143(n^2 + 2)$, which leads to $n = 1.541$.

measurements and dielectric measurements can therefore be used to study these types of rearrangement. Dielectric studies have the advantage over viscoelastic studies by virtue of the fact that a much wider range of frequencies can be used, ranging from about 10^4 s per cycle (10^{-4} Hz) up to optical frequencies of about 10^{14} Hz.

In section 7.2.1 the idea of a relaxation time is made explicit through the assumption that, in the simplest relaxation, the material relaxes to its equilibrium strain or stress on application of a stress or strain in such a way that the rate of change of strain or stress is proportional to the difference between the fully relaxed value and the value at any instant. A similar assumption is made in the present section for the relaxation of polarisation on application of an electric field.

In considering dielectric relaxation it is, however, necessary to remember that there are two different types of contribution to the polarisation of the dielectric, the deformational polarisation P_d and the orientational polarisation P_r, so that $P = P_d + P_r$. For the sudden application of a field E that then remains constant, the deformational polarisation can be considered to take place 'instantaneously', or more precisely in a time of the order of 10^{-14} s. When alternating fields are used the deformational polarisation can similarly be considered to follow the applied field exactly, provided that the frequency is below optical frequencies. If the limiting value of the dielectric constant at optical frequencies is called ε_∞, equation (9.5) shows that

$$P_d = \varepsilon_0(\varepsilon_\infty - 1)E \tag{9.19}$$

Assuming that, in any applied field with instantaneous value E, P_r responds in such a way that its rate of change is proportional to its deviation from the value that it would have in a static field of the same value E, it follows that

$$\frac{dP_r}{dt} = \frac{\varepsilon_0(\varepsilon_s - \varepsilon_\infty)E - P_r}{\tau} \tag{9.20}$$

where ε_s is the dielectric constant in a static field and τ is the *relaxation time*. Note that, in equation (9.20), it is necessary to subtract from the total polarisation $P = \varepsilon_0(\varepsilon_s - 1)E$ in a static field the polarisation $P_d = \varepsilon_0(\varepsilon_\infty - 1)E$ due to deformation in order to obtain the polarisation due to the orientation of dipoles.

Consider first the 'instantaneous' application at time $t = 0$ of a field E that then remains constant. Then it follows from equation (9.20) that

$$P_r = \varepsilon_0(\varepsilon_s - \varepsilon_\infty)(1 - e^{-t/\tau})E \tag{9.21}$$

Now consider an alternating applied field $E = E_0 e^{i\omega t}$. Because the deformational polarisation follows the field with no time lag, P_d is given, according to equation (9.19), by

$$P_d = \varepsilon_0(\varepsilon_\infty - 1)E_0 e^{i\omega t} \qquad (9.22)$$

P_r is now given, from equation (9.20), by

$$\frac{dP_r}{dt} = \frac{\varepsilon_0(\varepsilon_s - \varepsilon_\infty)E_0 e^{i\omega t} - P_r}{\tau} \qquad (9.23)$$

Assume that $P_r = P_{r,0} e^{i\omega t}$, where $P_{r,0}$ is complex to allow for a phase lag of P_r with respect to the field. It follows that

$$\frac{dP_r}{dt} = i\omega P_{r,0} e^{i\omega t} = \frac{\varepsilon_0(\varepsilon_s - \varepsilon_\infty)E_0 - P_{r,0}}{\tau} e^{i\omega t} \qquad (9.24)$$

or

$$P_{r,0} = \frac{\varepsilon_0(\varepsilon_s - \varepsilon_\infty)}{1 + i\omega\tau} E_0 \qquad (9.25)$$

The total complex polarisation P is thus given, from equations (9.19) and (9.25), by

$$P = P_d + P_r = [\varepsilon_0(\varepsilon_\infty - 1) + \varepsilon_0(\varepsilon_s - \varepsilon_\infty)/(1 + i\omega\tau)]E \qquad (9.26)$$

and the complex dielectric constant ε, from equation (9.5), by

$$\varepsilon = 1 + \frac{P}{\varepsilon_0 E} = 1 + (\varepsilon_\infty - 1) + \frac{\varepsilon_s - \varepsilon_\infty}{1 + i\omega\tau} = \varepsilon_\infty + \frac{\varepsilon_s - \varepsilon_\infty}{1 + i\omega\tau} \qquad (9.27)$$

The expression for ε given on the RHS of equation (9.27) is called the *Debye dispersion relation*. Writing the dielectric constant $\varepsilon = \varepsilon' - i\varepsilon''$, it follows immediately from equation (9.27) that

$$\varepsilon' = \varepsilon_\infty + \frac{\varepsilon_s - \varepsilon_\infty}{1 + \omega^2\tau^2} \quad \text{and} \quad \varepsilon'' = \frac{(\varepsilon_s - \varepsilon_\infty)\omega\tau}{1 + \omega^2\tau^2} \qquad (9.28a,b)$$

These equations are the dielectric equivalents of the equations developed in section 7.3.2 for the real and imaginary parts of the compliance or modulus. Just as the imaginary part of the compliance or modulus is a measure of the energy dissipation or 'loss' per cycle (see section 7.3.2), so is ε''. The variations of ε' and ε'' with ω are shown in fig. 9.2.

The terms *relaxed* and *unrelaxed dielectric constant* are used for the static and high-frequency values ε_s and ε_∞, respectively. These quantities

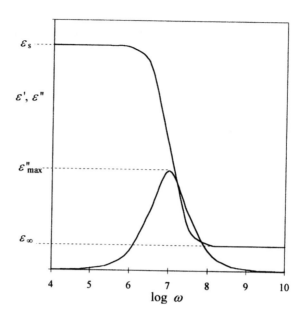

are the electrical analogues of the relaxed and unrelaxed moduli or compliances defined in section 7.3.2 and a *dielectric relaxation strength* is
defined as $\varepsilon_s - \varepsilon_\infty$, in exact analogy with the definition of mechanical
relaxation strength.

9.2.5 The dielectric constants and relaxations of polymers

As discussed in sections 9.2.2 and 9.2.3, the high-frequency, or optical,
dielectric constant depends only on the deformational polarisability of
the molecules and can be calculated from equation (9.8) (with $\alpha = \alpha_d$)
and values of the molar bond refractions, such as those given in table
9.1. Rearrangement of this equation shows, as expected, that the greater
the molar refraction the greater the high-frequency dielectric constant and
hence the refractive index. Table 9.1 then shows, for instance, that polymers with a large fraction of $C{=}C$ double bonds will generally have higher
refractive indices than will those with only single $C{-}C$ bonds and that
polymers containing chlorine atoms will have higher refractive indices than
will their analogues containing fluorine atoms. Some refractive indices of
polymers are given in table 9.3 and these trends can be observed by reference to it.

 If a polymer consisted entirely of non-polar groups, or contained polar
groups arranged in such a way that their dipoles cancelled out, it would
have a low dielectric constant at all frequencies, determined only by the

Table 9.3. *Approximate refractive indices for selected polymers at room temperature*

Polymer	Repeating unit	Refractive index
Polytetrafluoroethylene	$+(CF_2-CF_2)+$	1.35–1.38
Poly(vinlidene fluoride)	$+(CH_2-CF_2)+$	1.42
Poly(butyl acrylate)	$+(CH_2-CH(COO(CH_2)_3CH_3))+$	1.46
Polypropylene (atactic)	$+(CH_2-CH(CH_3))+$	1.47
Polyoxymethylene	$+(CH_2-O)+$	1.48
Cellulose acetate	See section 1.1	1.48–1.50
Poly(methyl methacrylate)	$+(CH_2-C(CH_3)(COOCH_3))+$	1.49
Poly(1,2-butadiene)	$+(CH_2-CH(CH=CH_2))+$	1.50
Polyethylene	$+(CH_2-CH_2)+$	1.51–1.55 (depends on density)
Polyacrylonitrile	$+(CH_2-CH(CN))+$	1.52
Poly(vinyl chloride)	$+(CH_2-CHCl)+$	1.54–1.55
Epoxy resins	Complex cross-linked ether	1.55–1.60
Polychloroprene	$+(CH_2-CCl(CH=CH_2))+$	1.55–1.56
Polystyrene	$+(CH_2-CH(C_6H_5))+$	1.59
Poly(ethylene terephthalate)	$+((CH_2)_2-(C=O)-(C_6H_4)-(C=O))+$	1.58–1.60
Poly(vinylidene chloride)	$+(CH_2-CCl_2)+$	1.60–163
Poly(vinyl carbazole)	$+(CH_2-CH)+$	1.68

deformational polarisation, and it would not undergo dielectric relaxation. Examples of polymers potentially in this class are polyethylene, in which the $>CH_2$ groups have very low dipole moments and are arranged predominantly in opposite directions along the chain, and polytetrafluoroethylene, in which the dipole moment of the $>CF_2$ group is moderately large but the molecule is helical and the dipole moments cancel out because they are approximately normal to the helix axis for each group. Such cancelling out is never exact, however, because of conformational defects. An additional factor is that oxidation, even if it does not significantly perturb the conformation, can introduce groups with different dipole moments from those of the groups replaced. The dipoles then no longer cancel out and both a higher dielectric constant and dielectric relaxation are observed.

If a polymer contains polar groups whose moments do not cancel out, the actual value of the dielectric constant depends very strongly on the molecular conformation, which in turn may depend on the dipoles, because strong repulsion between parallel dipoles will cause the conformation to change so that the dipoles are not parallel. The effect of the molecular conformation is incorporated into the correlation factor g introduced in equation (9.18). The value of g for an amorphous polymer depends on the angles between the dipoles within those parts of the molecule that can rotate independently when the electric field is applied. These angles and the sections of chain that can rotate independently are temperature- and frequency-dependent, particularly in the region of the glass transition, so it is not generally possible to calculate g. In practice, measurements of the static dielectric constant and the refractive index are used with equation (9.18) to determine g values and so obtain information about the dipole correlations and how they vary with temperature.

Matters are more complicated still for semicrystalline polymers, for which not only are the observed dielectric constants averages over those of the amorphous and crystalline parts but also the movements of the amorphous chains are to some extent restricted by their interactions with the crystallites, which causes changes in g and hence in the static dielectric constant of the amorphous material. These facts are illustrated by the temperature dependences of the dielectric constants of amorphous and 50% crystalline PET, shown in table 9.4.

Table 9.4. *Dielectric constants of amorphous and 50% crystalline poly(ethylene terephthalate)*[a]

Polymer state	T(K)	Dielectric constant	
		Unrelaxed, ε_∞	Relaxed, ε_s
Amorphous	193	3.09	3.80
	203	3.09	3.80
	213	3.12	3.80
	223	3.13	3.79
	233	3.14	3.78
	354	3.80	6.00
Crystalline	193	3.18	3.66
	203	3.18	3.67
	213	3.19	3.66
	373	3.71	4.44

[a]Adapted by permission from Boyd, R. H. and Liu, F. 'Dielectric spectroscopy of semicrystalline polymers' in *Dielectric Spectroscopy of Polymeric Materials: Fundamentals and Applications*, eds. James P. Runt and John J. Fitzgerald, American Chemical Society, Washington DC, 1997, Chap. 4, pp. 107–136, Table I, p. 117.

Table 9.4 shows that the unrelaxed values for these two samples are very similar and do not change much even at the glass transition (at about 354 K). The relaxed values are different for the two samples, particularly above the glass-transition temperature, where the relaxed dielectric constant of the amorphous sample is significantly greater than that of the crystalline sample. This temperature dependence arises from the fact that, at low temperature, the C=O groups tend to be *trans* to each other across the benzene ring, so that their dipoles almost cancel out, whereas at higher temperatures rotations around the ring—(C=O) bonds take place more freely in the amorphous regions so that the dipoles can orient more independently and no longer cancel out. The value of the relaxed constant for the crystalline sample is lower than that of the non-crystalline sample even at the lower temperatures because the rigid crystalline phase not only contributes little to the relaxed constant but also restricts the conformations allowed in the amorphous regions.

The width of the dielectric loss peak given by equation (9.28b) can be shown to be 1.14 decades (see problem 9.3). Experimentally, loss peaks are often much wider than this. A simple test of how well the Debye model fits in a particular case is to make a so-called *Cole–Cole plot*, in which ε'' is plotted against ε'. It is easy to show from equations (9.28) that the Debye model predicts that the points should lie on a semi-circle with centre at $[(\varepsilon_s + \varepsilon_\infty)/2, 0]$ and radius $(\varepsilon_s - \varepsilon_\infty)/2$. Figure 9.3 shows an example of such a plot. The experimental points lie within the semi-circle, corresponding to a lower maximum loss than predicted by the Debye model and also to a wider loss peak. A simple explanation for this would be that, in an amorphous polymer, the various dipoles are constrained in a wide range of different ways, each leading to a different relaxation time τ, so that the observed values of ε' and ε'' would be the averages of the values for each value of τ (see problem 9.4).

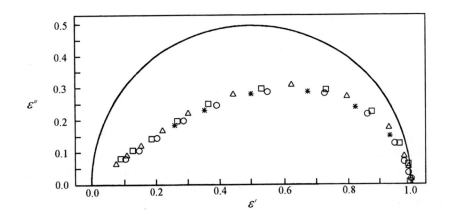

Fig. 9.3 A Cole–Cole plot for samples of poly(vinyl acetate) of various molar masses:
*, 11 000 g mol^{-1};
O, 140 000 g mol^{-1};
□, 500 000 g mol^{-1}; and
△, 1 500 000 g mol^{-1}. The dielectric constant has been normalised so that $\varepsilon_s - \varepsilon_\infty = 1$. (Adapted by permission of Carl Hanser Verlag.)

Various empirical formulae have been given for fitting data that deviate from the simple semi-circular Cole–Cole plot. The most general of these is the Havriliak and Negami formula

$$\frac{\varepsilon - \varepsilon_\infty}{\varepsilon_s - \varepsilon_\infty} = [1 + (i\omega\tau)^a]^{-b} \tag{9.29}$$

This equation has the merit that, on changing the values of a and b, it becomes the same as a number of the other formulae. In particular, when $a = b = 1$ it reduces, with a little rearrangement, to equation (9.27).

In section 5.7 various motions that can take place in polymers and can lead to NMR, mechanical or dielectric relaxation are considered mainly through their effects on the NMR spectrum. Mechanical relaxation is considered in detail in chapter 7. Although all the motions that are effective in mechanical relaxation can potentially also produce dielectric relaxation, the relative strengths of the effects due to various relaxation mechanisms can be very different for the two types of measurement. For example, if there is no change in dipole moment associated with a particular relaxation, the corresponding contribution to the dielectric relaxation strength is zero. Relaxations that would be inactive for this reason can be made active by the deliberate introduction of polar groups that do not significantly perturb the structure. An example is the replacement of a few $>CH_2$ groups per thousand carbon atoms in polyethylene by $>C\!=\!O$ groups.

As stated in section 5.7.2, measured mechanical and dielectric relaxation times are not expected to be equal to the fundamental relaxation times of the corresponding relaxing units in the polymer because of the effects of the surrounding medium. These effects are different for the two types of measurement, so it is unlikely that the measured relaxation times will be exactly the same. There are arguments that suggest that, in order to take this into account, the observed relaxation times for dielectric and mechanical relaxation should be multiplied by $\varepsilon_\infty/\varepsilon_s$ and J_u/J_r, respectively, before the comparison is made. The second of these ratios can sometimes be much larger than the first, but for sub-T_g relaxations neither ratio is likely to be greatly different from unity, so that the relaxation times determined from the two techniques would then not be expected to differ significantly for this reason. There are, however, other reasons why the two measured relaxation times need not be the same.

As discussed in sections 5.7.5 and 7.2.5 and above, a relaxation generally corresponds not to a single well-defined relaxation time but rather to a spread of relaxation times, partly as a result of there being different environments for the relaxing entity within the polymer. The g values for the different environments will be different, as will the contributions to mechanical strain. Suppose, for example, that the higher relaxation times correspond to larger g factors than do the lower relaxation times but that the lower relaxation times correspond to larger contributions to mechanical strain than do the higher relaxation times. The mean relaxation time observed in dielectric relaxation will then be higher than that observed in mechanical relaxation, leading to a different position of the peaks in a temperature scan. The glass transition does not correspond to a simple localised motion and has a particularly wide spread of relaxation times, so that its position in mechanical and dielectric spectra is not expected to be closely the same even allowing for differences of frequency.

Dielectric measurements can easily be made over a wide range of frequencies, allowing true relaxation strengths $\varepsilon_s - \varepsilon_\infty$ to be determined, but the corresponding measurement is not easy in mechanical studies. For reasons explained in section 7.6.3, the maximum value of $\tan\delta$ in an isochronal temperature scan is frequently used as a measure of relaxation strength for mechanical spectra. Such scans are frequently used to compare mechanical and dielectric relaxation phenomena.

Figure 9.4 shows a comparison of the dielectric and mechanical relaxation spectra of various forms of polyethylene. The most obvious feature is that the main relaxations, here the α, β and γ relaxations, occur at approximately the same temperatures in both spectra, although their relative relaxation strengths in the two spectra are different. This is a feature common to the spectra of many polymers. The peaks are not, however, in exactly the same positions in the two spectra for the same type of polyethylene. In addition to the possible reasons for this described above, the frequencies of measurement are different. The dielectric measurements were made at a much higher frequency than that used for the mechanical measurements, as is usual. Molecular motions are faster at higher temperatures, so this factor alone would lead to the expectation that the dielectric peaks would occur at a higher temperature than the mechanical peaks. The γ peak, which is assigned to a localised motion in the amorphous material and is in approximately the same place for all samples, behaves in accord with this expectation.

The β relaxation in polyethylene, which is most prominent in the low-crystallinity LDPE, is associated with the amorphous regions and almost certainly corresponds to what would be a glass transition in an amorphous polymer; a difference in its position in mechanical and dielectric spectra is therefore not surprising. The α relaxation, as discussed in section 7.6.3, is associated with helical 'jumps' in the crystalline regions and, provided that the lamellar thickness is reasonably uniform, might be expected to correspond to a fairly well-defined relaxation time and to a narrow peak in the relaxation spectrum. The dielectric peak is indeed quite narrow, because the rotation of the dipoles in the crystalline regions is the major contribu-

Fig. 9.4 Dielectric and mechanical relaxation in various forms of polyethylene. The vertical lines are simply a guide to the eye in comparing the positions of the major relaxations. See the text for discussion. (Adapted by permission from Ákadémiai Kiadó, Budapest.)

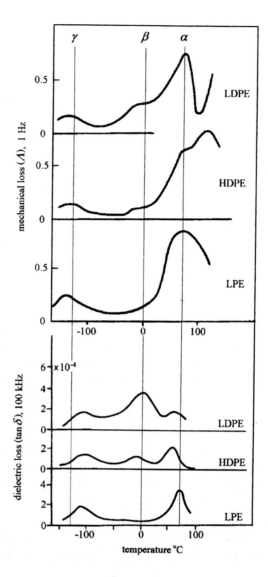

tor to the effect. As discussed in section 7.6.3, however, the mechanical relaxation is seen through the effect of the jumps on the freedom of motion of the amorphous interlamellar regions. The need for multiple jumps for the observation of mechanical relaxation leads to the mechanical relaxation having a higher mean relaxation time and a broader peak than does the dielectric relaxation.

The above discussion shows that measurements of dielectric relaxation provide very useful information about molecular motion to add to that provided by NMR and mechanical-relaxation spectroscopies. A particular advantage of dielectric measurements is that the relaxation strengths of crystalline and amorphous relaxations are directly proportional to the numbers of dipoles in each phase (although the g values and therefore the constants of proportionality are different for each phase). When constraints on the amorphous motions due to crystallites can be neglected this leads to a simpler dependence of dielectric relaxation strengths on crystallinity than that of mechanical relaxation strengths and hence to a simpler determination of the separate contributions of crystalline and amorphous regions. The problem is still not trivial, however, because the anisotropy of the crystallites and their dielectric constants leads to a non-linear relationship between dielectric relaxation strength and crystallinity. That the various techniques used to study relaxations in polymers are to some extent overlapping and to some extent complementary in terms of the information they provide, while at the same time each having its own difficulties, is a common feature of many studies on polymers and is due to their complicated morphologies.

9.3 Conducting polymers

9.3.1 Introduction

There are three principal mechanisms whereby a polymer may exhibit electrical conductivity:

(i)   metallic or other conducting particles may be incorporated into a non-conducting polymer;

(ii) the polymer may contain ions derived from small-molecule impurities, such as fragments of polymerisation catalysts, from ionisable groups along the chain or from salts specially introduced to provide conductivity for a specific use; and

(iii) the polymer may exhibit 'electronic' conductivity associated with the motion of electrons (more strictly, particle-like excitations of various kinds) along the polymer chains.

To produce conduction by the first method in an otherwise non-conducting polymer, the required concentration of metallic particles is likely to be greater than about 20% by volume, or 70% by weight. This high particle concentration is necessary in order to achieve continuity of conducting material from one side of the polymer sample to the other. Such high concentrations tend to destroy some of the desirable mechanical properties of the polymer, but composites of this kind are used in conducting paints and in anti-static applications. Conductivities of order $6 \times 10^3 \; \Omega^{-1} \; m^{-1}$ can be achieved in this way. Carbon black at concentrations above about

30% by weight can be used to achieve conductivities of about $100 \ \Omega^{-1} \ m^{-1}$ and has the advantages of not changing the density of the polymer much and being cheap. In such composite materials the polymer itself takes no part in the conduction process and acts merely as a support for the conducting material. In the rest of section 9.3 these materials are not considered further; instead, attention is directed towards the mechanisms (ii) and (iii) by which conduction takes place through the polymer itself.

9.3.2 Ionic conduction

Ionic conduction in a polymer below the glass-transition temperature is often due to the presence of small concentrations of ionisable impurity molecules. The degree of ionisation of the ion providers depends on the static dielectric constant ε_s of the polymer, because the energy required to separate two ions is inversely proportional to ε_s. It also depends on the temperature through the Boltzmann factor, which determines the relationship between the number of entities in a higher energy state, in this case the dissociated state, and a lower energy state, here the undissociated state. Consider a degree of dissociation f, where f measures the fraction of molecules dissociated. Only those ions that come closer than a certain maximum distance from each other will be able to recombine. This number will be equal to $CN_o^2 f^2$ per unit volume, where N_o is the total number of molecules per unit volume and C is a constant. If the degree of dissociation is small, the number of non-dissociated molecules can be considered equal to N_o. The equilibrium between the potentially recombinable ions, which are effectively excited molecules, and the neutral molecules is then given by

$$\frac{CN_o^2 f^2}{N_o} = \exp\left(\frac{-\Delta G}{kT}\right) \tag{9.30}$$

where ΔG is the difference between the Gibbs free energies per molecule for the dissociated and associated states. If the difference in entropy is ΔS and the difference in enthalpy is ΔH, then $\Delta G = \Delta H - T\Delta S$. If W_{ion} is the energy required to ionise the molecule in free space, then ΔH, which is the energy required to separate the ions in the dielectric, is equal to W_{ion}/ε_s. Substitution for ΔG into equation (9.30) thus leads to

$$\frac{CN_o^2 f^2}{N_o} = \exp\left(\frac{\Delta S}{k}\right) \exp\left(\frac{-W_{ion}}{\varepsilon_s kT}\right) \tag{9.31}$$

The concentration of ions, $N_o f$, is therefore given by

$$N_o f = C' N_o^{1/2} \exp\left(\frac{-W_{ion}}{2\varepsilon_s kT}\right) \tag{9.32}$$

where C' is a constant independent of temperature, concentration and dielectric constant. Assuming that each ion carries a single electronic charge e, the electrical current density j produced when an electric field E is applied to the polymer is thus given by

$$j = C' N_0^{1/2} e(u_+ + u_-) \exp\left(\frac{-W_{ion}}{2\varepsilon_s kT}\right) \tag{9.33}$$

where u_+ and u_- are the magnitudes of the mean drift velocities of the positive and negative ions, respectively, produced by the field E.

Mott and Gurney have given a simple model for deriving the values of u_+ and u_-, which is closely related to the site model applied to relaxation processes in section 5.7.3. In that model the sites, or potential wells, represent particular conformational states of the molecule. In the model of Mott and Gurney they represent different positions in the polymer where a positive or negative ion might be, which are spread uniformly throughout the polymer. For each type of ion the depths of the wells at all the possible sites are assumed to be the same and may be denoted by U_+ and U_- for the positive and negative ions, respectively. Let the separation between neighbouring potential wells for the two types of ion be d_+ and d_-, respectively. When the electric field E is applied the potential difference between points separated by $d_+/2$ along the field direction is equal to $Ed_+/2$, so that the barrier height for a positive ion attempting to move in the direction of the field is decreased by $eEd_+/2$ and the barrier height when it attempts to move in the opposite direction to the field is increased by this amount. The probabilities for jumps in the two directions in unit time will thus be $A \exp[-(U_+ \pm eEd_+/2)/(kT)]$ and the net probability of a jump in the direction of the field per unit time (the difference between these) will be $2A \exp[-U_+/(kT)] \sinh[eEd_+/(2kT)]$, where A is a constant. Because the distance moved per jump is d_+ the mean drift velocity, u_+, is given by

$$u_+ = 2 A d_+ \exp[-U_+/(kT)] \sinh[eEd_+/(2kT)] \tag{9.34}$$

with a similar expression for u_-.

Assuming that, to a first approximation, U_+ and U_- are equal to U, and d_+ and d_- are equal to d, substitution of the resulting expression for the drift velocities into equation (9.33) leads to

$$j = 4AC' N_0^{1/2} ed \exp\left(\frac{-[U + W_{ion}/(2\varepsilon_s)]}{kT}\right) \sinh\left(\frac{eEd}{2kT}\right) \tag{9.35}$$

In the limit of small fields, $\sinh[eEd/(2kT)] \approx eEd/(2kT)$ and normal ohmic conductivity is observed, with j proportional to E. As the temperature rises the conductivity $\sigma = j/E$ falls rapidly. The exponential factor dominates the temperature dependence, so that a plot of $\ln \sigma$ against $1/T$ is approximately straight with a slope of $-[U + W_{ion}/(2\varepsilon_s)]/k$. At high fields the conductivity is no longer ohmic and the current rises faster than linearly with increasing field. By fitting experimental data to equation (9.35), as shown in fig. 9.5, the assumed ionic mechanism for the observed conduction can be tested and values of the jump distance d can be obtained. These are found to be about 1–2 nm, which is similar to intermolecular spacings.

Figure 9.5 shows that, at 95 °C the current density j produced by a field of 60 MV m^{-1} is about 10^{-3} A m^{-2}, implying a conductivity of order 10^{-11} Ω^{-1} m^{-1}. PVC contains no ionisable groups and hence this is the conductivity due to impurities. Some polymers have intrinsic ionic conductivities several orders of magnitude higher than this because they contain ionisable groups. An example is nylon-6,6, in which significant dissociation of the >NH groups to give mobile protons takes place at elevated temperatures and the conductivity exceeds 10^{-8} Ω^{-1} m^{-1} above 100 °C. The conduction process in this and other polyamides may, however, be more complicated than that discussed above.

The temperature dependence of the conductivity of a glassy polymer generally changes significantly when its temperature is raised above the glass-transition temperature T_g. In the treatment just given the change

Fig. 9.5 The effect of electric field strength on conduction in PVC. (Reproduced by permission of the American Institute of Physics.)

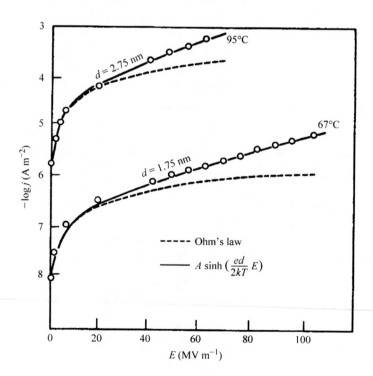

with temperature of the number of sites large enough to accept an ion following a jump was not taken into account. This approximation is not valid for temperatures above T_g because of the much faster rate of change of free volume of the polymer with temperature (see section 7.5.3), so that, in the vicinity of the glass transition, the conductivity j/E may increase faster than predicted by equation (9.35).

Polymers designed for use as electrolytes in batteries are used at temperatures well above T_g. In this region a different model for conduction is more applicable, namely one based on the fluctuations in the distribution of free volume that occur above T_g due to the localised movements of the polymer chain segments. It is assumed that any ion sits in a region of free volume and can move into an adjacent space only when a sufficiently large element of free volume becomes available immediately adjacent to it.

Let v_0 be the average size of an element of free volume and v' the minimum size of the element of free volume required for movement. The probability that an element of free volume permitting movement arises in unit time is then proportional to $\exp(-v'/v_0)$. At each opportunity to move the ion is accelerated in the direction of the applied electrical field and picks up a component of velocity $qE\tau$ in the direction of the field, where τ is the time taken for the move and q is the charge on the ion. The mean drift velocity in the field direction is thus proportional to $qE\tau n \exp(-v'/v_0)$, where n is the number of ions per unit volume. For given n the conductivity is thus proportional to $\exp(-v'/v_0)$. Strictly speaking the constant of proportionality depends on T, but this dependence is small compared with the dependence on T contained in the exponential. The quantity v_0 must clearly be proportional to the total fractional free volume V_f, so that the conductivity is proportional to $\exp(-c/V_f)$, where c is a constant for the particular polymer and ion. Substitution for V_f from equation (7.33) and rearranging leads finally to the temperature dependence of the conductivity:

$$\sigma = A \exp[-B/(T - T_g + V_g/\alpha)] = A \exp[-B/(T - T_{VFT})] \qquad (9.36)$$

where A depends only slightly on temperature and $B = c/\alpha$.

There are other theories for conduction in polymers above T_g, but the free-volume theory fits much experimental data well, although extraction of the fitting parameters is not easy. It also shows correctly the importance of choosing a polymer with a low T_g if it is to be used as an electrolyte, for example in a battery.

Increasing crystallinity causes a fall in the conductivity of a polymer because the mobility (drift velocity per unit field) of the ions in the crystalline regions is lower than that of the ions in the non-crystalline regions. When designing materials for high conductivity it is therefore desirable to choose polymers that have low crystallinities. The overall conductivity of a sample depends on the mobilities in both types of region and on the details of the morphology as well as the degree of crystallinity.

Some of the systems that have been studied extensively with a view to the eventual production of batteries with polymer electrolytes are based on salts such as $LiClO_4$, or $LiCF_3SO_3$ dissolved in high-M poly(ethylene oxide) (PEO) or polymers related to it but designed to avoid crystallinity. The lithium cations that result are responsible for the conductivity. One such system involves the polymer poly[*bis*(methoxy ethoxy ethoxy)-phosphazene], usually designated MEEP, which has short ethylene-oxide units as side-chains attached to the phosphorus atoms of a backbone with the repeat unit $-(N\!\!=\!\!P)-$, the full repeat unit being $-(N\!\!=\!\!P(O(CH_2)_2O(CH_2)_2OCH_3)_2)-$. This polymer has a glass-transition temperature of about 190 K, which is consistent with the requirement for a low T_g, and MEEP-based polymer electrolytes may have room-temperature conductivities of the order of 10^{-5} S cm^{-1}, which is unfortunately still too low for practical application. Useful conductivities can be obtained with PEO itself above the melting point and lithium–polymer batteries (LPBs) based on PEO and designed for operation at about 100 °C are now under development.

The chief advantages of polymer-based batteries are that they do not contain liquids that can leak, that they are mechanically relatively strong and that they can be easily shaped for various applications. Some other uses of polymer electrolytes are in displays, optical modulators and smart windows whose reflectance or transmission can be electrically controlled.

9.3.3 Electrical conduction in metals and semiconductors

The most obvious electronic conductors are metals, in which the current carriers can be regarded to a first approximation as free electrons. Before discussing electronic conductivity in polymers it is useful to consider briefly the mechanisms for conduction in metals and semiconductors.

The simplest metals are the alkali metals. Each alkali atom has an electronic structure consisting of a core of filled electronic sub-shells with one outer electron rather loosely bound to the atom. When the atoms come together to form a solid this outer electron comes under the influence of the positively charged cores of all the neighbouring atoms as well as its 'own' core. This results in the outer electrons of all the atoms becoming rather like a gas of electrons, with each electron moving in the relatively smooth potential due to the atomic cores and all the other electrons. To a first approximation the wave function of each electron can be treated as that of a particle moving in a uniform potential, i.e. as if it were in free space. The one-dimensional analogy for a metal then leads to the allowed electron wavelengths λ being given by the condition that there must be a whole number of wavelengths in the length L of the metal, i.e.

$$\lambda = L/n = Nd/n \tag{9.37}$$

where N is the number of atom cores within the length L, d is the separation between them and n is an integer. Equation (9.37) assumes either periodic boundary conditions or that the one-dimensional metal has been bent round to form a closed circuit. The momentum p of an electron with wavelength λ is given by de Broglie's relation $p = h/\lambda$ and the energy E is thus given by

$$E = \frac{p^2}{2m} = \frac{h^2}{2m\lambda^2} = \frac{n^2 h^2}{2mN^2 d^2} \tag{9.38}$$

where m is the mass of the electron. When N is large the separations between the values of E corresponding to adjacent values of n are very small and an *energy band* is thus formed.

It is important to remember, however, that the potential in which the electrons move is not completely smooth; it is actually periodic with period d, which means that waves with wavelength $2d/n'$, where n' is any integer, will be strongly reflected back along the one-dimensional metal because the waves reflected from different atomic cores will be exactly in phase. Waves of these wavelengths cannot propagate through the metal and must form standing waves with crests and troughs related to the positions of the atomic cores. Two such standing waves are possible, one with its crests and troughs at the cores and one with its crests and troughs between the cores. If there were no interaction of the electrons with the cores, these two waves would correspond to electrons with the same energy, but there is an interaction. The electron probability density is proportional to the square modulus of the wave function so that the electron is more likely to be at a crest or trough than between them, which means that the waves of the first type will have their potential energy lowered by the interaction and those of the second type will have their potential energy raised above the average potential for the metal. An *energy band gap* therefore opens up at every wavelength $2d/n'$.

An important question now arises: how many electron states are there within each of the regions between energy gaps? It is clear from equation (9.37) that $\lambda = 2d/n'$ whenever n is a whole multiple n' of $N/2$. There are, however, two waves moving in opposite directions for every value of λ and, when allowance is made also for the two possible directions of electron spin, it follows that there are $2N$ allowed electron states within each energy band between band gaps. Each alkali-metal atom supplies one electron to the total number of free electrons, so that the lowest band is half filled, with N electrons. The electrons take up the lowest energy levels consistent with minimising the Gibbs free energy. This results in the band being filled up to an energy level known as the *Fermi level*, or *Fermi energy*, E_F, with a slight blurring around this level, in the sense that there are some unfilled states just below it and some filled states just above it. That this will be the situation may be shown as follows.

For a half-filled band the maximum value of $n = N/4$. Substituting this value into equation (9.38) for E shows that $E_F = h^2/(32md^2)$. Setting $d = 1.4 \times 10^{-10}$ m gives $E_F = 7.7 \times 10^{-19}$ J or 4.8 eV. The value of kT at room temperature is about 0.025 eV, so that only electrons very near the Fermi level can be thermally excited into higher levels, giving an effective slight blurring of the Fermi level.

When an electric field is applied to the metal, electrons well below the Fermi level are unable to pick up energy from the field because all the states immediately above them are full. Near the Fermi level, however, there are empty states available and more of those corresponding to electrons travelling in the opposite direction to the field can be occupied at the expense of those corresponding to electrons travelling in the direction of the field. This will happen because the energies of the states corresponding to motion against the field will have their energy slightly lowered by the field and vice versa. The metal then carries an electron current in the opposite direction to the field, i.e. a conventional electrical current in the direction of the field.

If there were no mechanism to scatter electrons from states corresponding to travel in one direction to states corresponding to travel in the other direction, the electrons would continue to pick up energy from the electric field and the current would increase continuously. Such scattering mechanisms are always present in the form of phonons (thermal vibrations of the atoms), impurities, dislocations, etc., so that the current settles to a finite value. These scattering mechanisms determine the mobility of the electrons.

Even in a highly conducting material like copper the scattering is such that the electron mean free path is about 2 nm at room temperature. The energy gained by an electron travelling this distance parallel to an electric field of 1 MV m^{-1} is $2 \times 10^{-9} \times 10^6$ eV $= 2 \times 10^{-3}$ eV. The distribution of electrons among the various energy levels is therefore controlled mainly by thermal energies, which are about ten times this value, and the electric field produces only a small perturbation of this distribution to give a slight excess of electrons travelling in the direction opposite to the field.

The above discussion shows that the simplest form of electronic conduction can occur in a medium with loosely bound electrons only if there are states above the Fermi level that are thermally populated. The conductivity then depends on the number of these states and the mobility of the electrons, i.e. on the scattering mechanisms.

The energy levels in an insulator like pure silicon are quite different from those in a metal. Each silicon atom has four outer electrons and, in the solid, it is bound by single covalent bonds to four others in a three-dimensional lattice. Each bond involves a pair of electrons that are tightly bound to the atoms and their energy levels form a filled *valence band*. The energy needed to break a bond, i.e. to remove an electron and put it into the *conduction band* so that it can move freely through the material like the electrons in a metal, is about 1.1 eV and thus much greater than kT at room temperature. Pure silicon therefore has a very low conductivity, which rises with increasing temperature as the small number of thermally excited electrons in the conduction band increases. It is made more highly conducting (semiconducting) for use in electronic devices by the incorporation of 'impurity' atoms instead of silicon atoms at some of the sites, the process called *doping*.

If a silicon atom is replaced by a pentavalent atom there is one electron left over after the four bonds linking it to the lattice have been made and this electron is only loosely bound, giving rise to a *donor level* in the band gap just below the bottom of the conduction band, from which it can easily be excited thermally into the conduction band. Similarly, if a silicon atom is replaced by a trivalent atom, there is one electron too few for the four bonds required to link it to the lattice. It requires only a small amount of energy to transfer an electron from a silicon atom to this vacant site, which therefore gives rise to an *acceptor level* just above the top of the valence band. At ordinary temperatures thermal excitation fills most of these acceptor levels and leaves behind *holes* in the valence band, which can give rise to conduction by jumping from atom to atom.

9.3.4 Electronic conduction in polymers

Polymers appear at first sight to be most unlikely materials to exhibit electronic conduction. A polymer such as polyethylene contains only fully saturated chemical bonds, each with two electrons closely bound to particular atoms, which thus constitute a filled valence band; there are no free electrons. Some polymers, however, contain unsaturated bonds in the backbone. An important example is polyacetylene, $-(CH{=}CH)_{\overline{n}}$, which has alternating single and double bonds in the backbone. Examples of other polymers with such alternation are given in fig. 9.6. Where a number of double bonds alternate in this way the phenomenon of *resonance* usually occurs and all the bonds tend to become similar. Notice, however, that, for all the molecules shown in fig. 9.6, except for *trans*-polyacetylene (*t*-PA),

Fig. 9.6 Some polymers with alternation (conjugation) of single and double backbone bonds. (Adapted by permission of E. M. Conwell and H. A. Mizes.)

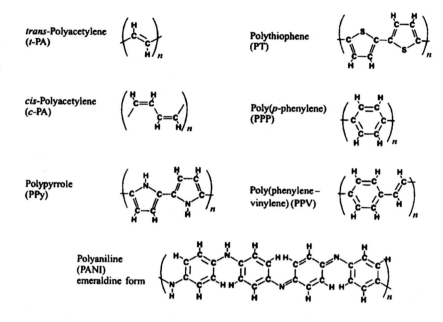

interchange of double and single bonds would produce different structures with higher energies. For t-PA the two structures have the same energy and the ground state of this molecule is *doubly degenerate*.

Benzene is the extreme example of a molecule that at first sight might be expected to exhibit bond alternation, but in this molecule all the C—C bonds become completely equivalent. Only three of the four electrons on each carbon atom are involved in sp^2 *hybrid atomic orbitals*[†] bonding to the hydrogen atom and the two adjacent carbon atoms, in so-called σ *bonds*. The fourth electron takes up a p orbital, which is largely situated at right angles to the plane of the molecule (fig. 9.7(a)). These p orbitals are fairly extended in space and overlap to form a pair of π-*electron clouds* on each side of the plane of the molecule, as illustrated in fig. 9.7(b).

A similar phenomenon takes place for the molecule of 1,3-butadiene, $CH_2{=}CH{-}CH{=}CH_2$, with 'sausage-shaped' π-electron clouds on each side of the plane of the molecule (see fig. 9.7(c)), but here the central C—C bond remains slightly different from the other two. The tendency for all bonds to become equal in such a structure is expected to be particularly strong for the planar zigzag form of t-PA, for which the choice of which alternate bonds are single and which double seems arbitrary. If the bonds

---

[†] Atomic orbitals show the forms of the electron-probability-density clouds for the outermost electrons of the atoms. Where clouds on adjacent atoms overlap, bonds can be formed by electrons with paired spins.

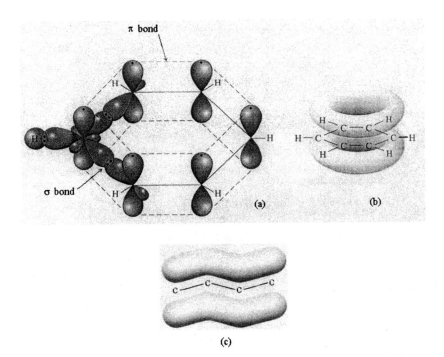

Fig. 9.7 (a) Molecular orbitals for benzene, showing the p orbitals on all carbon atoms and the σ bonds made by one C atom to the adjacent hydrogen and carbon atoms; (b) the π electron density clouds of benzene formed by overlap of the p orbitals; and (c) the π electron density clouds above and below the plane of the molecule for *trans*-butadiene ((a), (b) Adapted by permission of W. H. Freeman and Company. (c) Adapted by permission of John Wiley & Sons, Inc.)

did become equal, which they do not, for reasons explained below, the π-electron clouds would form a half-filled electron band, giving the possibility of metallic conduction along the chain. Even if this did happen, conduction could not occur unless there were levels close to the Fermi level that could be thermally populated and unless the mobility were sufficiently high, i.e. the scattering were sufficiently low.

It is easy to show (see problem 9.5) that, for a chain of N atoms, the separation between the Fermi level and the next highest level is given approximately by $h^2/(4mNd^2)$ for large N. For a polymer chain of 1000 carbon atoms 1.4×10^{-10} m apart this evaluates to the equivalent of 0.04 eV, which is comparable to thermal energies, so that the first condition is satisfied. The second condition is harder to satisfy, because any deviation of the chain from the planar zigzag conformation will cause scattering. It can, however, be satisfied in highly crystalline material, wherein conductivities as high as those in metals might then be expected. The conductivity of pure *t*-PA is nevertheless only about $10^{-5} \; \Omega^{-1} \; cm^{-1}$. One of the reasons for this, as has already been hinted at, is that, in *long* chains, the backbone bonds are not in fact equivalent, as was first shown by Peierls and by Fröhlich: alternate bonds are actually either shorter or longer than the average length d. The reason for this alternation is as follows.

If an alternating arrangement of longer and shorter bonds is formed, so that there are pairs of atom cores separated by a larger external distance d_1 than their internal separation d_2, where $d_1 + d_2 = d$, π-electron standing waves of wavelength $4d$ can now have maxima of probability density either at the core pairs or halfway between them. The former not only have a lower energy than the latter because they interact more strongly with the cores, giving a band gap, but also have lower energy than they would have with equal separations d of the cores. The σ-bonding electrons actually have a higher energy in the alternating state than they do in the non-alternating state, but the lowering of the energy of the π-electron states more than compensates for this.

The bond alternation means that the repeat unit is twice the length that it would be if the bonds were equivalent, so that there are two nearly free electrons per repeat unit and the band gap therefore appears at the Fermi level. The magnitude of this energy gap for t-PA turns out be about 1.5 eV. There can thus be no metallic-type conductivity and *intrinsic* t-PA is a semiconductor. In the other polymers of the type shown in fig. 9.6 the backbone bonds are intrinsically non-equivalent and a band gap therefore exists for them too, so that they also cannot exhibit metallic-type conductivity.

As described in the previous section, the conductivity of intrinsic semiconductors such as silicon and germanium can be increased by doping. It turns out that the conductivity of polyacetylene and other polymers with alternating double and single bonds in the backbone can also be dramatically increased by doping, by up to ten orders of magnitude for some forms of t-PA, so that its conductivity can become comparable to that of copper. The concentrations of dopant required are a few per cent, whereas those used in semiconductor devices may be of the order of a few parts per million or less, and the mechanisms by which high conductivity is achieved in polymers are quite different from those in doped semiconductors. The impurity atoms or molecules do not substitute for atoms of the polymer chain but are incorporated *interstitially*, i.e. they take up locations between the chains, which leads to the production of radical cations or anions on the polymer chains each associated with an inert counter-ion. Polymeric cation-rich materials are called *p-doped* and polymeric anion-rich materials are called *n-doped* in analogy with the nomenclature used for extrinsic semiconductors. Examples of substances used for doping are iodine, which leads to p-doped materials with mainly I_3^- counter-ions, and alkali metals, which leads to n-doped materials with Na^+, K^+, etc. counter-ions.

A complete understanding of the conduction processes both in intrinsic and in doped polymers has not yet been obtained. It is clear that at least two types of processes are required: charge transport along the chains and charge transport between chains. Transport along the chains may be possible because of the formation of various kinds of *pseudo-particles* such as *solitons* (or *kink solitons*) and *polarons*, which are localised but mobile excitations.

In a polymer such as *t*-PA a soliton is most easily visualised as a point at which the single–double-bond alternation shifts by one bond, leaving two adjacent single bonds, as shown in fig. 9.8(b). This means that the carbon atom between these bonds effectively carries an electron not involved in forming a bond. If one of the two pairs of electrons forming a double bond to a carbon atom adjacent to this central atom 'flips over' so that it converts one of the two single bonds originally adjacent to each other into a double bond, the carbon atom carrying an electron that is not bonded is now *two* carbon atoms along the chain from the original one.

A proper quantum-mechanical treatment shows that each soliton is not in fact concentrated at a single carbon atom but is spread out over about ten atoms on either side of the central atom, i.e. the reversal of the bond-alternation pattern takes place over about 20 carbon atoms. The wave function has a finite amplitude only on alternate carbon atoms, as expected, and thus has a wavelength equal to four times the average separation of the atoms, i.e. to the gap wavelength. An *anti-soliton* state involving the alternate set of carbon atoms can also exist. Because the soliton and the anti-soliton have the same energy they have the free-electron energy corresponding to their wavelength, i.e. the energy at the centre of the gap. The soliton as described so far is a neutral *quasi-particle*, but it

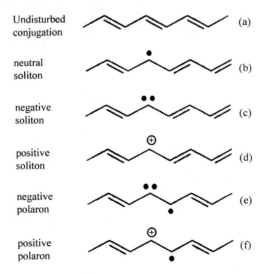

Fig. 9.8 A schematic diagram of some excitations in *trans*-polyacetylene: ● indicates electrons not involved in bonding and ⊕ indicates a missing electron. (Adapted by permission of John Wiley & Sons Limited.)

carries the spin appropriate to a single localised electron, i.e. $\frac{1}{2}$. It is also possible for *charged solitons* to be formed, as shown in figs 9.8(c) and (d). If an electron is added to a soliton the spins of the pair of electrons now involved must be opposed, so that a *negatively charged soliton* with zero spin is formed. Similarly, if the soliton loses its electron, a *positively charged soliton* with zero spin is formed.

Isolated solitons cannot be generated in *t*-PA because the bond alternation of large sections of chain would have to change. Soliton–anti-soliton pairs can, however, be formed by the conversion of a double bond into a single bond, either leaving an unbonded electron on each of two adjacent carbon atoms or leaving an electron pair on one of them. The first of these corresponds to the formation of a neutral-soliton–neutral-anti-soliton pair and the second to a positive-soliton–negative-anti-soliton pair (see fig. 9.9). The neutral pairs quickly recombine and disappear because they have a higher energy than the unperturbed state but the charged pairs can be separated if they are formed in the presence of an electric field. Such pairs can be formed by absorption of light of the correct wavelength and this leads to the possibility of *photoconductivity*.

When the polymer is doped further possibilities arise, including the formation of *polarons* (see figs 9.8(e) and (f)). A *negatively charged polaron* is formed when an electron from a donor is added to the chain. The chain can distort locally to create an attractive potential well in which the electron is trapped. In a similar way, an electron can be lost by the chain to an acceptor, so that a *positively charged polaron* is formed. The polaron state has a width slightly greater than that of the soliton and has spin $\frac{1}{2}$. The electron added to form the negative polaron occupies a state in the band-gap just below the top of the gap and the positive polaron state lies in the band-gap just above the bottom of the gap. Soliton pairs cannot form in polymers that do not have degenerate ground states because, as fig. 9.9

Fig. 9.9 A charged soliton–anti-soliton pair in *t*-PA: (a) undisturbed conjugation, (b) the pair as formed and (c) the pair after being driven apart by the electric field. Note that the bond alternation in the region between the two solitons in (c) is the reverse of that in (a).

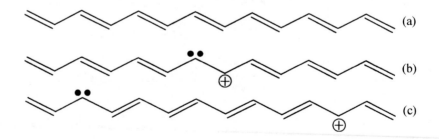

shows, this would involve interchanging double and single bonds for the section of chain between the solitons and, for these polymers, this increases the energy of the chain. Polarons of a somewhat more complicated type than those of *t*-PA can, however, form.

Although the charged quasi-particles just described may be quite mobile along intact segments of polymer chain, it is necessary for them to move from one chain to another for macroscopic conductivity to occur. Various mechanisms have been proposed to describe how this takes place. The problem is complicated because it is necessary for both the charge and the distortion of the chain corresponding to the quasi-particle to move from one chain to the next and different mechanisms may operate at different degrees of doping. As an example of a possible mechanism, a soliton–anti-soliton pair with a fairly short region of reversed alternation between them might 'hop' from one chain to an adjacent one, since only a small rearrangement of the chains would be required. It is now believed, however, that, even in polyacetylene, polarons may play a more important role in conduction than do solitons.

The most successful theory for samples with moderate concentrations of dopant involves the *variable-range-hopping* mechanism. In this theory the excitations are localised by pinning to the donor or acceptor ions and have discrete energies in a range that includes the Fermi energy. The range of energies of the localised states is assumed to be $> kT$ in the temperature range considered. At the higher temperatures in this range hopping takes place between sites that are close together, but at lower temperatures it can take place only between states separated by low energies, even though the nearest such states may be some distance away. The theory leads to a temperature dependence of the conductivity of the form

$$\sigma = CT^{1/2}\exp[-(T_\text{o}/T)^{1/4}] \tag{9.39}$$

which has been found to fit well the conductivity of some *t*-PA samples in the range 10–300 K (fig. 9.10). Various hopping models have been considered for other moderately doped polymers, leading to different temperature dependences. A complication of any attempt to produce a theoretical description is the non-uniformity of doping, which can lead to some regions becoming more highly conducting than others, whereupon it may happen that the macroscopic conductivity is limited by the rate of hopping or tunnelling between the more highly conducting regions.

At high concentrations of dopant the counter-ions tend to take up ordered arrangements, so it is probable that a full theoretical treatment of conduction in the metallic-like state that can then be formed will have to take into account the complete three-dimensional nature of the polymer–dopant complex. An important way of trying to understand the mechanism for

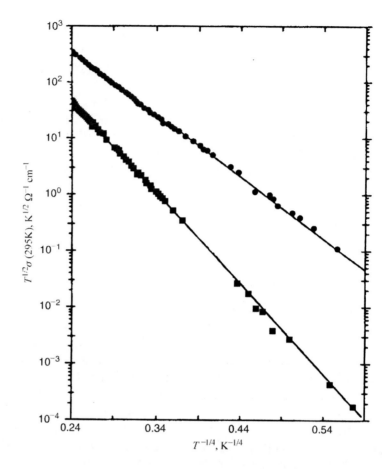

Fig. 9.10 The temperature dependence of the conductivity of samples of *t*-PA doped with iodine $\left[CH(I_3)_x\right]_n$. The lines are fits to the variable-range-hopping model: •, $x = 0.017$; and ■, $x = 0.042$. (Adapted by permission of the American Physical Society.)

conduction is to study the magnetic susceptibility of the material. Charged solitons do not possess spin and hence have no associated magnetic susceptibility, whereas polarons have spin $\frac{1}{2}$, like electrons, and should therefore give rise to the same type of Pauli paramagnetism as that of free electrons. Figure 9.11 shows a comparison of the conductivity and magnetic susceptibility of *t*-PA doped with AsF_6^-. Although the conductivity rises steeply for dopant concentrations above 0.001, the susceptibility does not rise significantly until the concentration is above about 0.06, whereupon it then increases almost discontinuously. This transition is called the *semiconductor–metal transition*. One explanation that has been suggested for it is that the regular spacing of the counter-ions and the pinning of the excitations to them can lead to the production of soliton or polaron lattices and, under suitable conditions, to soliton or polaron bands. At high doping polarons may become more favoured energetically than solitons and a half-filled polaron band can then develop, leading to metallic-like conductivity.

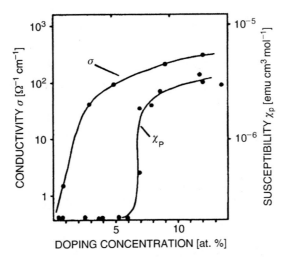

Fig. 9.11. The Pauli susceptibility χ_P and electrical conductivity σ plotted against the doping concentration in t-PA. (Reproduced by permission of John Wiley & Sons Limited.)

Whatever the precise mechanisms of conduction, the macroscopic conductivity observed in any conducting polymer will depend strongly on the morphology of the sample and on whether it is oriented and, if so, to what degree, because these factors will influence both intra- and inter-chain mobility. For these reasons the conductivities of samples prepared under very similar conditions may differ considerably. The above account has merely attempted to indicate some of the types of process that may be involved in conduction in these relatively new materials, for which it is certain that no single mechanism can explain all the experimental results. Even when a particular theoretical description fits the conductivity, it cannot be accepted as the correct mechanism unless it can also account for the observations of magnetic susceptibility, thermo-electric power, photoconductivity and other properties.

9.4 Optical properties of polymers

9.4.1 Introduction

Many polymers in everyday use contain fillers and colouring agents that render them opaque. The optical properties of the base polymer are thus obscured. On the other hand the clarity of optical transmission of many polymers and the fact that they are almost colourless, coupled with their low density and excellent mechanical properties, are the reasons for their use to replace glass in many applications. In other applications glass would be totally inappropriate because of its greater brittleness. Other polymers, such as the ubiquitous polyethylene, are often colourless but translucent rather than transparent.

9.4.2 Transparency and colourlessness

The transparency and colourlessness of a substance arise from the fact that it does not absorb or scatter any significant amount of visible electromagnetic radiation. The reason why many polymers behave in this way is that their molecules consist of atoms linked together either by single bonds, or by isolated double bonds or benzene rings. Such molecules give rise to absorption of electromagnetic radiation in the infrared region, by the mechanisms described in section 2.6.1, and to absorption in the ultraviolet region through electronic transitions, which generally require photons of energies greater than those of visible light. The polymer then appears colourless and transparent, because all wavelengths of visible light can pass through it unmodified, unless the light is appreciably scattered, when it will appear to be colourless and translucent, i.e. it will appear somewhat 'milky'.

Significant scattering may occur if the polymer contains structures of a similar size to the wavelength of light. Such structures occur when the polymer is appreciably crystalline, as discussed in chapters 2, 3 and 5. The high degree of transparency of many polymers in their unoriented state is good evidence that they do not contain such structures. This means that they cannot be significantly crystalline and that the non-crystalline material of which they are composed is actually amorphous at least on a scale greater than about 0.1 μm. Conversely, if the structure of a polymer is such that its chains cannot take up any regular conformation, it is expected to be amorphous and transparent; poly(methyl methacrylate) (PMMA or Perspex) is a good example. In the commercial form the polymer is almost atactic and, because of the relatively large side group —C=O.COCH$_3$, it cannot form straight chains. Its use as a glass substitute in situations where stress is expected is well known.

Polymers may absorb significant amounts of light at some or all visible wavelengths if they contain significant lengths of chain in which there are adjacent double bonds. Such structures arise in poly(vinyl chloride), for instance, if it does not contain stabiliser to prevent thermal or photochemical degradation. In the degradation a molecule of hydrogen chloride is eliminated at an initiation site along the chain, probably a branch point or an allylic end group, —CH=CH—CH$_2$Cl, and a double bond or additional double bond is formed. This acts as a propagation point for sequential loss of hydrogen chloride by an 'unzipping' process and conjugated polyene sequences $-(CH=CH)_n$ of various lengths are formed. Those with about seven or more double bonds have electronic absorption spectra in or near the visible region, giving the commercially unacceptable problem of discoloration to a yellowish brown or even almost black colour, depending on the degree of degradation.

The reason for the lowering of the electronic transition energy in the polyene sequences is the delocalisation of the bonding electrons within each conjugated sequence, as described in section 9.3.4. The extreme limit of conjugated sequences is found in the polyacetylenes, for which n is equal to the degree of polymerisation. It is explained in section 9.3.4, however, that t-PA does not have metallic properties as might be expected, but is a semiconductor with a band gap of about 1.5 eV. This means that it can absorb strongly at wavelengths lower than $ch/(1.5e)$, or about 825 nm, which includes the visible region. Materials with strong absorption at a particular wavelength reflect strongly at that wavelength. Suitably prepared t-PA samples can therefore have a metallic-like appearance.

9.4.3   The refractive index

The use of *scalar* molar bond refractions for the calculation of refractive indices is described in section 9.2.3. This allows the refractive index of a randomly oriented sample to be calculated and also permits an understanding of the wide range of refractive indices that polymers can have, as shown in table 9.3.

Polymers are not, however, always randomly oriented and the phenomenon of orientation is discussed in the next two chapters, chapters 10 and 11. One of the ways used to obtain information about the degree of orientation of the polymer is to measure either the refractive indices of the polymer for light polarised in different directions or its birefringences, i.e. the differences in refractive index for light polarised in different directions. In chapter 11 the theory of the method is described in terms of the polarisability of structural units of the polymer. This polarisability is a second-rank tensor like the molecular polarisability referred to in the earlier sections of this chapter and, insofar as the assumption of additivity in section 9.2.3 holds, it is in fact the sum of the polarisability tensors of all the bonds in the unit. Since, however, the whole basis of the method is that the structural units are anisotropic, the tensors must be added correctly, taking account of the relative orientations of the bonds, unlike the treatment used to calculate the refractive index of PVC in example 9.1, where scalar bond refractions are used.

It is assumed that the orientation of every bond in the unit is known with respect to a set of axes $OX_1X_2X_3$ fixed in the unit and that the components of the bond polarisability with respect to axes fixed within each bond are also known. Equations (A.7) of the appendix are then used to refer the tensor components for each bond to

$OX_1X_2X_3$ so that corresponding components for all bonds can be added. It is usual to assume that the polarisability tensor for a bond is cylindrically symmetric around the bond axis. This means (see the appendix) that, if the bond axis for a particular bond in the unit is chosen as the Ox_3 axis of a set of axes $Ox_1x_2x_3$ the tensor takes the form $\alpha_{11} = \alpha_{22} = \alpha_t$ and $\alpha_{33} = \alpha_p$, with all other components zero, where α_t represents the component transverse to the bond and α_p represents the component parallel to the bond. Values of α_t and α_p are given for various types of bond in table 9.5.

Let the bond make the polar and azimuthal angles θ and ϕ with respect to the axes $OX_1X_2X_3$ of the unit and choose the Ox_1 axis so that it lies in the OX_1X_2 plane. This can always be done because of the cylindrical symmetry of the tensor. The direction-cosine matrix linking the sets of axes $Ox_1x_2x_3$ and $OX_1X_2X_3$ is then

$$
\begin{array}{cccc}
 & x_1 & x_2 & x_3 \\
\begin{array}{c} X_1 \\ X_2 \\ X_3 \end{array} &
\left(\begin{array}{c} -\sin\phi \\ \cos\phi \\ 0 \end{array} \right. &
\begin{array}{c} -\cos\theta\cos\phi \\ -\cos\theta\sin\phi \\ \sin\theta \end{array} &
\left. \begin{array}{c} \sin\theta\cos\phi \\ \sin\theta\sin\phi \\ \cos\theta \end{array} \right)
\end{array}
$$

Table 9.5. *Polarisabilities of selected bonds*[a]

Bond	α_p (10^{-41} F m^2)	α_t (10^{-41} F m^2)
C—C	10.8	2.8
C=C	32.3	11.9
C≡C	39.4	14.1
C—C aromatic	25.0	5.3
C(aromatic)—C(aliphatic)	15.6	3.3
C—N	15.4	2.4
C—O	16.2	1.9
C=O	22.3	11.1
C—H	9.1	6.7
N—H	6.5	9.3
O—H	5.3	8.9
C—F	16.7	4.5
C—Cl	40.8	23.1
C—Br	56.1	32.0

[a]Adapted by permission of Oxford University Press from *Chemical Crystallography* by C. W. Bunn, 2nd Edn, Clarendon Press, Oxford, 1961/3. © Oxford University Press 1961.

Example 9.2

Calculate the contributions of each C—C bond of a planar zigzag chain backbone to the components of the polarisability of the chain, α_{33} parallel to the chain, α_{11} in the plane of the chain and α_{22} perpendicular to the plane of the chain.

Solution

From table 9.5, $\alpha_p = 10.8 \times 10^{-41}$ F m^{-2} and $\alpha_t = 2.8 \times 10^{-41}$ F m^{-2}. For tetrahedral bonding of the backbone, $\theta = (180° - 109.5°)/2 = 35.25°$ and $\phi = 0$, because the bond lies in the OX_1X_3 plane. Thus, from equations (9.40),

$$\alpha_{33} = (2.8 \sin^2 35.25° + 10.8 \cos^2 35.25°) \times 10^{-41} \text{ F m}^{-2} = 8.14 \times 10^{-41} \text{ F m}^{-2}$$
$$\alpha_{11} = (2.8 \cos^2 35.25° + 10.8 \sin^2 35.25°) \times 10^{-41} \text{ F m}^{-2} = 5.46 \times 10^{-41} \text{ F m}^{-2}$$

and $\alpha_{22} = \alpha_t = 2.8 \times 10^{-41}$ F m^{-2}.

---

and use of equation (A.7) of the appendix then leads to the following expressions for the components of the tensor for the bond with respect to $OX_1X_2X_3$:

$$\alpha_{11} = \alpha_t(\sin^2 \phi + \cos^2 \theta \cos^2 \phi) + \alpha_p \sin^2 \theta \cos^2 \phi \qquad (9.40a)$$

$$\alpha_{22} = \alpha_t(\cos^2 \phi + \cos^2 \theta \sin^2 \phi) + \alpha_p \sin^2 \theta \sin^2 \phi \qquad (9.40b)$$

$$\alpha_{33} = \alpha_t \sin^2 \theta + \alpha_p \cos^2 \theta \qquad (9.40c)$$

The off-diagonal components α_{12} etc. are not zero, but, if it is assumed that the axes $OX_1X_2X_3$ are symmetry axes of the unit, these off-diagonal components will sum to zero when the components for all the bonds in the unit are added to obtain the components α_{11}, α_{12}, etc. of the polarisability tensor for the unit. The assumption that the axes $OX_1X_2X_3$ are symmetry axes for the unit means that the unit must have at least orthotropic (statistical orthorhombic) symmetry. Units of lower symmetry are not considered here.

Equations (9.40) can be used to evaluate the refractive indices for the crystals of any polymer with orthorhombic symmetry, provided that the co-ordinates of all the atoms within the crystal structure are known, so that all the bond orientations can be calculated. The value of α_{11} for the crystal is substituted into the Lorentz–Lorenz equation (9.9) in place of α and the value of n obtained is assumed to be equal to the refractive index for light polarised with the electric vector parallel to OX_1. Similar calculations are performed to calculate the indices for light polarised parallel to OX_2 and

OX_3. This use of the Lorentz–Lorenz equation for an anisotropic medium cannot be rigorously justified but usually leads to results that are in reasonable agreement with experiment.

The calculation of the principal refractive indices for non-orthorhombic crystals is a little more complicated because the axes of the indicatrix, or refractive-index ellipsoid (see section 2.8.1), cannot be predicted in advance of the calculation. It is therefore necessary to calculate the values of all six independent components of the polarisability tensor of the crystal with respect to arbitrarily chosen axes and then to find the principal axes of the resulting tensor.

9.5   Further reading

(1) *Electrical Properties of Polymers*, by A. R. Blythe, CUP, 1979. This book, which is at a level similar to or slightly more advanced than the present book, gives a good introduction to its subject but it is now becoming a little out of date, particularly regarding conducting polymers, and some of the equipment described for making measurements has been superseded.

(2) *Anelastic and Dielectric Effects in Polymer Solids*, by N. G. McCrum, B. E. Read and G. Williams. See section 7.7.

(3) *Polymers*, by D. Walton and P. Lorimer. See section 1.6.2.

(4) *Polymers: Chemistry and Physics of Modern Materials*, by J. M. G. Cowie. See section 1.6.2.

(5) *One-dimensional Metals: Physics and Materials Science*, by S. Roth, VCH, Weinheim, 1995. This book considers not only polymers but also other essentially one-dimensional conducting materials in a rather unconventional style.

9.6   Problems

9.1. Show that the field at the centre of a spherical hole in a uniformly polarised dielectric due to the apparent surface charges is $P/(3\varepsilon_0)$, where P is the polarisation.

9.2. Calculate the refractive index of polyoxymethylene (POM), repeat unit $-(CH_2-O)-$, using the bond refractions given in table 9.1. Assume that the density of POM is 1.25 Mg m^{-3}.

9.3. Show that, according to the Debye model, the width of the dielectric loss peak ε'' plotted against frequency is 1.14 decades measured at half the peak height.

9.4. By adding the contributions to the dielectric constant ε of three equally strong Debye relaxations with relaxation times $\tau/4$, τ and 4τ, show that a distribution of relaxation times leads to points on a Cole–Cole plot lying inside the semi-circle. (Hint: use a spreadsheet. For all relaxations take $\tau = 10^{-7}$ s, $\varepsilon_s = 9$ and $\varepsilon_\infty = 1$ and consider values of $\log_{10} \omega$ lying between 3.5 and 10.5 at intervals of no more than 0.25, with ω in rad s^{-1}.)

9.5. Show that, for a linear chain of N atoms, each with one free electron, the separation between the Fermi level and the next highest level is given by $h^2/(4mNd^2)$, where m is the mass of the electron, d is the separation between the atom cores and N is assumed to be $\gg 1$.

9.6. Starting from the Clausius–Mosotti equation (9.7) and assuming that it can be applied to each principal direction Ox_i in an orthorhombic crystal, show that the Lorentz–Lorenz equation can be written in a form that leads to the equation $n_i = \sqrt{(1 + 2x_i)/(1 - x_i)}$, where n_i is the refractive index for light polarised parallel to Ox_i and $x_i = \alpha_i/(3\varepsilon_0 V)$, with α_i equal to the polarisability per unit cell for light polarised parallel to Ox_i and V the volume of the unit cell.

9.7. Assuming that the backbone of the polyethylene molecule in the crystal has the CCC angle 112° and that the HCH angle is tetrahedral, calculate the polarisability of each molecule per translational repeat length in the three symmetry directions. Use the result in problem 9.6 to deduce the refractive indices for light polarised parallel and perpendicular to the chain axis in the crystal. (Data for the polyethylene unit cell are given in section 4.4.1.)

Chapter 10

Oriented polymers I – production and characterisation

10.1 Introduction – the meaning and importance of orientation

Consider a volume within a sample of polymer that is large compared with the size of individual chain segments, crystallites or spherulites, but small compared with the sample as a whole. If equal numbers of the axes of the chain segments or of the crystallites within this region point in every direction in space, the region is *isotropic* on a macroscopic scale. If all such regions within the sample are isotropic, the sample itself is said to be isotropic, *randomly oriented* or *unoriented*. If any region of the sample is not isotropic it is said to be *oriented* or *partially oriented*. The sample as a whole may be *homogeneously oriented* if all regions of it are oriented in the same way, or it may be *inhomogeneously oriented*. The various types and degrees of orientation that are possible are considered in detail in subsequent sections.

It is important to realise that, even in a sample that is randomly oriented on a *macroscopic* scale, there may be regions, such as crystallites, in which the molecular chains are oriented on a *microscopic* scale.

Polymers are not the only materials that can exhibit orientation. Any polycrystalline material, such as a metal, can exhibit orientation and this can confer desirable or undesirable properties on the material. The effects are, however, often much greater for polymers than they are for other materials. Table 10.1 compares the effects of the forging of steel and the extrusion of polypropylene. In addition to the other changes shown in table 10.1, the elongation that takes place before breaking occurs, often called the *elongation to break*, is substantially reduced for the polymer.

Other desirable changes in properties may be produced by orientation, for example polystyrene sheet that is biaxially oriented (see section 10.2.2) can be highly flexible, whereas unoriented sheet is brittle; oriented sheet exhibits substantial increases in *impact strength*, in *tensile-yield strength* and in *resistance to stress crazing*. It should not, however, be assumed that such improvements will always result from orientation.

Table 10.1. *Effects of forging on the properties of steel and polypropylene*[b]

Sample	Extrusion ratio	Longitudinal yield strength (10^7 N m^{-2})	Longitudinal tensile strength (10^7 N m^{-2})	Elongation at breakage (%)
Carbon steel: ingot	1	29	56	19
Carbon steel: forged rod	12	34	56	27
Polypropylene: billet	1	1.0	2.2	200
Polypropylene: extruded rod	5.5	[a]	22	24

[a]No drop in yield is observed.
[b]Adapted by permission of Kluwer Academic Publishers from L. Holliday and I. M. Ward 'A general introduction to the structure and properties of oriented polymers' in *Structure and Properties of Oriented Polymers*, Ed. I. M. Ward, Applied Science Publishers, London, 1975, Chap. 1, pp. 1–35.

Undesirable properties due to orientation may include anisotropy in properties and dimensional instabilities at elevated temperatures in thermoplastic materials, e.g. *shrinkage* of textile fibres, due to randomisation of the orientation of the amorphous regions on heating above the glass transition temperature.

Natural polymers often owe their important properties to the fact that they are highly oriented, e.g. *cellulose*, in the form of wood and cotton, and silk fibre, which is a *protein*.

10.2   The production of orientation in synthetic polymers

There are many methods by which orientation can be produced in synthetic polymers, but they fall into two groups. In one group the orientation is introduced after the polymer has first solidified in an unoriented state or a state of low orientation, a simple example of this being the *drawing* (stretching) of a cast film. In the other group the orientation is achieved by solidification from a fluid state in which the molecules are sometimes partially aligned. Further orientation is then often produced by subsequent drawing. Both undesirable, incidental, orientation and deliberate orientation can arise in ways belonging to either of these groups. The first of these is considered briefly below and the second in rather more detail. Information about both types of orientation can in principle be obtained by the methods described in the later sections of this chapter.

In section 8.2.2 the phenomenon of cold drawing and the associated natural draw ratio are described. This process is referred to again in section 10.2.2 and discussed further in section 10.2.4.

10.2.1   Undesirable or incidental orientation

Undesirable or incidental orientation may result from

(a) flow and deformation in the melt, for example in extrusion dies and moulds, which leads to shearing of the melt and hence to molecular orientation; and

(b) deformation in the solid state. In cold forming, or solid-state forming, the polymer is forced into shape at a temperature below the softening point and the product is stable at room temperature but not at elevated temperatures.

10.2.2   Deliberate orientation by processing in the solid state

There are essentially four types of process that may be used:
  i.   *drawing* (stretching), including *die drawing*;
  ii. *extrusion*;
  iii. *blowing*; and
  iv. *rolling*.

Two or more of these methods may sometimes be used together, either simultaneously or sequentially.

To understand the conditions for orientation to be achieved by any of these processes it is necessary to consider the forms that may be taken by the load–strain curves for a polymer. These are considered briefly in section 6.1 and fig. 6.2 is reproduced here as fig. 10.1. Most thermoplastic polymers can exhibit behaviour ranging from (a) to (e) under the appropriate testing conditions, e.g. the correct temperature and strain-rate and whether the stress is applied in tension or compression. The behaviour illustrated is as follows (for a sample initially of uniform cross-section).

(a) *Low extensibility* followed by *brittle fracture*: if the load is removed before fracture occurs, the strain disappears almost completely and virtually no orientation results.

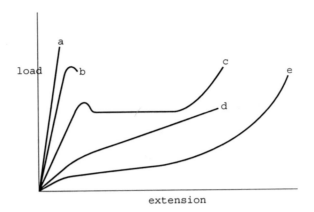

Fig. 10.1 Possible forms of the load–extension curve for a polymer: (a) low extensibility followed by brittle fracture, (b) localised yielding following by fracture, (c) necking and cold drawing, (d) homogeneous deformation with indistinct yield and (e) rubber-like behaviour.

(b) *Strain-softening* behaviour: there is a definite drop in load due to localised yielding, followed by *fracture* in the deformation zone. Very little orientation is obtained outside the deformation zone.

(c) *Cold-drawing* behaviour: *yield* is followed by a drop in load; the load then remains constant for increasing extension and the sample exhibits *necking* (see figs 8.3 and 10.2). When the neck has moved through all the material the load increases rapidly for further extension, with breaking occurring after little further extension. The drawn material is stable when the load is removed and is highly oriented.

(d) *Homogeneous deformation* throughout, with an *indistinct yield point*. Some orientation is obtained once the strain is above the yield point.

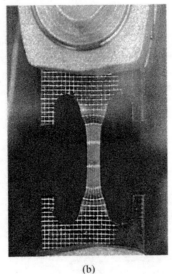

(a)   (b)

Fig. 10.2 The appearance of a sample undergoing necking when drawn: (a) the sample before drawing, with a reference grid of squares marked on it; and (b) after drawing. The central section has increased in length by a factor of about five, i.e. the draw ratio is 5. (Courtesy of Dr J. Sweeney and Dr P. Caton-Rose.)

(e) *Rubber-like behaviour* above the glass transition temperature, T_g. If the load is removed at the stretching temperature the strain disappears, leaving no residual orientation, *but* fast cooling to below T_g in the stretched state results in the orientation being *frozen in*.

Some of the processes of yield and fracture are discussed more fully in chapter 8. Only behaviour of the types shown in (c), (d) and (e) leads to significant orientation and, whichever of the processes (i)–(iv) is used to produce orientation, one of these types of behaviour is required. The individual processes are described briefly in the remainder of this section.

i. *Drawing*

Drawing can be done in essentially three different ways.

(i)   A single piece of material, in the form of a fibre or sheet, is drawn in a tensile testing machine. This process is often used for experimental studies.

(ii) As a continuous process: a continuous filament or sheet passes through two sets of rolls (fig. 10.3). The *stretching rolls* rotate at a higher speed than the *feed rolls* and the ratio of the speeds controls the *draw ratio*. In the commercial drawing of fibres, the *neck* is localised by a heated roller (called a *pin*) over which drawing takes place and the fibres may be *heat set* (crystallised) by passing over a heated *plate* before being wound up.

(iii) *Die drawing* – see below under extrusion.

Fibres can exhibit only *uniaxial orientation* in which the chains orient towards the fibre axis. Sheets can be oriented in a more complex way, because the chains may not only be oriented towards the draw direction but also may have a distribution of orientations of the chain axes that is

Fig. 10.3 Continuous drawing of a sheet. The feed rolls cause the sheet to move at speed V_1 at the stretching rolls cause it to move at the higher speed V_2. (Reprinted by permission of Kluwer Academic Publishers.)

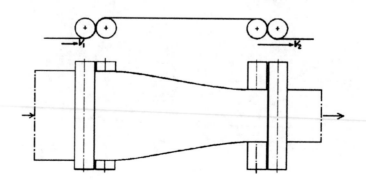

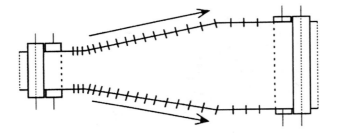

Fig. 10.4 A schematic diagram of a tentering frame. The short lines represent the clips that move apart longitudinally as well as laterally. The mechanism carrying the clips is not shown.

not cylindrically symmetric around the draw direction. This is called *biaxial orientation*.

Biaxial orientation can be produced by drawing a broad sheet at constant width, which is achieved by placing the feed and stretching rolls close together, or by using a *tentering frame*, in which the edges of the sheet are gripped by clips that move in such a way as to provide both lateral and longitudinal stretching at the same time (see fig. 10.4). Tentering frames are very complicated pieces of machinery, so the process is often implemented in two stages. The first stage is longitudinal stretching, which is followed by lateral stretching on a machine in which the clips do not accelerate longitudinally but only move apart laterally as the sheet passes through.

For any kind of drawing the resulting extension ratios λ_i (defined in section 6.3.2) are called the *draw ratios*.

ii. *Extrusion*

In the process of extrusion a rod of polymer is pushed through a *die* to reduce its diameter and increase its length. Either hydrostatic pressure or a piston (ram) may be used. Extrusion can be combined with *blowing* (see the next section), for the production of tubular films.

A related process is *die-drawing*. In this process the sample is *pulled*, not *pushed*, through the die. The important difference here is that the type of stress applied is quite different. In extrusion the polymer is under compressive stress, whereas in die-drawing it is largely under tensile or shear stress, with plane sections normal to the draw direction often remaining almost plane, giving rise to *plug flow*. This type of stress is necessary for some polymers in order to allow them to draw and orient.

Figure 10.5 shows the difference between extrusion and die drawing.

iii. *Blowing*

This process involves the extrusion of molten, fairly viscous material through an *annular die* and is described in section 1.5.3. The film produced acquires a biaxial type of orientation.

Fig. 10.5 A schematic
diagram illustrating the
difference between
extrusion and die-drawing.

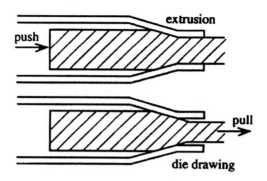

iv. *Rolling*

In this process the polymer is passed between rolls separated by less than
its thickness, which causes it to expand longitudinally and possibly also
laterally and leads to biaxial orientation.

10.2.3   Deliberate orientation by processing in the fluid state

There are essentially four types of process that may be used for producing
fibres from polymer chains processed in the fluid state:

(i)   melt spinning,
(ii) solution spinning of 'conventional' polymers,
(iii) gel spinning of 'conventional' polymers and
(iv) solution spinning of liquid-crystal polymers.

(i) *Melt spinning*

Both conventional and liquid-crystal polymers can be processed by
melt spinning. The principal requirement is that the polymer must not
degrade before becoming molten. Examples are nylon-6, nylon-6,6 and
poly(ethylene terephthalate) ('polyester', 'terylene', or 'dacron') for conventional polymers and the 'Vectra' series (see section 12.4.3) among the
thermotropic liquid-crystal polymers.

The essential melt-spinning process is the production of polymer filaments by allowing molten polymer to pass downwards through small holes
in a *spinneret*. The originally liquid filaments eventually solidify and are
drawn off by take-up rolls situated some distance below the spinneret.
Between the rolls and the spinneret the filaments may pass through tubes
where they are subjected to heat or other treatment. The precise mechanism by which orientation is produced in the fibres depends on the polymer
and the details of processing and is not always easy to determine. It can
take place through the crystallisation or glass formation of molecules

already oriented in the molten thread-line or by subsequent drawing after solidification, or by a mixture of the two processes. It has, however, been shown that at the very high speeds at which modern spinning lines operate the melt can become oriented before solidification occurs even with conventional polymers.

(ii) *Solution spinning of 'conventional' polymers*

Many conventional polymers cannot be melt-spun because they degrade before melting. Examples are those used to produce the acrylic fibres, such as 'Orlon' and 'Courtelle', which are based on copolymers of polyacrylonitrile, $-\!(CH_2CH(CN))_{\overline{n}}$, and the cellulosic fibres (see section 1.2).

There are two methods of solution spinning, 'dry' and 'wet'. Wet spinning is carried out by passing the solution emerging from the spinneret into a coagulating bath containing a liquid with which the solvent is miscible but which is not itself a solvent for the polymer. There may but need not be an air gap between the spinneret and the coagulating bath. In dry spinning the jet issuing from the spinneret passes through a column of heated air in which the solvent evaporates. The as-spun fibres are not usually particularly oriented and are subsequently stretched.

Most of the work on the production of very highly aligned fibres by solution spinning has been done on polyethylene. The important feature of the method is that the chains must be extended in the solution by an *elongational flow* of the solution, i.e. a flow in which there is acceleration of the solution along the direction of flow. In one method a fibre is pulled from a flowing solution in the opposite direction to the flow. The solution at the tip of the growing fibre is essentially static and accelerates away from the tip, thus producing the elongational flow and the orientation of the polymer molecules. The temperature of the solution is kept higher than the normal static-solution crystallisation temperature.

(iii) *Gel spinning of 'conventional' polymers*

It was found in the early 1980s that fibres can also be produced by continuous drawing of ultra-high molecular weight polyethylene (UHMW-PE, $M_w > 5 \times 10^5 \, g \, mol^{-1}$) from a solution within the annular region between a fixed cylinder and an inner rotating cylinder. The tip of the fibre becomes attached to the rotating cylinder by reason of a layer of polymer adsorbed onto the surface in the form of a gel. With optimised conditions, this method showed that fibres of very high modulus and strength could be produced from UHMW-PE. The rate of production is, however, low, so the method cannot be adopted for commercial production. The processes now called gel spinning are quite different and at first sight are very similar to wet solution spinning.

Gel spinning was first performed with UHMW-PE. The polymer is dissolved in a solvent such as decalin at a concentration of a few per cent and hot extruded into a cooling bath containing a liquid, such as water, with which the solvent is immiscible. This results in the formation of gel-like filaments, i.e. filaments of polymer incorporating solvent, which can subsequently be drawn, with or without prior removal of solvent, to very high draw ratios. This yields fibres of very high modulus and strength. Gel spinning can also be used to improve the properties of fibres made from other flexible polymers, such as high-molar-mass polypropylene.

(iv) *Solution spinning of liquid-crystal polymers*
This topic is discussed in chapter 12, section 12.4.5.

10.2.4   Cold drawing and the natural draw ratio

There is evidence, particularly from studies on non-crystalline polymers, that the natural draw ratio observed in a particular experiment depends on the degree to which the polymer is already oriented prior to drawing. As described in section 10.2.3, non-crystalline melt-spun fibres are oriented glasses. If they are subsequently heated above the glass transition they shrink to give an unoriented rubber-like material, in which entanglements act as effective cross-link points. If the ratio of the length of the fibre before shrinkage to its length after shrinkage is called λ_1 and the natural draw ratio observed when the fibre is cold-drawn rather than shrunk is called λ_2, then there is evidence that $\lambda_1\lambda_2$ is approximately constant for a given grade of polymer. This quantity represents the effective draw ratio in drawing from an originally unoriented state to the final drawn state and appears to be a kind of limiting draw ratio that cannot be exceeded even in the two-stage orienting process.

The simplest explanation is that there is a rubber-like network present and that this has a maximum extensibility due to the degree of entanglement, which is constant for a given grade of polymer and depends on its molar mass and method of polymerisation. This limiting extensibility is not to be confused with the limit of applicability of the affine rubber model for predicting orientation distributions discussed in section 11.2.1 because the limiting extension can involve non-affine deformation.

10.3   The mathematical description of molecular orientation

In general it is not meaningful to talk about the orientation of a polymer molecule as a whole; it is therefore useful at the outset to refer to the orientation of *structural units*. Such a unit may be a crystallite, a regular molecular segment between points of conformational disorder, or a particular part of a molecule, e.g. a benzene ring. The polymer is imagined to be composed of a large number of identical structural units, and a set of rectangular Cartesian axes $Ox_1x_2x_3$ is imagined to be fixed identically within each unit. These axes may be related to the crystallographic axes for crystallites or they may be, e.g., the axes shown in fig. 10.6 for a benzene ring.

The polymer sample is assumed to have at least *orthotropic symmetry*, i.e. it contains three mutually perpendicular directions such that if it is rotated through 180° about any one of these directions its macroscopic properties are unchanged. Axes $OX_1X_2X_3$ are chosen parallel to these three symmetry directions of the sample. The orientation of a particular structural unit can then be specified in terms of three *Euler angles*, θ, ϕ and ψ, as shown in fig. 10.7.

The discussion of the most general form of biaxial orientation involves fairly complicated mathematics. Further discussion in this section and in the sections on the characterisation of orientation is therefore largely restricted to a special simple type of distribution of orientations of the structural units. This distribution is the *simplest type of uniaxial orientation*, for which the following conditions apply.

(i) ϕ is random around OX_3. This is the *definition* of *uniaxial orientation*, which is sometimes called *cylindrical symmetry* or *transverse isotropy*.

(ii) ψ is random, i.e. there is no preferred orientation of the structural units around their Ox_3 axes. This additional condition *defines* the *simplest type of uniaxial orientation*.

For the simplest type of uniaxial orientation the distribution function reduces to $N(\theta)$, where $N(\theta)\,d\omega$ is the fraction of units for which Ox_3 lies within any small solid angle $d\omega$ at angle θ to OX_3. By *characterising the distribution* is meant finding out as much as possible about $N(\theta)$ for the various types of structural unit that may be present in the polymer.

Most experimental methods give only limited information about $N(\theta)$. Many give only $\langle \cos^2 \theta \rangle$, the average value of $\cos^2 \theta$, or give only $\langle \cos^2 \theta \rangle$

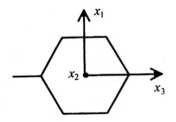

Fig. 10.6 A possible choice of axes within the benzene-ring unit.

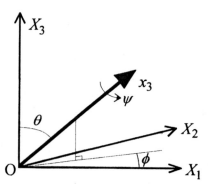

Fig. 10.7 The Euler angles used for specifying the orientation of the set of axes $Ox_1x_2x_3$ of a unit with respect to the set of axes $OX_1X_2X_3$ fixed in the sample. For the simplest type of uniaxial orientation ϕ and ψ take random values.

and $\langle \cos^4 \theta \rangle$, the average value of $\cos^4 \theta$, where the averages are taken over all the structural units of a given type within the sample. Methods that can give $\langle \cos^4 \theta \rangle$ are mathematically more difficult to describe and are therefore considered rather briefly here; detailed discussion is limited to methods that can give only $\langle \cos^2 \theta \rangle$.

The value of $\langle \cos^2 \theta \rangle$ varies from $\frac{1}{3}$ for random orientation to 1 for complete orientation; similarly, $\langle \cos^4 \theta \rangle$ varies from $\frac{1}{5}$ to 1. The averages $\langle P_2(\cos \theta) \rangle$ and $\langle P_4(\cos \theta) \rangle$ of the functions

$$P_2(\cos \theta) = \tfrac{1}{2}(3\cos^2 \theta - 1) \tag{10.1a}$$

$$P_4(\cos \theta) = \tfrac{1}{8}(3 - 30\cos^2 \theta + 35\cos^4 \theta) \tag{10.1b}$$

which are the second- and fourth-order *Legendre polynomials* in $\cos\theta$, are often used instead of $\langle \cos^2 \theta \rangle$ and $\langle \cos^4 \theta \rangle$. These averages have the simple property that they each range from 0 for random orientation to 1 for complete orientation of the Ox_3 axes parallel to the OX_3 axis. The Legendre polynomials also have important mathematical properties that are used in later sections. $\langle P_2(\cos\theta) \rangle$ is sometimes called the *Hermans orientation function*.

Some methods, as described in the following section, can give the complete orientation distribution function, $N(\theta)$. It is often useful to calculate from this the values of $\langle P_2(\cos\theta) \rangle$ and $\langle P_4(\cos\theta) \rangle$ for comparison with values obtained by other methods. They can be obtained from the equation

$$\langle P_l(\cos \theta) \rangle = \frac{\int_0^{\pi/2} P_l(\cos \theta) N(\theta) \sin \theta \, d\theta}{\int_0^{\pi/2} N(\theta) \sin \theta \, d\theta} \tag{10.2}$$

with $l = 2$ or 4. The $\sin \theta$ in each integral arises because $N(\theta)$ is the fraction of units oriented per unit solid angle at θ to the draw direction and the total solid angle in a range $d\theta$ is $2\pi \sin \theta \, d\theta$ when all angles ϕ around OX_3 are considered.

Example 10.1

Show that, for a randomly oriented sample, $\langle \cos^2 \theta \rangle = \frac{1}{3}$, where θ is the angle between the axis of any unit in the sample and a specific direction in the sample.

Solution

For a random sample $N(\theta)$ is constant and independent of θ. Thus, by analogy with equation (10.2),

$$\langle \cos^2 \theta \rangle = \int_0^{\pi/2} \cos^2 \theta \sin \theta \, d\theta \bigg/ \int_0^{\pi/2} \sin \theta \, d\theta = \left[\frac{1}{3}\cos^3 \theta\right]_{\pi/2}^0 \bigg/ \left[\cos \theta\right]_{\pi/2}^0 = \frac{1}{3}.$$

10.4 Experimental methods for investigating the degree of orientation

Many methods have been developed for characterising molecular orientation: they include measurement of the following:

(i)   optical refractive indices or birefringence – can give $\langle \cos^2 \theta \rangle$ or $\langle P_2(\cos \theta) \rangle$ only;

(ii) UV, visible or infrared dichroism – can give $\langle \cos^2 \theta \rangle$ or $\langle P_2(\cos \theta) \rangle$ only;

(iii) polarised fluorescence – can give $\langle P_2(\cos \theta) \rangle$ and $\langle P_4(\cos \theta) \rangle$;

(iv) polarised Raman scattering – can give $\langle P_2(\cos \theta) \rangle$ and $\langle P_4(\cos \theta) \rangle$;

(v)   the angular dependence of X-ray scattering – can give the complete distribution of orientations for crystallites; and

(vi) the angular dependence of broad-line NMR spectra – can give $\langle P_2(\cos \theta) \rangle$ and $\langle P_4(\cos \theta) \rangle$ and possibly higher averages. Modern 2-D NMR methods can give the complete distribution of orientations.

The theory behind methods (i)–(v) is described in the following sections. Method (vi) is rather complicated and is not discussed further.

10.4.1 Measurement of optical refractive indices or birefringence

Methods of measuring birefringence are discussed briefly in section 2.8.1. Before discussing the use of refractive indices or birefringence for studying molecular orientation it is necessary to draw attention to the fact that *form birefringence* can occur even in systems in which there is no *molecular* orientation. It is then necessary for two different phases with different refractive indices to be present and for one phase to be present as anisotropically *shaped* regions that are aligned within a matrix of the other phase. Samples of this type often occur for block copolymer systems (see section 12.3.2). In oriented semicrystalline samples of homo-polymers such aligned arrays can in principle occur. Although it is often assumed that form birefringence is unimportant for oriented polymer samples, it may in fact contribute up to 10% of the total birefringence in oriented crystalline polymers such as polyethylene. In the remainder of the present section it will be neglected.

As discussed in section 9.2.1, the refractive index of a substance depends on the *polarisabilities* of its basic units. The *Lorentz–Lorenz equation* derived there as equation (9.9) links the refractive index n and the *polarisability* α of the structural units in any isotropic medium. It can be written in the slightly simpler form

$$\alpha = \frac{3\varepsilon_0(n^2 - 1)}{N_0(n^2 + 2)} \tag{10.3}$$

by substituting $\varepsilon = n^2$ into equation (9.7) and rearranging slightly.

N_0 is the number of structural units per unit volume and the polarisability α relates the electric dipole μ induced in a unit to the applied electric field E. For isotropic units, as assumed in section 9.2.1,

$$\mu = \alpha E \tag{10.4}$$

If the units are anisotropic, α becomes a *second-rank tensor* $[\alpha_{ij}]$ (see the appendix) in this equation and it is then assumed that the Lorentz–Lorenz equation can be modified to read

$$\langle \alpha_{ii} \rangle = \frac{3\varepsilon_0(n_i^2 - 1)}{N_0(n_i^2 + 2)} \tag{10.5}$$

where $i = 1, 2$ or 3, the average is taken over all units and n_i is the principal refractive index for light polarised parallel to Ox_i. This generalisation of the Lorentz–Lorenz equation cannot be rigorously justified, but it has been found to apply well in practice.

The refractive index n_i thus depends on the average value of α_{ii}, which depends on the anisotropy of the polarisability and the degree of orientation of the sample. It is shown below that, for uniaxial orientation with respect to OX_3 (for which $n_1 = n_2 = n_t$) and a cylindrical polarisability tensor with low asymmetry,

$$\langle P_2(\cos \theta) \rangle = \Delta n / \Delta n_{\max} \tag{10.6}$$

where $\Delta n = n_3 - n_t$ is the *birefringence* of the sample and $\Delta n_{\max}$ is the maximum birefringence for a fully uniaxially oriented sample. As indicated above, only $\langle P_2(\cos \theta) \rangle$ or $\langle \cos^2 \theta \rangle$ can be obtained and, if the structural units are of different types, say crystallites and amorphous

Example 10.2

A specially oriented (hypothetical) sample of a polymer has one half of its structural units oriented with their axes parallel to the draw direction and the other half oriented with their axes uniformly distributed in the plane normal to the draw direction. If the birefringence of this sample is 0.02, calculate the birefringence Δn_{max} of a fully uniaxially oriented sample.

Solution

$$\langle \cos^2 \theta \rangle = (\tfrac{1}{2} \times 1) + (\tfrac{1}{2} \times 0) = \tfrac{1}{2}$$
$$\langle P_2(\cos \theta) \rangle = \tfrac{1}{2}(3\langle \cos^2 \theta \rangle - 1) = \tfrac{1}{4}$$

Thus, from equation (10.6), $\Delta n_{max} = 4 \times 0.02 = 0.08$.

chain segments, a weighted average of their orientations is obtained (see section 10.5).

Figure 10.8 shows Δn plotted against the draw ratio for some PVC samples drawn at various temperatures.

Equation (10.6) is derived as follows. Assume that the principal axes of the polarisability tensor coincide with the axes $Ox_1 x_2 x_3$ of the structural unit and that, with respect to these axes, the polarisability tensor $[\alpha_{ij}]$ takes the form

$$[\alpha_{ij}] = \begin{bmatrix} \alpha_1 & 0 & 0 \\ 0 & \alpha_2 & 0 \\ 0 & 0 & \alpha_3 \end{bmatrix} \tag{10.7}$$

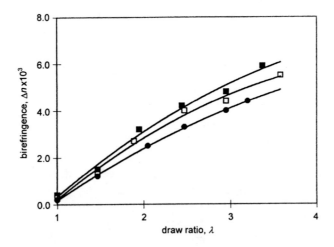

Fig. 10.8 Birefringences for three sets of uniaxially drawn PVC samples plotted against the draw ratio: ■, draw temperature 90 °C; □, draw temperature 100 °C; and ●, draw temperature 110 °C. (Reprinted with permission of Elsevier Science.)

with $\alpha_1 = \alpha_2 = \alpha_t$ and $\alpha_3 - \alpha_t = \Delta\alpha$, i.e. assume that $[\alpha_{ij}]$ has cylindrical symmetry around Ox_3. $\Delta\alpha$ is then the *anisotropy* of $[\alpha_{ij}]$. Further, define the *mean value* α_o of $[\alpha_{ij}]$ as

$$\alpha_o = \tfrac{1}{3}(\alpha_1 + \alpha_2 + \alpha_3) = \tfrac{1}{3}(2\alpha_t + \alpha_3) \tag{10.8}$$

Equation (A.7) of the appendix then shows that, with respect to the symmetry axes $OX_1X_2X_3$ of the sample,

$$\alpha_{33} = a_{31}^2\alpha_1 + a_{32}^2\alpha_2 + a_{33}^2\alpha_3 \tag{10.9}$$

where (a_{ij}) is the direction-cosine matrix for rotation from unit axes to sample axes. Thus

$$\langle\alpha_{33}\rangle = \langle a_{31}^2\rangle\alpha_1 + \langle a_{32}^2\rangle\alpha_2 + \langle a_{33}^2\rangle\alpha_3 \tag{10.10}$$

If there is no preferred orientation around Ox_3, as is the case in the simplest form of uniaxial orientation, $\langle a_{31}^2\rangle = \langle a_{32}^2\rangle$. Equation (A.5) then shows that

$$\langle a_{31}^2\rangle = \langle a_{32}^2\rangle = \tfrac{1}{2}(1 - \langle a_{33}^2\rangle) \tag{10.11}$$

However, $\langle a_{33}^2\rangle = \langle\cos^2\theta\rangle$, $\alpha_1 = \alpha_2 = \alpha_t$ and $\alpha_3 - \alpha_t = \Delta\alpha$, so that equation (10.10) becomes

$$\langle\alpha_{33}\rangle = \alpha_t + \langle\cos^2\theta\rangle\Delta\alpha \tag{10.12}$$

It may be shown in a similar way that

$$\langle\alpha_{11}\rangle = (1 - \langle a_{13}^2\rangle)\alpha_t + \langle a_{13}^2\rangle\alpha_3 \tag{10.13}$$

However, a_{13} is the direction cosine of OX_1 with respect to Ox_3, so that (see fig. 10.7) $a_{13}^2 = \sin^2\theta\cos^2\phi$. Because ϕ is assumed random, i.e. independent of θ, averages can be taken separately over functions of θ and ϕ, so that $\langle\sin^2\theta\cos^2\phi\rangle = \langle\sin^2\theta\rangle\langle\cos^2\phi\rangle$. Also, because ϕ is random, $\langle\cos^2\phi\rangle = \tfrac{1}{2}$, so that $\langle a_{13}^2\rangle = \tfrac{1}{2}\langle\sin^2\theta\rangle = \tfrac{1}{2}(1 - \cos^2\theta)$. Remembering again that $\alpha_1 = \alpha_2 = \alpha_t$ and $\alpha_3 - \alpha_t = \Delta\alpha$, equation (10.13) now leads to

$$\langle\alpha_{11}\rangle = \alpha_t + \tfrac{1}{2}(1 - \langle\cos^2\theta\rangle)\Delta\alpha \tag{10.14}$$

Equations (10.12) and (10.14) are closely related to equations (9.40). If α_t, α_3 and $\langle\cos^2\theta\rangle$ are known, they can be used to calculate n_t $(= n_1 = n_2)$ and n_3 from the modified Lorentz–Lorenz equation (10.5). More usually the measured values of n_1 and n_3 are used to calculate $\langle\cos^2\theta\rangle$. It follows from equations (10.12) and (10.14) that

$$\frac{\langle\alpha_{33}\rangle - \langle\alpha_{11}\rangle}{\langle\alpha_{33}\rangle + 2\langle\alpha_{11}\rangle} = \left(\frac{\Delta\alpha}{6\alpha_o}\right)(3\langle\cos^2\theta\rangle - 1) = \left(\frac{\Delta\alpha}{3\alpha_o}\right)\langle P_2(\cos\theta)\rangle \tag{10.15}$$

Substituting into equation (10.15) the values of $\langle \alpha_{33} \rangle$ and $\langle \alpha_{11} \rangle$ in terms of n_3 and $n_1 (= n_t)$ from the modified Lorentz–Lorenz equations (10.5) gives

$$\left(\frac{\Delta\alpha}{3\alpha_0}\right)\langle P_2(\cos\theta)\rangle = \frac{n_3^2 - n_t^2}{n_t^2 n_3^2 + n_t^2 - 2} = \frac{(n_3 - n_t)(n_3 + n_t)}{n_t^2 n_3^2 + n_t^2 - 2} \qquad (10.16)$$

If the birefringence $\Delta n = n_3 - n_t$ is small the substitution $n_3 = n_t = n_0$ can be made in all terms except $n_3 - n_t$ on the RHS of this equation, which then reduces to

$$\left(\frac{\Delta\alpha}{3\alpha_0}\right)\langle P_2(\cos\theta)\rangle = \frac{2n_0\Delta n}{n_0^4 + n_0^2 - 2} = \frac{2n_0\Delta n}{(n_0^2 - 1)(n_0^2 + 2)} \qquad (10.17)$$

Setting $\Delta n = \Delta n_{max}$ when $\langle P_2(\cos\theta)\rangle = 1$, i.e. for full orientation, leads immediately to equation (10.6).

Equation (10.6) states that the birefringence is directly proportional to $\langle P_2(\cos\theta)\rangle$. This is true only for small birefringences. If the birefringence is larger the more correct equation (10.16) should be used, but equation (10.6) is often still used as an approximation. In either case, the quantities Δn_{max} or $\Delta\alpha/(3\alpha_0)$ must be determined from measurements on a very highly aligned sample or by calculating the refractive indices of such a sample by the method explained in section 9.4.3 for calculating the refractive indices of a crystal.

A biaxially oriented sample has three independent principal refractive indices, so the method can be adapted to give information about the degree of orientation with respect to the three principal axes of the sample.

10.4.2 Measurement of infrared dichroism

As explained in section 2.6.1, any infrared-absorption peak is associated with a particular vibrational mode of the polymer chain. For each of these modes there is a particular direction within the polymer chain, called the *infrared-transition dipole* or *transition-moment axis*, which is the direction of the absorption dipole μ. Infrared radiation is absorbed by a particular chain of the polymer only if two conditions are satisfied: the frequency of the radiation must correspond to the frequency of the vibration and there must be a component of the electric vector E of the incident radiation parallel to the transition-dipole axis. If the molecules become oriented, so do the dipole axes; hence the absorption of the sample will depend on the polarisation of the radiation.

The experiment involves the measurement of the transmittance or absorbance of a sample, usually in the form of a thin film, for infrared radiation polarised parallel or perpendicular to the draw direction for a particular

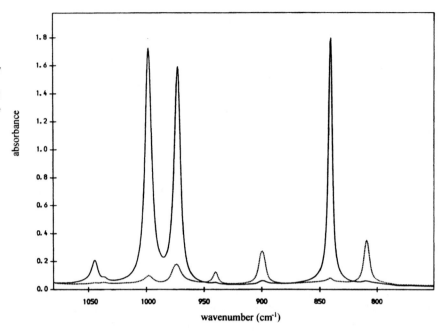

absorption peak in the spectrum. Figure 10.9 shows the spectra of an oriented sample of polypropylene for two different polarisation directions.

Provided that the attenuation of the infrared beam is due only to absorption, not to reflection or scattering, the Beer–Lambert law discussed in section 2.6.4 states that

$$I = I_o \times 10^{-kct} \tag{10.18}$$

where k is the extinction coefficient, which is proportional to the average fraction of the incident intensity per unit area absorbed by a single absorber (its absorption cross-section), c is the concentration of absorbers and t is the thickness of the sample.

For an unoriented sample, the absorbance A (see section 2.6.4) is given by

$$A = \log_{10}(I_o/I) = kct \tag{10.19}$$

For an oriented sample, absorbances $A_\parallel$ and $A_\perp$ are defined for the two polarisation directions parallel and perpendicular to the draw direction, so that

$$A_\parallel = k_\parallel ct \qquad A_\perp = k_\perp ct \tag{10.20}$$

where $k_\parallel$ and $k_\perp$ are proportional to the average fractions of the incident energy per unit area absorbed by a single absorber for radiation polarised parallel and perpendicular to the draw direction, respectively.

Absorption occurs because there is an oscillating dipole μ associated with the vibration. Thus $A_\parallel \propto \langle \mu_\parallel^2 \rangle$ and $A_\perp \propto \langle \mu_\perp^2 \rangle$, where $\mu_\parallel$ is the component of μ parallel to the draw direction and $\mu_\perp$ is the component perpendicular to the draw direction in the plane of the sample, i.e. parallel to OX_1 (see fig. 10.10).

Thus

$$A_\parallel = C \langle \cos^2 \theta_\mu \rangle \tag{10.21}$$

and, assuming uniaxial symmetry,

$$A_\perp = C \langle \sin^2 \theta_\mu \cos^2 \phi_\mu \rangle = \tfrac{1}{2} C \langle \sin^2 \theta_\mu \rangle \tag{10.22}$$

where

$$C = A_\parallel + 2A_\perp = 3A_{iso} \tag{10.23}$$

and

$$\langle \cos^2 \theta_\mu \rangle = \frac{A_\parallel}{A_\parallel + 2A_\perp} \tag{10.24}$$

Here A_{iso} is the absorbance of the same thickness of isotropic polymer and the simplification in equation (10.22) is due to the random orientation around OX_3, as explained in deducing equation (10.14) in section 10.4.1. The ratio $A_\parallel/A_\perp$ is sometimes called the *dichroic ratio*, D. It then follows that

$$\langle \cos^2 \theta_\mu \rangle = D/(D+2) \qquad \langle P_2(\cos \theta_\mu) \rangle = (D-1)/(D+2) \tag{10.25}$$

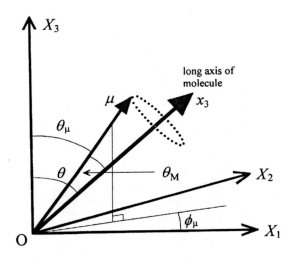

Fig. 10.10 Orientation of the infrared transition dipole μ with respect to the long axis of the polymer molecule and the set of axes $OX_1X_2X_3$ fixed in the sample. The axes OX_1 and OX_3 are assumed to lie in the plane of the sample.

The value of $\langle P_2(\cos\theta)\rangle$ or $\langle\cos^2\theta\rangle$ can be calculated from $\langle\cos^2\theta_\mu\rangle$ provided that the angle θ_M that the infrared dipole $\pmb{\mu}$ makes with the chain-axis direction is known. It can be shown that, for any value of l and for any uniaxial distribution of $\pmb{\mu}$ around the long axis of the molecule

$$\langle P_l(\cos\theta_\mu)\rangle = \langle P_l(\cos\theta)\rangle\langle P_l(\cos\theta_M)\rangle \tag{10.26}$$

provided that there is no preferred orientation of $\pmb{\mu}$ around the long axis of the molecule, as is assumed for the simplest type of uniaxial orientation. Provided that θ_M is the same for all units, equation (10.26) becomes

$$\langle P_2(\cos\theta)\rangle = \langle P_2(\cos\theta_\mu)\rangle / P_2(\cos\theta_M) \tag{10.27}$$

The quantity $\langle P_2(\cos\theta_\mu)\rangle$ is easily calculated from $\langle\cos^2\theta_\mu\rangle$, using equation (10.1a). The angle θ_M might not be precisely known, in which case only a quantity proportional to $\langle P_2(\cos\theta)\rangle$, i.e. $\langle P_2(\cos\theta_\mu)\rangle$, can be determined, not its absolute value.

In the simplified treatment given, the effect of reflection at the surfaces of the sample and internal-field effects have not been taken into account. Although these must be taken into account in more accurate treatments, their effects are generally small. It has also been assumed that absorption peaks due to different modes do not overlap. If they do, they must be 'resolved' to find their separate absorbances. There are various methods available for doing this.

Figure 10.11 shows some results obtained for oriented PET. The 875 cm^{-1} peak is the out-of-plane, in-phase H-bending mode of the benzene ring, sometimes called the 'umbrella mode', for which the dipole is expected to be perpendicular to the plane of the ring and approximately perpendicular to the chain axis. We thus expect $P_2(\cos\theta_M)$, which is the value of $\langle P_2(\cos\theta_\mu)\rangle$ for full alignment, to be approximately equal to -0.5; the results are in good agreement with this.

Although the infrared method can give only $\langle P_2(\cos\theta)\rangle$ or $\langle\cos^2\theta_\mu\rangle$, it has the important advantage over refractive-index measurements that various absorption peaks can be used to give information about different parts of the molecule or different types of units, e.g. crystallites or amorphous chain segments.

Infrared measurements can be used for characterising biaxial samples, but this involves making some measurements with samples tilted so that the incident IR beam is not normal to the sample surface. Three different values of the absorbance or extinction coefficient can then be determined, which provide information about the degree of orientation with respect to the three principal axes of the sample.

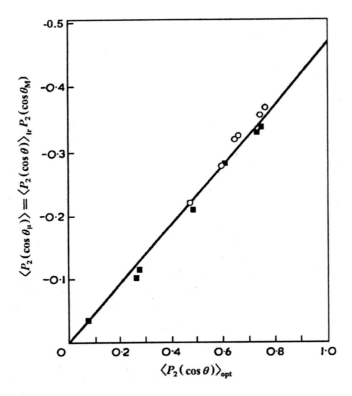

Fig. 10.11 $\langle P_2(\cos\theta_\mu)\rangle$ plotted against $\langle P_2(\cos\theta)\rangle_{opt}$, the value of $\langle P_2(\cos\theta)\rangle$ determined by refractive-index measurements, for the 875-cm^{-1} infrared vibration for two different series of uniaxially drawn samples of poly(ethylene terephthalate). $\langle P_2(\cos\theta)\rangle_{ir}$ signifies the value of $\langle P_2(\cos\theta)\rangle$ obtained from infrared measurements, which should ideally be the same as $\langle P_2(\cos\theta)\rangle_{opt}$. (Reprinted with permission of Elsevier Science.)

Example 10.3

A uniaxially oriented polymer sample transmits 21% or 90% of infrared radiation corresponding to a particular absorption for radiation polarised parallel or perpendicular to the draw direction, respectively. If the transition dipole is parallel to the chain axis, calculate the value of $\langle P_2(\cos\theta)\rangle$ for the chains.

Solution

$$A_\parallel = \log(I_o/I_\parallel) = \log(100/21) = 0.6778$$

$$A_\perp = \log(100/90) = 0.0458, \text{ so that}$$

$$C\langle\cos^2\theta\rangle = 0.6778 \quad \text{and} \quad \tfrac{1}{2}C\langle\sin^2\theta\rangle = 0.0458$$

Hence $C = 0.6778 + (2 \times 0.0458) = 0.7694$, leading to

$$\langle\cos^2\theta\rangle = 0.6778/0.7694 = 0.8809$$

and

$$\langle P_2(\cos\theta)\rangle = \tfrac{1}{2}(3\langle\cos^2\theta\rangle - 1) = 0.82$$

10.4.3 Polarised fluorescence

Many substances exhibit fluorescence. If they are exposed to electromagnetic radiation within a particular wavelength range, often in the
ultraviolet part of the spectrum, they absorb some of the radiation and
a short time later emit radiation of a longer wavelength. Some polymers
are naturally fluorescent, but the majority are not, so that, in the
polarised-fluorescence method for characterising orientation, fluorescent
probe molecules are usually dissolved in the polymer. The molecules
chosen are usually long and thin and are expected to mimic the orientation of the surrounding polymer chains. The probe molecules do not
usually penetrate the crystalline parts of a semicrystalline polymer, so
that the method is one of the few that provide information specifically
about the orientation of the amorphous parts of the polymer. This is one
of the principal reasons for its use.

The principle of the method is similar to that of the infrared method
except that the polymer is illuminated by polarised radiation of a suitable
wavelength to induce fluorescence and the emitted radiation is examined
using an analyser. Information is thus obtained about molecular orientation with respect to *two* directions, the transmission directions of the polar-
izer and analyser.

A fluorescent molecule absorbs light of a certain wavelength, or range
of wavelengths, provided that there is a component of the electric vector
of the incident light parallel to an axis fixed in the molecule called the
absorption axis. In an analogous way to the absorption of infrared
radiation discussed in section 10.4.2, the rate of absorption is proportional to $\cos^2 \alpha$, where α is the angle between the polarisation vector of
the incident light and the absorption axis. After a short time, the
molecule emits light of a longer wavelength or wavelengths. This light
is emitted from an electric dipole parallel to an *emission axis* fixed in the
molecule. The intensity, I, observed through an analyser is thus
proportional both to the rate of absorption and to $\cos^2 \beta$, where β is
the angle between the emission axis and the polarisation axis of the
analyser. Thus

$$I \propto \cos^2 \alpha \cos^2 \beta \qquad\qquad (10.28)$$

The emission and absorption axes usually either coincide or are separated by a small angle. If they coincide and both the exciting and the
analysed fluorescent light are polarised parallel to the draw direction in a
uniaxially oriented polymer, then $\alpha = \beta = \theta_A$, where θ_A is the angle
between the absorption axis and the draw direction. The intensity observed
will thus be proportional to $\langle \cos^4 \theta_A \rangle$. It is easy to show that, if the incident

light is polarised parallel to the draw direction and the transmission direction of the analyser is perpendicular to the draw direction, then

$$I \propto \langle \tfrac{1}{2}\cos^2 \theta_A \sin^2 \theta_A \rangle = \tfrac{1}{2}(\langle \cos^2 \theta_A \rangle - \langle \cos^4 \theta_A \rangle) \qquad (10.29)$$

with the same constant of proportionality, so that $\langle \cos^2 \theta_A \rangle$ and $\langle \cos^4 \theta_A \rangle$ can be determined by making both measurements.

Because the angle θ_A refers to the absorption/emission axis of the fluorescent molecule, the method does not characterise the orientation of the polymer molecules directly when probe molecules are used rather than fluorescence of the polymer itself. Some relationship between the orientations of the probes and the polymer chains must then be assumed if information about the orientation distribution of the chains is to be deduced.

Fig. 10.12 shows values of $\langle \cos^2 \theta_A \rangle$ for a particular probe molecule called VPBO dissolved in essentially amorphous PET plotted against the birefringence. It is seen that, although the values of $\langle \cos^2 \theta_A \rangle$ increase with birefringence, they are not proportional to the birefringence, which would be expected if the probe molecules mimicked the orientation of the amorphous material exactly.

Biaxially oriented samples can be characterised by making tilted-film measurements somewhat similar to those made for the infrared characterisation of biaxial samples.

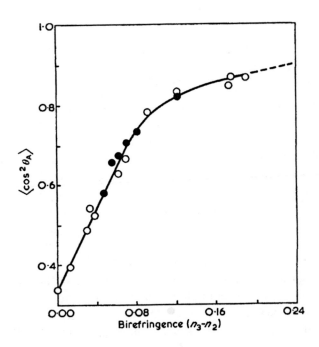

Fig. 10.12 $\langle \cos^2 \theta_A \rangle$ for VPBO molecules dissolved in poly(ethylene terephthalate) plotted against the birefringence of the sample. (Reprinted with permission of Elsevier Science.)

10.4.4   Raman spectroscopy

The general principles involved in Raman spectroscopy are described in section 2.6. This method for characterising orientation is very similar to the fluorescence method, in that it involves both incident and emerging radiation, but the Raman process is a scattering process; the re-radiation is essentially instantaneous. It is also similar to the infrared method in that it is associated with molecular vibrations and it has the same advantages of potential selectivity to a part of the polymer molecule or a phase of the polymer.

The theory of the method is rather complicated, because the amplitude of Raman scattering is described by a second-rank tensor, so it will not be discussed here. Just like for fluorescence, $\langle P_2(\cos\theta)\rangle$ and $\langle P_4(\cos\theta)\rangle$, or $\langle\cos^2\theta\rangle$ and $\langle\cos^4\theta\rangle$, can, at least in principle, be obtained for the simplest type of uniaxial orientation distribution and these values now refer directly to the molecules of the polymer itself. In practice it is often necessary to make various simplifying assumptions. For biaxially oriented samples several other averages can be obtained.

Figure 10.13 shows the C—Cl stretching region of the Raman spectra of two different oriented samples of PVC obtained with various combinations of polarisation directions of the incident and scattered light. As expected, the spectra in parts (a) and (d) of figure 10.13 are very similar for the sample with lower birefringence, as are the spectra in (b) and (c). For the sample with higher birefringence all four spectra are, however, quite different from each other.

10.4.5   Wide-angle X-ray scattering

This method is based on the fact that the intensity of a particular Bragg 'reflection' (see section 2.5.1) depends on the total volume of the crystallites that are oriented at the Bragg angle, θ_B. An X-ray system is set up so that the correct angle $2\theta_B$ for a particular set of crystal planes is formed between the direction of the incident X-ray beam and the direction of detection. Diffraction can then be observed only from crystallites for which the normals to this set of planes are parallel to the bisector of the incident and diffracted X-ray beams. The sample is rotated so that its draw direction always lies in the scattering plane and makes the angle θ (not to be confused with θ_B) with this bisector, e.g. as shown in fig. 10.14. The observed intensity is then proportional to the volume of the crystallites that have the normal for this set of planes lying within some small solid angle $d\omega$ making the angle θ with the draw direction. The value of $d\omega$ is the same for all values of θ and depends on instrumental parameters. It is therefore possible

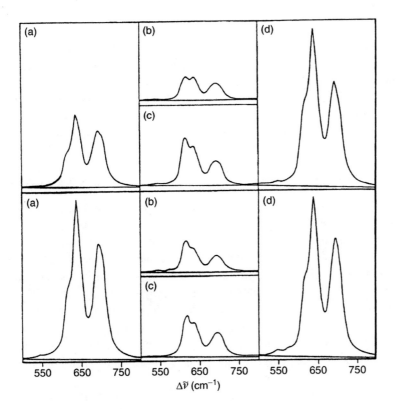

Fig. 10.13 Raman spectra of two uniaxially oriented samples of PVC for various combinations of polarisation directions of incident and scattered light. The upper set is for a sample with $\Delta n = 5.49 \times 10^{-3}$; the lower set is for a sample with $\Delta n = 1.58 \times 10^{-3}$: (a) S_{33}, (b) S_{32}, (c) S_{12} and (d) S_{11}, where S_{ij} is the spectrum obtained with the incident light polarised parallel to Ox_j and the polarisation of the analyser parallel to Ox_i. (© 1978 John Wiley & Sons, Inc.)

to obtain the complete angular distribution for this set of planes. For biaxially oriented samples it is necessary to combine the results of measurements with more than one set of planes in order to determine the complete orientation-distribution function for the crystallites.

Other possible arrangements than the one shown in fig. 10.14 can be used. That arrangement would require corrections for the variation of

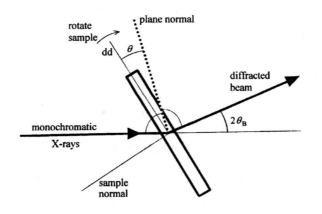

Fig. 10.14 The characterisation of molecular orientation by means of wide-angle X-ray scattering. The angles marked but not labelled are equal. See the text for explanation.

effective sample thickness and absorption with variation of θ that other arrangements might not require. If the sample shown in fig. 10.14 is a thin uniaxially oriented sheet viewed edge-on, rotation around the sample normal could be used, with the sample oriented so that the plane normal being studied is in the plane of the sample. This keeps a constant thickness of sample in the beam. Several other corrections that cannot be considered here must be applied when very accurate results are required.

10.5   The combination of methods for two-phase systems

The degrees of molecular orientation of the non-crystalline and crystalline components of a semicrystalline polymer are not necessarily the same. To the extent that the polymer can be imagined to consist of two completely distinct phases, the birefringence of the sample will depend on the concentrations and orientations of the different phases. In the same way, the birefringence of a segregated block copolymer (see section 12.3) will depend on the concentration and orientation of the two blocks. In both cases there may also be a form birefringence. It is usual to assume the additivity of the birefringences of the various components, although this is not really correct; it is really the polarisabilities that should be added, but this leads to a rather complicated equation.

Assuming additivity of birefringences, it follows that for a two-phase system,

$$\Delta n = \Delta n_f + f_1 \Delta n_1 + f_2 \Delta n_2 \tag{10.30}$$

where Δn is the birefringence of the sample, f_1 and f_2 are the volume fractions of phases 1 and 2, Δn_1 and Δn_2 are the corresponding birefringences and Δn_f is the form birefringence. Now $\Delta n_i = \Delta n_{i,max} \langle P_2(\cos\theta) \rangle_i$, where i is 1 or 2, so that

$$\Delta n = \Delta n_f + f_1 \Delta n_{1,max} \langle P_2(\cos\theta) \rangle_1 + f_2 \Delta n_{2,max} \langle P_2(\cos\theta) \rangle_2 \tag{10.31}$$

Equation (10.31) shows that if the maximum birefringences $\Delta n_{i,max}$ and the fractions f_i are known (for a two-phase system $f_2 = 1 - f_1$) and the form birefringence can be neglected or determined, knowledge of one of the $\langle P_2(\cos\theta) \rangle_i$ allows the other to be determined from a measurement of the birefringence of the sample. For a semicrystalline polymer the fractional crystallinity can be determined from density or X-ray-diffraction measurements, whereas for a block copolymer the composition will usually be known. X-ray measurements can give the value of $\langle P_2(\cos\theta) \rangle$ for the crystalline phase, so that equation (10.31) then gives the corresponding value for the amorphous material. In the case of a block copolymer, in

---

Example 10.4

A particular sample of a semicrystalline polymer with the simplest form of uniaxial orientation is found to have a birefringence of 0.042. Density measurements show that it has a volume crystallinity of 0.45 and X-ray measurements show that $\langle P_2(\cos\theta)\rangle$ for the crystalline phase is 0.91. Assuming that Δn_{max} is 0.050 for the crystalline phase and 0.045 for the amorphous phase, estimate $\langle P_2(\cos\theta)\rangle_a$ for the amorphous phase.

Solution

Equation (10.31) gives

$$0.042 = 0.45 \times 0.050 \times 0.91 + (1 - 0.45) \times 0.045\langle P_2(\cos\theta)\rangle_a$$

Hence $\langle P_2(\cos\theta)\rangle_a = 0.87$.

---

which both components may be non-crystalline, a method such as measuring infrared dichroism might be used to give $\langle P_2(\cos\theta)\rangle$ for one phase.

When the two phases are sufficiently different, methods such as measurement of infrared dichroism or Raman scattering can sometimes be used to characterise the orientation of the two phases independently, which will usually give more accurate results. Nevertheless, birefringence measurements are often used as a cross-check on the results, because all methods of characterising orientation are subject to errors that are sometimes difficult to quantify and it is usually desirable to obtain as much data as possible by various methods. In practice, the maximum birefringences of the various phases are not always well known, so large uncertainties can arise from using equation (10.31).

10.6 Methods of representing types of orientation

10.6.1 Triangle diagrams

When biaxially oriented materials are studied, the results for the second-order orientation averages are often expressed as $\langle\cos^2\theta_1\rangle$, $\langle\cos^2\theta_2\rangle$ and $\langle\cos^2\theta_3\rangle$, where the angles θ_1, θ_2 and θ_3 are the angles between a specific axis in the unit and the OX_1, OX_2 and OX_3 axes, respectively. In accordance with the usual properties of direction cosines, these three averages must add up to 1, so that only two are independent. A helpful way to display these averages for any set of samples is by means of the *equilateral triangle diagram*, shown in fig. 10.15. In this diagram the perpendicular from each vertex of the triangle onto the opposite side is assumed to be of unit length. The point representing any sample with averages $\langle\cos^2\theta_i\rangle$ for

Fig. 10.15 A triangle
diagram for showing
values of $\langle \cos^2 \theta_i \rangle$: □, a
series of samples with unit
axes uniaxially oriented
towards OX_3; ■, a series of
samples with unit axes
oriented preferentially
towards the OX_3X_1 plane,
with no preference for OX_3
rather than OX_1; and ●, a
series of samples with unit
axes oriented towards the
OX_1X_2 plane, with a
preference for OX_2 rather
than OX_1.

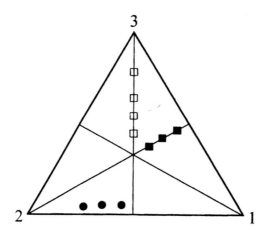

$i = 1, 2$ and 3 is then placed so that the perpendicular distance of the point
from the side opposite the vertex i is equal to $\langle \cos^2 \theta_i \rangle$ for all three values of
i. (Readers might like to prove that this can be done!) In such a diagram a
randomly oriented sample is represented by the point of intersection of the
three perpendiculars from each vertex to the opposite side. A uniaxial
sample with symmetry axis Ox_i is represented by a point on the perpendicular through the vertex i. If the units are oriented towards, rather than
away from, Ox_i then the point lies between the triple intersection and i.
Figure 10.15 shows a triangle diagram for various hypothetical sets of
samples.

10.6.2   Pole figures

The results obtained for biaxially oriented samples by methods that can
give the full orientation distribution, such as WAXS, are often expressed in
terms of *pole figures*. If a particular crystallite is imagined to be at the
centre of a large sphere, the point of intersection of the sphere with the
normal to a particular set of crystal planes is called the *pole* for those
planes. For an oriented sample placed at the centre of the sphere, the
density of poles at various points of the sphere gives a complete description
of the orientation distribution. It is inconvenient to work with spheres,
however, and the pole figure is a two-dimensional representation of the
three-dimensional spherical description. Because of the symmetry of the
sample it is necessary to represent only one eighth of the sphere, but it is
usual to represent one half. The pole figure is usually obtained by imagin-
ing every pole in the hemisphere to be joined by a straight line to the
central point on the surface of the other hemisphere. The pole is then

represented by the point in the common diametral plane to the two hemi-spheres where this line cuts it (see fig. 10.16). Such a pole figure is a *stereographic projection.*

If a particular distribution of orientations corresponded to a discrete set of poles the description just given would be completely correct. Of course, the more realistic case is a distribution of poles with different densities for different parts of the sphere. The pole figure is then represented by a series of contour lines of constant pole density. These must first be imagined drawn *on the sphere* before being projected onto the diameter, because the densities of projected poles for different regions of the sphere are not proportional to the actual densities. Figure 10.16(b) shows how a pole figure might look for a biaxial distribution in which the chains are oriented preferentially both towards the OX_3 direction and towards the OX_3X_1 plane.

The stereographic projection is only one of many projections that can be used for mapping a distribution on the surface of a sphere onto a plane. Other projections are sometimes used for representing X-ray data.

10.6.3   Limitations of the representations

Neither a point on a single triangle diagram nor even a single pole figure characterises fully the distribution of orientations of a sample. A pole figure shows the complete distribution of orientations, with respect to the sample axes, of a single axis within the structural units. At least one further pole figure for a different axis in the units is required for a full characterisation of the orientation, because the first figure gives no infor-mation about the orientation of the other axes of the units around the one chosen, which is frequently not random even when the chosen axis is the chain axis. A point on a single triangle diagram may be thought of as representing just one average of the data that could be more fully repre-

Fig. 10.16 The stereographic-projection pole figure. (a) The construction of a pole figure centred on OX_3. The circle represents the section of the sphere on which the figure is based, formed by its intersection with the OX_3X_1 plane. P is the pole for a specific axis in a given unit that happens to lie in this plane and P' is the corresponding point in the pole figure plotted on the diametral plane OX_1X_2. Points for poles lying off the OX_3X_1 plane are constructed in a similar way. (b) A schematic pole figure centred on OX_3 for a hypothetical distribution of unit axes oriented preferentially towards the OX_3 axis and preferentially towards the OX_3X_1 plane. The numbers inside the circle give the relative densities of poles for the contours shown.

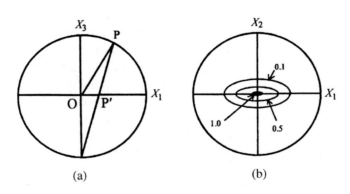

(a)   (b)

sented on a single pole figure. As explained in previous sections, some methods of studying orientation can give only this second-order average, but there is an infinite number of possible distributions corresponding to any second-order average. Just as for the pole figure, a second diagram is required for another axis in the unit if any information about orientation around the first chosen axis is to be conveyed.

10.7   Further reading

(1) *Structure and Properties of Oriented Polymers*, edited by I. M. Ward, 2nd Edn, Chapman and Hall, London, 1997. This is in general an advanced monograph, but some sections are less difficult than others. The book concentrates on methods of studying oriented polymers and the results obtained, but also includes some material on the production of orientation.

10.8   Problems

10.1. A hypothetical uniaxial distribution of orientations has 20% of the axes of the units oriented randomly, 50% oriented on a cone of semi-angle $\alpha = 30°$ and the remainder parallel to the axis of the cone. Calculate $\langle \cos^2 \theta \rangle$, $\langle \cos^4 \theta \rangle$, $\langle P_2(\cos \theta) \rangle$ and $\langle P_4(\cos \theta) \rangle$, where θ is the angle between the axis of a unit and the cone axis.

10.2. The orientation-distribution function $N(\theta)$ for a particular set of crystal planes in a uniaxially oriented polymer is found from WAXS measurements to be given approximately by $N(\theta) = A(1 + \cos \theta)$. Calculate $\langle P_2(\cos \theta) \rangle$ for this distribution.

10.3. A uniaxially oriented fibre of a certain polymer has a birefringence of 0.03 and the birefringence for a fully oriented fibre of the polymer is 0.05. Assuming that, to a good approximation, all the polymer chains in the partially oriented fibre make the same angle θ with the fibre axis and that the simplest theory of birefringence applies, calculate θ.

10.4. A particular sample of a uniaxially oriented polymer is composed of structural units each of which is transversely isotropic with respect to an axis Oz within the unit. The value of $\langle \cos^2 \theta \rangle$ is 0.65, where θ is the angle between the Oz axis of a typical unit and the draw direction. There are 4.2×10^{27} structural units per m^3 in the polymer and the polarisabilities of an individual unit for light polarised parallel and perpendicular to Oz are 2.6×10^{-39} and 1.9×10^{-39} F m^2, respectively. Calculate the polarisabilities of the sample for light polarised parallel and perpendicular to the draw direction and hence the birefringence of the sample.

10.5. A particular polymer sample has all chain axes oriented closely parallel to the draw direction. Assuming that the chains have no preferred orientation around their axes, calculate the dichroic ratio expected for an infrared-absorption peak for which the corresponding absorption dipole makes an angle of $25°$ with the chain axis.

10.6. The dipole moment associated with a particular infrared absorption in a polymer makes an angle of $20°$ with the chain axis. A uniaxially drawn sample of the polymer is found to transmit 50% of incident infrared radiation at the wavelength corresponding to the absorption when the incident beam is polarised parallel to the draw direction and 75% when the polariser is rotated through $90°$ from this position. Making the usual simplifying assumptions (what are they?) calculate $\langle \cos^2 \theta \rangle$ for the sample, where θ is the angle between a typical chain axis and the draw direction.

10.7. The molecular chains in a particular sample of polymer film are almost completely aligned in the simplest type of uniaxial orientation. The fractional transmissions of this sample for radiation of a particular infrared wavelength incident normally are found to be 0.20 and 0.95 for radiation polarised parallel and perpendicular to the alignment axis, respectively. For a second film of the same polymer of lower orientation the corresponding fractional transmissions are 0.30 and 0.80. Making suitable assumptions, what can be deduced about the degree of orientation of the chains in the second film?

10.8. It is known that the dipole moment associated with a particular infrared absorption in a polymer makes an angle of $15°$ with the chain axis. A particular uniaxially drawn sample of the polymer is found to transmit 25% of incident infrared radiation at the wavelength corresponding to this absorption when the incident beam is polarised parallel to the draw direction. If the sample has a birefringence 0.85 times that of a perfectly oriented sample, estimate the corresponding percentage infrared transmission for an incident beam polarised perpendicular to the draw direction. State clearly the assumptions made.

10.9. A highly uniaxially oriented polymer sample in the form of a thin film has mean optical polarisabilities of 2.3×10^{-39} and 1.7×10^{-39} F m^2 per structural unit for light polarised parallel and perpendicular to the draw direction, respectively; it transmits 50% of infrared radiation at a particular wavelength λ when the radiation is polarised parallel to the draw direction and transmits almost all radiation at this wavelength when the radiation is polarised perpendicular to the draw direction. Calculate (a) the mean optical polarisability per structural unit and (b) the infrared transmission at the wavelength λ for an isotropic sample of the polymer. In (b) assume that the thickness of the isotropic sample is the same as that of the highly drawn sample. What simplifying approximations do you need to make?

10.10. A uniaxially oriented sample of poly(ethylene terephthalate) is found to have a volume crystallinity of 0.35 and a birefringence of 0.109. X-ray diffraction and polarised fluorescence measurements show that the values of $\langle P_2(\cos\theta)\rangle$ for the crystalline and amorphous phases are 0.91 and 0.85, respectively. The birefringence of a fully oriented crystalline sample is calculated to be 0.132. Given that the densities of the crystalline and amorphous components are 145.5 and 133.5 kg m^{-3}, respectively, check whether these data are consistent with the assumption that the contribution that a given chain segment makes to the birefringence depends only on its orientation and not on whether it is in a crystalline or amorphous region.

10.11. Two sets A and B of oriented thin film polymer samples have the following orientation averages for their chain axes.

Sample	A1	A2	A3	A4	B1	B2	B3	B4
$\langle\cos^2\theta_3\rangle$	0.40	0.50	0.76	0.92	0.38	0.42	0.45	0.48
$\langle\cos^2\theta_2\rangle$	0.30	0.25	0.12	0.04	0.24	0.16	0.10	0.04

For each set of samples the axes OX_1 and OX_3 lie in the plane of the film. Plot the two sets of data on an equilateral triangle plot and describe the corresponding orientation distributions qualitatively.

Chapter 11
Oriented polymers II – models and properties

11.1 Introduction

In this chapter an attempt to answer the following questions is made.

(i) What kind of theoretical models, or *deformation schemes*, can be envisaged for predicting the distribution of molecular orientations in a drawn polymer?

(ii) How do the predictions of these models agree with experimental results for the orientation averages such as $\langle P_2(\cos\theta)\rangle$?

(iii) How can experimental or theoretical values of the orientation averages be used to predict the properties of the drawn polymer and with how much success?

These questions form the topics of discussion for sections 11.2, 11.3 and 11.4, respectively. In extremely highly drawn material, a great deal of modification of the original structure of the polymer must have taken place and most of the chains are essentially parallel to the draw direction, so that $\langle\cos^2\theta\rangle$ and $\langle P_2(\cos\theta)\rangle$ are both close to 1. It is difficult to describe theoretically and in detail how molecular orientation takes place under these circumstances and thus to predict the small departures from unity for such materials. Discussion of the prediction and use of orientation averages is therefore limited to materials of moderate draw ratios. Models used for interpreting the properties of highly oriented polymers are discussed in sections 11.5 and 11.6.

The deformation schemes to be described in the following section are expected to be valid for single-phase, i.e. non-crystalline, polymers or for one of the two phases present in semicrystalline polymers. In such two-phase systems the prediction of the properties involves combining those of the two oriented phases in the appropriate way. The models considered in section 11.5 show how this can be done.

11.2 Models for molecular orientation

Two principal *deformation schemes* have been used in attempts to predict the distribution of orientations:

(i)   the *affine rubber* deformation scheme, or model; and
(ii) the *aggregate* or *pseudo-affine* deformation scheme or model.

11.2.1 The affine rubber deformation scheme

In this scheme it is assumed that, at the temperature of drawing or other deformation, the polymer is in the rubbery state, either as a true rubber containing chemical cross-links or as a polymer containing tangled amorphous chains, in which the entanglements can behave like cross-link points. Once deformation has been accomplished, the polymer is cooled down into the glassy state before the stress is removed, so that the orientation is 'frozen in'.

The simplest version of the rubber model makes the assumption of an *affine deformation*: when the polymer is stretched the cross-link points move exactly as they would if they were points in a completely homogeneous medium deformed to the same macroscopic deformation (see section 6.4.4, fig. 6.12). The following *additional assumptions* are also made in the simplest form of the theory.

(i)   Each chain (i.e. the portion of a polymer molecule between two successive cross-link points) consists of the same number n of freely jointed random links, each of length l.
(ii) The length of the end-to-end vector r_o of each chain in the unstrained state is equal to the RMS vector length of the free chain, which is shown in section 3.3.3 to be $n^{1/2}l$.
(iii) There is no change in volume on deformation, so that $\lambda_1\lambda_2\lambda_3 = 1$, where the quantities λ_i are the extension ratios defined in section 6.3.2. (It is useful to remember that this really means that fractional changes in volume are not actually zero but are very small compared with fractional changes in linear dimensions.)

For simplicity, only uniaxially oriented samples are considered here, which means that $\lambda_1 = \lambda_2 = \sqrt{1/\lambda_3}$ and hence only λ_3, which will be called λ, the *draw ratio*, need be specified.

When the polymer is drawn, two effects contribute to the orientation of the random links:

(i)   the end-to-end vector of each chain orients towards the draw direction; and
(ii) the random links of each chain orient towards the end-to-end vector of the chain and therefore towards the draw direction, as illustrated in fig. 11.1.

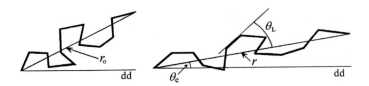

Fig. 11.1 The orientation of a chain and its links before and after drawing. dd represents the draw direction. For exact application of equation (11.3) there must be a large number of links per chain. A small number are shown here for clarity.

The calculation of $N(\theta)$, the distribution of orientations of the random links with respect to the draw direction, therefore involves three steps (see fig. 11.1):

(i) the calculation of $N(\theta_e)$, the distribution function for the end-to-end vectors with respect to the draw direction after drawing

(ii) the calculation of $N(\theta_L, r)$, the distribution function for the links of a chain of stretched length r with respect to the end-to-end vector r, where r depends on θ_e; and

(iii) the combining of these two functions to give $N(\theta)$.

It is a straightforward piece of geometry (see problem 11.1) to show that

$$N(\theta_e) = \frac{\lambda^3}{4\pi[\lambda^3 - (\lambda^3 - 1)\cos^2\theta_e]^{3/2}} \tag{11.1}$$

The calculation of the distribution function $N(\theta_L, r)$ for the random links with respect to the end-to-end vector of the chain of stretched length r is somewhat more difficult. Provided that n is sufficiently

---

Example 11.1

The end-to-end vector of a particular chain in an undrawn polymer makes an angle of 45° with the direction in which the polymer is subsequently drawn to $\lambda = 6$. Calculate the angle θ that the chain makes with the draw direction after drawing.

Solution

Imagine a unit cube of the sample before drawing. The chain under consideration can be thought of as one of the face diagonals. After stretching, the length of the cube in the stretching direction is $\lambda = 6$ and each of the other sides has length $\lambda_t = 1/\sqrt{6}$ ('constant volume'). The new angle θ that the chain makes with the draw direction is thus $\tan^{-1}(\lambda_t/\lambda) = \tan^{-1}[(1/\sqrt{6})/6] = 3.9°$.

large the usual statistical methods can be used, by assuming that the distribution that actually exists is the most probable one, i.e. the one that can be obtained in the largest number of different ways subject to the constraints that, if there are n_L links inclined at θ_L to the end-to-end vector r, then:

$$\sum_L n_L = n \quad \text{and} \quad l \sum_L n_L \cos \theta_L = r \qquad (11.2a,b)$$

This is a familiar type of problem in statistical mechanics. The solution is

$$N(\theta_L, r) = \frac{n\beta \exp(\beta \cos \theta_L)}{4\pi \sinh \beta} \qquad (11.3)$$

where β satisfies the equation

$$\coth \beta - 1/\beta = r/(nl) \qquad (11.4)$$

β is said to be the *inverse Langevin function* of $r/(nl)$. It is important to note that the form of $N(\theta_L, r)$ depends on r, which in turn depends on θ_e.

It is very difficult to combine equations (11.1) and (11.3) to give $N(\theta)$, so it is usual to calculate only the values of $\langle P_2(\cos \theta) \rangle$ and possibly $\langle P_4(\cos \theta) \rangle$ in the following stages.

(a) Find the values of $\langle P_l(\cos \theta_L) \rangle$ for $N(\theta_L, r)$ for a given r, i.e. a given θ_e.
(b) Use

$$\langle P_l(\cos \theta) \rangle_r = \langle P_l(\cos \theta_L) \rangle P_l(\cos \theta_e) \qquad (11.5)$$

to calculate $\langle P_l(\cos \theta) \rangle_r$, the average value of $P_l(\cos \theta)$ for all those random links which belong to chains of length r, for which the end-to-end vectors lie on a cone of angle θ_e.
(c) Calculate $\langle P_l(\cos \theta) \rangle$ for all random links of all chains by averaging over all values of r or, equivalently, over all values of θ_e, using the expression for $N(\theta_e)$ from equation (11.1).

The result will be quoted only for the second-order average $\langle P_2(\cos \theta) \rangle$, which is given by

$$\langle P_2(\cos \theta) \rangle = \frac{1}{5n} \left(\lambda^2 - \frac{1}{\lambda} \right) + \text{terms in } \frac{1}{n^2} \text{ etc.} \qquad (11.6)$$

Using only the first term is essentially equivalent to the Gaussian approximation of rubber elasticity theory developed in section 6.4.4. The form of $\langle P_2(\cos \theta) \rangle$ versus λ for the rubber model is then as shown

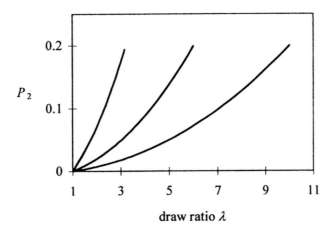

Fig. 11.2 $P_2 = \langle P_2(\cos\theta)\rangle$ plotted against the draw ratio λ according to the simplest form of the affine rubber deformation scheme (the first term of equation (11.6)). Curves from left to right are for $n = 10$, 36 and 100, respectively, and each curve is plotted up to $\lambda = \sqrt{n}$.

in fig. 11.2. The relationship between $\langle P_4(\cos\theta)\rangle$ and λ is similar but it takes much smaller values for a given value of λ.

An important consequence of the affine assumption is that the rubber model in this simple form is applicable only up to $\lambda_{max} = \sqrt{n}$, because at this draw ratio the chains parallel to the draw direction in the undeformed material are fully extended in the deformed material and cannot extend further (see equation (3.5) and the sentence following it). Further extension of the material as a whole could thus take place only non-affinely. Equation (11.6) shows that $\langle P_2(\cos\theta)\rangle \approx \lambda^2/(5n)$ for $\lambda \geq 3$, so that the maximum value of $\langle P_2(\cos\theta)\rangle$ to which the simple theory applies is approximately 0.2, as fig. 11.2 shows.

Example 11.2

The birefringence Δn for a uniaxially oriented sample with draw ratio $\lambda = 3.5$ was found to be 8.1×10^{-3}. Assuming that the orientation of the polymer molecules is adequately described by equation (11.6) using only the first term on the RHS, deduce Δn for a similarly drawn sample with $\lambda = 2.0$.

Solution

For small birefringence, as here, $\langle P_2(\cos\theta)\rangle = \Delta n/\Delta n_{max}$. Equation (11.6) then shows that, if the subscripts 1 and 2 refer to the first and second samples, respectively, then

$$\frac{(\Delta n)_2}{(\Delta n)_1} = \frac{\lambda_2^2 - 1/\lambda_2}{\lambda_1^2 - 1/\lambda_1} = \frac{2^2 - 1/2}{3.5^2 - 1/3.5} = \frac{3.5}{11.96} = 0.293$$

Thus $(\Delta n)_2 = 0.293(\Delta n)_1 = 2.37 \times 10^{-3}$.

11.2.2   The aggregate or pseudo-affine deformation scheme

The basic assumptions of this scheme are the following.

(i)   The polymer behaves like an aggregate of anisotropic units.

(ii)   The properties of the units are those of the completely oriented polymer.

(iii)   In the simplest treatment, which applies to the simplest type of uni-axially oriented samples, each unit is also assumed to be *transversely isotropic* with respect to an axis within it, often called its 'unique axis'.

(iv)   In the isotropic medium, the unique axes of the units are distributed randomly in space.

(v)   On drawing, each unit rotates in exactly the same way as would the end-to-end vector of a chain in the affine rubber model if it were originally parallel to the axis of the unit considered.

(vi)   The units do not change in length or properties on drawing, unlike the chains in a rubber.

The last assumption is the *pseudo-affine* assumption; it should be seen immediately that this makes the model somewhat unphysical. It is there-fore sometimes called the *needles in plasticine model*, in which each unit corresponds to a needle and the plasticine corresponds to the surrounding polymer. In spite of this unreality, the model often predicts well the orien-tation distribution of the crystallites for a drawn semicrystalline polymer.

The distribution of orientations of the unique axes of the units for this scheme is given simply by equation (11.1) with θ instead of θ_e, i.e.

$$N(\theta) = \frac{\lambda^3}{4\pi[\lambda^3 - (\lambda^3 - 1)\cos^2\theta]^{3/2}} \tag{11.7}$$

and the corresponding expression for $\langle P_2(\cos\theta)\rangle$ becomes

$$\langle P_2(\cos\theta)\rangle = \frac{1}{2}\left(\frac{2\lambda^3 + 1}{\lambda^3 - 1} - \frac{3\lambda^{1/2}\cos^{-1}\lambda^{-3/2}}{(\lambda^3 - 1)^{3/2}}\right) \tag{11.8}$$

When $\langle P_2(\cos\theta)\rangle$ is calculated from equation (11.8) it is found to depend on λ as shown in fig. 11.3. The variation of $\langle P_4(\cos\theta)\rangle$ with λ is quite similar except that it is slightly concave upwards below $\lambda \approx 2$ and $\langle P_4(\cos\theta)\rangle$ has a somewhat lower value than $\langle P_2(\cos\theta)\rangle$ for a given λ, as shown in fig. 11.3. Unlike the curves for the affine rubber deformation scheme, which are different for different values of n, these curves have no free parameters, so that they are the same for all polymers. The shapes contrast strongly with those for the rubber model, being concave to the abscissa, whereas the latter are convex. The curves for the affine rubber

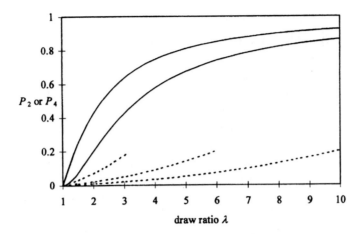

Fig. 11.3 The variation of $P_2 = \langle P_2(\cos\theta) \rangle$ and $P_4 = \langle P_4(\cos\theta) \rangle$ with λ according to the pseudo-affine deformation scheme. Full curves, P_2 (upper) and P_4; dashed curves are for affine deformation as in fig. 11.2.

model from fig. 11.2 are included in fig. 11.3 for comparison, from which it is clear that the values of $\langle P_2(\cos\theta) \rangle$ for a given λ are always very much smaller for the affine model than they are for the pseudo-affine model.

Example 11.3

Find the ratio of the numbers of units oriented within a small angle $\delta\theta$ to the draw direction at the angles $\theta = 10°$ and $\theta = 20°$ for a uniaxially oriented sample of draw ratio $\lambda = 5$ to which the pseudo-affine model applies.

Solution

The total solid angle between two cones with semi-angles θ and $\theta + \delta\theta$ is $2\pi \sin\theta\, \delta\theta$, so that the number δN of units oriented within the angular spread $\delta\theta$ at θ is $2\pi N(\theta) \sin\theta\, \delta\theta$.

The required ratio is thus, from equation (11.7),

$$\frac{\sin 10° \, (125 - 124\cos^2 20°)^{3/2}}{\sin 20° \, (125 - 124\cos^2 10°)^{3/2}}$$

Substitution of the values of the angles leads to the ratio 3.00. The reader should check that, although $\lambda^3 \gg 1$, λ^3 *cannot* be substituted for $\lambda^3 - 1$. Explain why (consider what happens as $\theta \to 0$).

11.3 Comparison between theory and experiment

11.3.1 Introduction

The simplest tests that can be made of the predictions of the deformation schemes considered in the previous two sections are comparisons between the values of $\langle P_2(\cos\theta)\rangle$ predicted by the schemes and values obtained from refractive-index or birefringence measurements. In the next two sections comparisons of this kind are made. It is clear from the proof given in the previous chapter that the reason why the birefringence depends on $\langle P_2(\cos\theta)\rangle$ but not on any higher-order averages is that the refractive index depends on a second-rank tensor, the polarisability tensor. Equations (A.7) show that the rotation equations for a *second*-rank tensor involve products of *two* direction cosines and, even for general biaxial orientation, these products can be expressed in terms of *second-order* generalised spherical harmonics, of which the Legendre polynomial $\langle P_2(\cos\theta)\rangle$ is the simplest and the only one required for the simplest uniaxial orientation.

Equations (A.7) also show that, in general, the prediction of a property that depends on a tensor of rank *l* will require knowledge of orientation averages of order *l*. The elastic constants of a material are fourth-rank tensor properties; thus the prediction of their values for a drawn polymer involves the use of both second- and fourth-order averages, in the simplest case $\langle P_2(\cos\theta)\rangle$ and $\langle P_4(\cos\theta)\rangle$, and thus provides a more severe test of the models for the development of orientation. The elastic constants are considered in section 11.4.

11.3.2 The affine rubber model and 'frozen-in' orientation

When a polymer is spun to form a fibre the molecular network in the melt is extended and subsequently 'frozen in' the extended state. In commercial fibre production, heat setting or subsequent hot drawing above the glass-transition temperature may also be used, but for fibres that have only been spun and drawn simultaneously the distribution of orientations in the fibre should conform to the predictions of the rubber model.

The effective draw ratio, λ, of the fibre can be determined from the shrinkage, S, when the fibre is re-heated to a temperature above the glass-transition temperature T_g. S is defined by

$$S = \frac{\text{length before shrinkage} - \text{length after shrinkage}}{\text{length before shrinkage}} = \frac{\lambda - 1}{\lambda}$$

(11.9)

Thus

$$\lambda = 1/(1 - S)$$

(11.10)

It follows from equation (11.6) and the fact that the birefringence, Δn, is approximately proportional to $\langle P_2(\cos\theta)\rangle$ that a plot of Δn against $[1/(1-S)^2 - (1-S)]$ should be a straight line. Figure 11.4 shows data

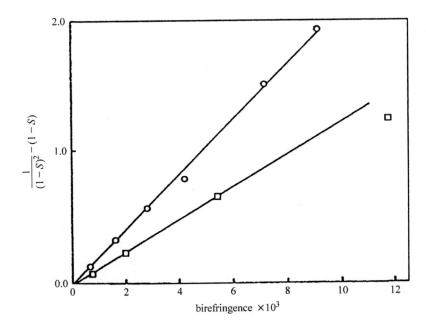

Fig. 11.4 Data obtained for shrinkage and birefringence for spun and drawn PET fibres of molar masses $\bigcirc$, 1.8×10^4 g mol^{-1}; and $\square$, 2.8×10^4 g mol^{-1}. (Reproduced by permission of the Royal Society of Chemistry.)

for PET fibres of two different molar masses. The fibres were obtained by spinning only, with different wind-up rates, giving different effective draw ratios. Good straight lines are found for each molar mass. The difference in slopes is due to differences in the degree of entanglement, i.e. different values of n, the polymer of higher molar mass having a larger number of entanglements and thus a lower value of n.

11.3.3   The affine rubber model and the stress–optical coefficient

For the polymer considered in the previous section the birefringence measurements and the stretching or shrinkage took place at different times; the birefringence was measured in the 'frozen-in' state of orientation. It is, however, possible to measure the birefringence of a real rubber when it is still under stress at a temperature above its glass-transition temperature. This provides a simultaneous test of the predictions of the rubber deformation theory for both orientation and stress.

It is useful to collect together three equations from previous sections, namely equation (10.3):

$$\alpha_o = \frac{3\varepsilon_o(n_o^2 - 1)}{N_o(n_o^2 + 2)}$$

equation (10.17):

$$\left(\frac{\Delta\alpha}{3\alpha_o}\right)\langle P_2(\cos\theta)\rangle = \frac{2n_o\,\Delta n}{(n_o^4 + n_o^2 - 2)}$$

and equation (11.6), keeping only the first term on the RHS:

$$\langle P_2(\cos\theta)\rangle = \frac{1}{5n}\left(\lambda^2 - \frac{1}{\lambda}\right)$$

In the first two of these equations the subscript on the refractive index n_o indicates that it refers to the mean refractive index and also distinguishes it from n in the third equation, which denotes the number of random links per chain. If the equivalent random link is taken as the structural unit, N_o is now the number of random links per unit volume and α_o and $\Delta\alpha$ are the mean value and the anisotropy of the polarisability of a random link.

The second of these equations can be rewritten

$$\Delta n = \left(\frac{\Delta\alpha}{3\alpha_o}\right)\langle P_2(\cos\theta)\rangle\frac{(n_o^2 - 1)(n_o^2 + 2)}{2n_o} \tag{11.11}$$

Substituting the expressions for α_o and $\langle P_2(\cos\theta)\rangle$ from the other two equations leads to

$$\Delta n = \left(\frac{\Delta\alpha}{90\varepsilon_o}\right)N\left(\lambda^2 - \frac{1}{\lambda}\right)\frac{(n_o^2 + 2)^2}{n_o} \tag{11.12}$$

where the substitution $N_o/n = N$ has been made, so that N is the number of *chains* per unit volume. Equation (6.49) shows that the true stress σ on the rubber is given, in the Gaussian approximation, by $\sigma = NkT(\lambda^2 - 1/\lambda)$, so that

$$\frac{\Delta n}{\sigma} = \frac{\Delta\alpha}{90\varepsilon_o kT}\frac{(n_o^2 + 2)^2}{n_o} \tag{11.13}$$

This equation shows that the ratio of the birefringence to the true stress should be independent of stress. The expression on the RHS of equation (11.13) is known as the *stress–optical coefficient*. A test of equation (11.13) can be made by plotting Δn against σ, when a straight line should be obtained. Such plots for a vulcanised natural rubber at various temperatures are shown in fig. 11.5. The hysteresis shown in the curves for the lower temperatures is interpreted as being due to stress crystallisation, with the crystallites produced being oriented in the stretching direction and

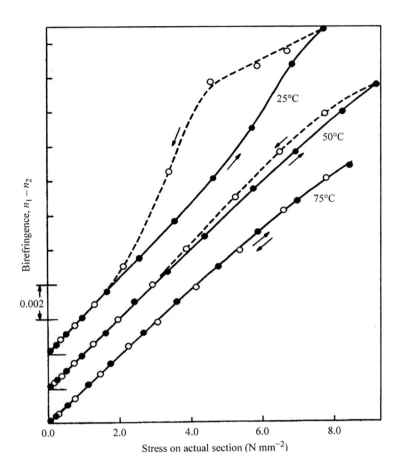

Fig. 11.5 The relation between birefringence and stress for natural rubber at various temperatures: ●, strain increasing; and ○, strain decreasing. Data for the various temperatures are offset along the birefringence axis. (Reproduced by permission of the Royal Society of Chemistry.)

giving an excess contribution to the birefringence. At the higher temperatures this effect is smaller or absent and the plot is linear to much higher stresses, confirming the applicability of the rubber model for predicting orientation and stress.

An estimate of the number of monomer units per equivalent random link can be obtained by dividing the value of $\Delta\alpha$ calculated from the stress–optical coefficient by the anisotropy of the polarisability of the monomer unit calculated from bond polarisabilities. This number can more interestingly be expressed in terms of the number of single bonds in the equivalent random link and is found to be about 5 for natural rubber, about 10 for gutta percha and about 18 for polyethylene. (For the last two the values are extrapolated from measurements at elevated temperature.) The number for polyethylene is considerably higher than the value of 3 suggested by the assumption of totally free rotation around the backbone bonds (see section 3.3.3 and problem 3.7).

Fig. 11.6 Birefringence plotted against draw ratio for a series of drawn samples of low-density polyethylene. The broken curve shows values calculated according to the pseudo-affine deformation scheme. (Adapted by permission of I. M. Ward.)

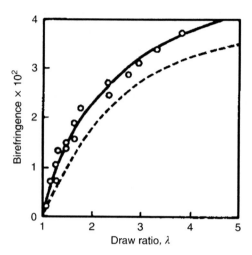

11.3.4   The pseudo-affine aggregate model

This model should apply best to glassy or semicrystalline polymers, e.g. polyethylene. Figure 11.6 shows the experimental birefringence Δn plotted against λ for a series of drawn samples of low-density polyethylene and Δn calculated according to the pseudo-affine deformation scheme. The general features of the curve are fitted, in particular the concavity relative to the λ-axis, but the fit is not quantitative.

11.4   Comparison between predicted and observed elastic properties

11.4.1   Introduction

The prediction of the elastic constants of an oriented polymer provides a further, more stringent, test for the deformation models because these properties depend in a more complex way on the details of the distribution of orientations.

It is shown in the appendix that the state of stress within a homogeneously stressed material can be represented by a symmetric second-rank tensor $[\sigma_{ij}]$ with six independent components. For a cube with faces perpendicular to $OX_1X_2X_3$ the component σ_{ii} represents the outwardly directed force per unit area applied in the direction parallel to Ox_i on the face perpendicular to Ox_i. It thus represents a *normal stress*. The component σ_{ij} for $i \neq j$ represents the force per unit area applied in the Ox_j direction to the face perpendicular to Ox_i, so that it represents a *shear stress*. It is also shown in the appendix that the state of strain in the material can be described by a symmetric second-rank tensor $[\varepsilon_{ij}]$, where the component ε_{ii} represents a *tensile strain* parallel to Ox_i and the components ε_{ij} for $i \neq j$ represent *shear strains*.

The relationship between the stresses and strains is given by

$$\sigma_{ij} = c_{ijkl}\varepsilon_{kl} \quad \text{or} \quad \varepsilon_{ij} = s_{ijkl}\sigma_{kl} \qquad (11.14a,b)$$

where the Einstein convention for summation over repeated subscripts is used (see the appendix). The quantities c_{ijkl} and s_{ijkl} are components of the *elastic-stiffness* and *elastic-compliance* tensors, respectively, which are fourth-rank tensors. (Note the misleading, but standard, notation: c for stiffness, s for compliance.) Either the compliances or the stiffnesses may be meant when the term *elastic constants* is used. The terms *modulus and moduli* are sometimes used rather than stiffness(es). For a material with orthotropic (statistical orthorhombic) symmetry, such as a biaxially oriented polymer sample, only nine of the 81 components of either the stiffness tensor or the compliance tensor are independent. For a uniaxial sample only four are independent. The non-zero values for reference axes $OX_1X_2X_3$ chosen parallel to the symmetry axes of the sample and their relationships are listed in table A.1 of the appendix.

Example 11.4

Deduce the relationship between Young's modulus E_i measured in the OX_i direction and the appropriate compliance.

Solution
Equation (11.14b) shows that if the only non-zero stress is $\sigma_{ii} = \sigma$, as it is when determining Young's modulus in the OX_i direction, then $\varepsilon_{ii} = s_{iiii}\sigma$, so that $E_i = \sigma/\varepsilon_{ii} = 1/s_{iiii}$.
Note that, if the value of E is required for a direction OA not parallel to one of the axes $OX_1X_2X_3$ with respect to which the compliance tensor is given, the tensor components must first be expressed with respect to a set of axes of which one is parallel to OA.

11.4.2   The elastic constants and the Ward aggregate model

In the simplest theory for calculating the elastic constants, the sample is assumed to consist of an *aggregate* of anisotropic units whose elastic constants are known. It is then assumed that the elastic constants of the aggregate are obtained by averaging those of the units, taking into account the distribution of orientations of the units. Because the elastic constants are tensor properties, the averaging involves expressing the components of the tensor for each unit with respect to axes fixed in the sample and then averaging these values over all units to find the corresponding components of the tensor for the aggregate. Equation (A.7) of the appendix shows that because the elastic constants are fourth-rank tensors, the results involve averages over four direction cosines multiplied together. For the simplest uniaxial orientation these averages can be expressed in terms of $\langle \cos^2 \theta \rangle$ and $\langle \cos^4 \theta \rangle$ or $\langle P_2(\cos \theta) \rangle$ and $\langle P_4(\cos \theta) \rangle$.

If it is assumed that the *strain* is uniform throughout the sample, an average called the *Voigt average* is obtained. If it is assumed that the *stress* is uniform throughout the sample, an average called the *Reuss average* is obtained. It follows from equation (11.14a) that the Voigt average is obtained directly by averaging the stiffnesses c_{ijkl}, because for any value of i and j the value of ε_{ij} is assumed to be the same for all units, so that the average value of σ_{ij} is simply $\langle c_{ijkl} \rangle \varepsilon_{kl}$, again using the Einstein notation. Similarly, the Reuss average is obtained directly by averaging the compliances s_{ijkl}.

It is, however, possible to calculate a complete set of corresponding compliances from a complete set of stiffnesses and vice versa, so that it is possible to derive a set of Voigt average compliances or Reuss average stiffnesses, but only if the complete set of the first type is known. It is usual in practice to compare experimental data with the results obtained for both types of average. It can be shown theoretically that, for a random aggregate, the true values of the elastic constants must lie between those predicted by the Reuss and Voigt assumptions. It cannot be shown that this must be true for non-random aggregates, but it is often found to be so in practice.

As an example, it can be shown that the Reuss average compliance s_{3333} for a uniaxial sample consisting of an aggregate of transversely isotropic units is given in terms of the compliances s^o_{ijkl} of the unit by

$$s_{3333} = s^o_{1111} \langle \sin^4 \theta \rangle + s^o_{3333} \langle \cos^4 \theta \rangle + 2(s^o_{1133} + 2s^o_{2323})\langle \sin^2 \theta \cos^2 \theta \rangle$$

$$(11.15)$$

where the OX_3 axis in the sample coincides with the draw direction and the Ox_3 axis corresponds to the unique axis of the unit. The expressions for the remaining s_{ijkl} are rather more complicated and are given in the book cited as (1) in section 11.7. The RHSs of equations such as (11.15) can easily be converted into expressions involving only $\langle \cos^2 \theta \rangle$ and $\langle \cos^4 \theta \rangle$ or $\langle P_2(\cos \theta) \rangle$ and $\langle P_4(\cos \theta) \rangle$ (see problem 11.5).

Figure 11.7 shows a comparison of the elastic moduli (stiffnesses) determined experimentally with those calculated using values of $\langle P_2(\cos \theta) \rangle$ and $\langle P_4(\cos \theta) \rangle$ predicted by the pseudo-affine model for a series of samples of

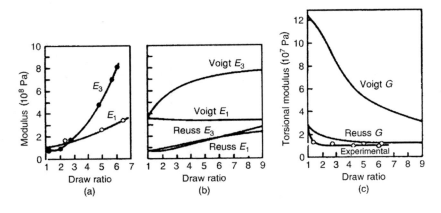

Fig. 11.7 A comparison between experimental moduli for low-density polyethylene filaments and values calculated using the pseudo-affine aggregate model: E_3, extensional modulus; E_1, transverse modulus; and G, torsional modulus; (a) experimental, (b) calculated and (c) experimental and calculated. (Reprinted by permission of Kluwer Academic Publishers.)

low-density polyethylene. The E's are Young's modulus and G is the torsional, or rigidity, modulus. The agreement is at best qualitative and the Reuss averages agree better with the experimental results than do the Voigt averages. For other polymers the Voigt average fits better than does the Reuss average; for yet others an average of the two is quite good.

There are two possible sources for the discrepancies between measured and predicted elastic constants when the orientation averages used are determined from models for molecular orientation: the models may be incorrect and neither Voigt nor Reuss averaging may be appropriate. In order to examine the second of these possibilities more directly it is necessary to determine the orientation averages experimentally.

As an example, fig. 11.8 shows a comparison of experimental and predicted moduli for cold-drawn low-density polyethylene. The theoretical curves are calculated using the Reuss averaging scheme and the values of $\langle P_2(\cos\theta) \rangle$ and $\langle P_4(\cos\theta) \rangle$ are derived from NMR data for the crystalline regions only. The fit is now quite good.

The results described and many others show that the development of optical and low-strain mechanical anisotropy in moderately oriented polymers can be understood at least semi-quantitatively in terms of the theoretical models that have been considered so far. For polymers of higher orientation somewhat different approaches are required. Some of these are considered in the remaining sections of this chapter and in section 12.4.7, where the properties of highly aligned liquid-crystal polymers are described.

11.5 Takayanagi composite models

In a polymer that contains highly oriented crystalline material and also non-crystalline material, which is usually less highly oriented, a good first approximation is to treat the two phases as totally distinct, each having its

Fig. 11.8 A comparison between experimental moduli for low-density polyethylene filaments and values calculated using the aggregate model with values of $\langle P_2(\cos\theta)\rangle$ and $\langle P_4(\cos\theta)\rangle$ derived from NMR measurements for the crystalline material. (Reproduced by permission of IOP Publishing Limited.)

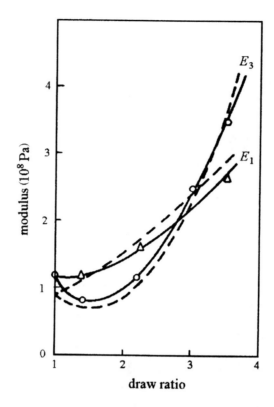

own elastic properties. If these properties are known and a suitable model is chosen to describe how the different phases are arranged, it is possible to calculate the elastic properties of the material.

As an example of this approach, consider first for simplicity the two physically rather unlikely models shown in fig. 11.9. The model shown in part (a) has alternating layers of crystalline and non-crystalline ('amorphous') material, as shown on the LHS. If a tensile stress is applied parallel to the long axis of the sample, the stress on each crystalline or amorphous portion is the same, so that it does not matter how they are distributed along the length of the sample. The elastic compliance $S = 1/E$ of the sample can therefore be calculated from the simplified model on the RHS, which has the same total fraction a of amorphous material. If E_a and E_c are the moduli of the amorphous and crystalline regions, respectively, the total strain in the sample on applying stress σ will be $a\sigma/E_a + (1 - a)\sigma/E_c$, so that

$$1/E = a/E_a + (1 - a)/E_c \tag{11.16}$$

For the model shown on the LHS of fig. 11.9(b) the fractions of the stress carried by the amorphous and crystalline regions will be proportional to

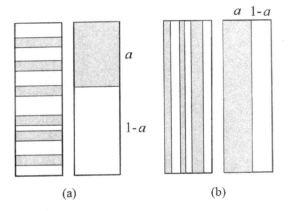

Fig. 11.9 The simplest composite models for a drawn fibre, with the draw direction vertical. In both (a) and (b) the left-hand figure shows the postulated structure, whereas the right-hand figure shows the equivalent for calculation of the elastic properties. (a) Alternating layers of crystalline and non-crystalline (shaded) material. (b) Columns of crystalline and non-crystalline (shaded) material.

their respective cross-sectional areas, so that the simplified model on the RHS, which has the same total fraction a of amorphous material, can be used in calculating the modulus for the sample. If the strain in the sample (and thus both in the crystalline part and in the amorphous part), is e, the total stress σ will be $aE_a e + (1 - a)E_c e$, so that

$$E = aE_a + (1 - a)E_c \qquad (11.17)$$

Equation (11.16) shows that, if $E_a \ll E_c$, then $E \approx E_a/a$ unless the fraction of amorphous material is very small, so that E thus depends mainly on the fraction of amorphous material and its modulus. Equation (11.17) shows, on the other hand, that, if $E_a \gg E_c$, then $E \approx (1 - a)E_c$ and thus depends mainly on the fraction of crystalline material and its modulus.

So far only the modulus for stretching parallel to the long axis of the sample has been considered. If, however the sample (a) were stretched perpendicular to that direction, equation (11.17) would apply, rather than equation (11.16). It is thus possible to account for the fact that the modulus of a drawn semicrystalline polymer is sometimes found to be greater when it is measured in the direction perpendicular to the draw direction than when it is measured in the direction parallel to the draw direction (see example 11.5). Such an anomaly is likely to disappear at low temperatures when the modulus of the amorphous material rises because it is no longer in the rubbery state.

Example 11.5 draws attention to the fact that the crystalline and non-crystalline regions, particularly the crystalline regions, are likely to be highly anisotropic and that this must be taken into account in applying the models. A further complication is that, for many polymer samples, the structures illustrated schematically in fig. 11.9 are too simplified. Structures such as those shown in fig. 11.10, which allow for amorphous material both

Fig. 11.10 The series–parallel and parallel–series versions, (b) and (c), respectively, for the Takayanagi model (a), which allows for non-crystalline (shaded) material being partly in series and partly in parallel with crystalline material.

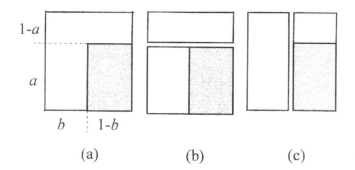

(a)   (b)   (c)

Example 11.5

A certain oriented polymer has a structure like that in fig. 11.9(a), with $a = 0.5$, and the moduli for the crystalline and amorphous regions are as follows, where the subscripts $\parallel$ and $\perp$ denote values measured parallel and perpendicular to the draw direction, respectively:

$E_{c\parallel} = 10^{11}$, $E_{c\perp} = 10^{9}$, $E_{a\parallel} = 10^{7}$ and $E_{a\perp} = 10^{7}$, all measured in pascals. Show that the modulus measured in the perpendicular direction is greater than that measured in the parallel direction.

Solution

For the parallel direction equation (11.16) gives $E_{\parallel} = 1/(0.5/10^{7} + 0.5/10^{11}) = 2 \times 10^{7}$ Pa. For the perpendicular direction equation (11.17) gives $E_{\perp} = 0.5(10^{7} + 10^{9}) = 5 \times 10^{8}$ Pa.

---

in series and parallel with crystalline material, must then be considered. It is left as an exercise for the reader to derive the appropriate equivalent of equation (11.16) or (11.17) for these models (see problem 11.7).

11.6 Highly oriented polymers and ultimate moduli

11.6.1 Ultimate moduli

Highly oriented polymers can exhibit very high tensile moduli and high tensile strengths. Many polymers, both conventional and liquid-crystalline, have been produced in highly oriented form to make use of these properties, so it is important to consider what controls the ultimate modulus available for a particular polymer. It is clear that the highest modulus will be obtained if all the polymer chains are parallel to the tensile axis. The modulus of the material will then depend on the force required to produce unit extension of a chain and on the number of chains passing through unit cross-sectional area normal to the tensile axis. Because the force required to open bond angles is low compared with that to stretch bonds, polymers with most of their backbone bonds parallel or nearly parallel to the chain axis when the chain is fully extended will be preferred when one is attempting to produce high-modulus polymers. If, however, the chains have many or large side groups they will not be able to pack closely together, so that the number of chains crossing unit area will be low. These considerations suggest immediately why some of the liquid-crystal polymers (LCPs) to be considered in the next chapter should be capable of producing high modulus materials.

It turns out that a rough guide to the relative ultimate moduli of various polymers can be obtained by simply calculating the 'effective' density that each would have if the masses of all atoms not forming part of the backbone were set equal to zero. Molecular-mechanics calculations of the ultimate theoretical moduli of various polymers show that polyethylene fits well into the trend suggested by this rule, whereas some other polymers depart significantly from it, including some of the LCPs. Diamond can be considered as the ultimate stiff polymer, because it consists entirely of carbon atoms bonded together tetrahedrally; it has a modulus of approximately 1200 GPa. The effective density of polyethylene is approximately a quarter that of diamond, giving an expected ultimate modulus of about 300 GPa, similar to the highest experimental values obtained at low temperatures for highly oriented linear polyethylene. Molecular mechanics calculations suggest a slightly higher value of about 350 GPa.

The reason for the lower experimental value lies, of course, in the facts that not all chain segments are parallel to the direction of alignment and that the average number of chains crossing unit area normal to the alignment direction is less than the theoretical number. In the next section models for highly oriented polyethylene are described. A model that has been found particularly useful for understanding the moduli of oriented LCPs is considered in section 12.4.7.

It should be noted that the moduli of many highly oriented polymers fall significantly as the temperature is raised and the highest moduli are obtained only at very low temperatures. The reason is that thermal motions involve torsional oscillations around single bonds as well as angle-bending and bond-stretching vibrations and may involve conformational changes at higher temperatures. If the molecule is in its most extended conformation, such torsional oscillations or conformational changes can only shorten it and give a partially entropic character to the tensile restoring force. Entropic restoring forces are weak and, as a result, the modulus falls with increasing temperature.

11.6.2   Models for highly oriented polyethylene

Very highly oriented polyethylene has been produced, usually in the form of fibres, by a number of different methods, including solution spinning and multi-stage drawing. Two principal models have been put forward to explain the increase in modulus with second draw ratio when already highly oriented fibres are subjected to a second drawing process. These models are the *crystalline fibril model* due to Arridge and the *crystalline bridge model* due to Gibson, Davies and Ward. Both models are related to the Takayanagi models described in section 11.5.

In the fibril model the original highly oriented fibre is assumed to consist of a set of crystalline fibrils, oriented with the chain axes parallel to the draw direction, embedded in a matrix of non-crystalline material. The fibrils are very long compared with their cross-section but still finite in length, i.e. they have a high aspect ratio. Equation (11.17) for the simple parallel Takayanagi model of fig. 11.9(b) is modified to allow for the fact that the average stress in the fibrils for a given overall strain of the fibre is less than it would be if the fibrils were of effectively infinite length, as is assumed in the simple model. This arises because of the rather low shear modulus of the matrix (non-crystalline) material, which allows the stress near the ends of the fibres to relax by causing the matrix to shear, a phenomenon known as *shear lag*. This effect is taken account of in a modified form of equation (11.17) by multiplying the first term on the RHS by a quantity less than unity that can be calculated from the theory of composite materials. It is assumed that when the fibre is subjected to the second drawing process, the degree of crystallinity does not change, but the aspect ratio of the fibrils increases according to an affine deformation, so that the shear lag effect is reduced and the modulus is therefore increased.

In the bridge model it is assumed that the highly oriented fibre consists of blocks of fully oriented crystalline material with the chain axes parallel to the draw direction, separated by non-crystalline material through which pass 'bridges' of fully oriented crystalline material, as shown schematically in fig. 11.11. The modulus of the material is calculated on the basis of a modification of the Takayanagi parallel–series model of fig. 11.10(c), in which the crystalline chains involved in the bridges are in parallel with the series combination of the rest of the crystalline material and the non-crystalline material. The modification to the model is similar to that for the fibril model and takes account of the finite length of the bridging molecules by again introducing the effect of shear lag. The bridges are assumed to be sited randomly, so the probability of a crystalline sequence crossing a disordered region (and hence the total amount of material contained in bridging molecules) can be calculated from the average crystal length

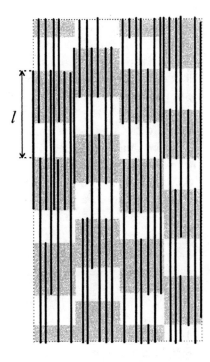

Fig. 11.11 A schematic diagram of the crystalline-bridge model. The grey area represent blocks of crystalline material in which the chains are all parallel to the draw direction, which is assumed vertical in the diagram. The thick vertical lines represent chains or groups of chains that pass through the non-crystalline regions to link two or more crystalline blocks. Within the crystal blocks these bridging molecules are indistinguishable from the other molecules making up the block and are only shown differently so that their paths through the blocks can be seen.

deduced from WAXS and the long period, shown as l in the diagram, from SAXS.

11.7 Further reading

(1) *An Introduction to the Mechanical Properties of Solid Polymers*, by I. M. Ward and D. W. Hadley. See section 6.6. Chapters 7 and 8 are particularly relevant to this chapter. The second book in section 6.6(1) is also relevant.

(2) *Structure and Properties of Oriented Polymers*, edited by I. M. Ward. See section 10.7.

11.8 Problems

11.1. Prove equation (11.1).

11.2. Show that the differential with respect to θ of the function

$$\frac{-\cos\theta}{[\lambda^3 - (\lambda^3 - 1)\cos^2\theta]^{1/2}}$$

is

$$\frac{\lambda^3 \sin\theta}{[\lambda^3 - (\lambda^3 - 1)\cos^2\theta]^{3/2}}$$

Hence deduce an expression for the fraction of units making angles less than a given value θ_0 with the draw direction for a material that fits the pseudo-affine deformation law of equation (11.7) and evaluate the result for $\lambda = 5$ and $\theta_0 = 10°$ and $20°$.

11.3. The variation of birefringence with draw ratio for a set of uniaxially drawn samples of a certain polymer is found to be consistent with the simplest version of the affine rubber model when the draw ratio is less than 3.5. If the birefringence is 7.65×10^{-3} for draw ratio 3.0, calculate its value for a sample of draw ratio 1.5. If the birefringence for a very highly oriented sample is 0.045, what is the effective number of random links per chain?

11.4. The birefringence Δn of a certain uniaxially drawn polymer is found to vary with the draw ratio λ as follows.

λ	1.5	2.0	2.5	3.0	3.5	4.0
$10^3 \Delta n$	2.5	5.5	9.1	13.3	17.9	22.8

Deduce whether the affine rubber or the pseudo-affine aggregate model better describes these data. Given that the birefringence observed for a very highly uniaxially oriented sample of the polymer is 0.078, deduce what numerical information you can about a sample with draw ratio 3.0.

11.5. Express the RHS of equation (11.15) entirely in terms of $\langle \cos^2 \theta \rangle$ and $\langle \cos^4 \theta \rangle$ and the compliances of the unit. Show that the new equation for s_{3333} gives the correct result for a fully uniaxially oriented sample. Calculate from the equation the extensional compliance for a randomly oriented sample for which the units have the following compliances: $s^0_{1111} = 8.5$, $s^0_{3333} = 1.2$, $s^0_{1133} = -0.45$ and $s^0_{2323} = 12.0$, all in units of 10^{-10} Pa^{-1}.

11.6. The independent non-zero components of the elastic compliance tensor s_{ijkl} for a particular sample of a uniaxially oriented polymer are as follows: $s_{1111} = 8.9$, $s_{3333} = 1.1$, $s_{1122} = -3.9$, $s_{1133} = -0.47$ and $s_{2323} = 14.0$, all in units of 10^{-10} Pa^{-1}. Calculate the value of the Young modulus for the following directions in the polymer: (i) the OX_3 direction and (ii) any direction making an angle of $45°$ with OX_3.

11.7. Deduce the equations corresponding to equations (11.16) and (11.17) for the Takayanagi models shown in figs 11.10(b) and (c). A polymer that is 75% crystalline by volume has amorphous and crystalline moduli E_a and E_c equal to 1.0×10^9 and 1.0×10^{11} Pa, respectively, parallel to the draw direction. Calculate the modulus parallel to the draw direction for the simple series and simple parallel Takayanagi models and for the series–parallel and parallel–series models, assuming that in the last two models $a = b = 0.5$.

Chapter 12
Polymer blends, copolymers and liquid-crystal polymers

12.1 Introduction

Although the properties of the polymers in the three groups considered in this chapter are in some ways very different, they share the possibility, under suitable circumstances, of self-organisation into supermolecular structures that are of different kinds from the crystallites and spherulites typical of the semicrystalline polymers discussed in the earlier chapters. These kinds of organisation give rise to properties that are of great interest both scientifically and commercially and it is the aim of this chapter to indicate some of the factors that control the various kinds of organisation and to give some indication of the wide variety of structures and properties that can be obtained.

Two or more existing polymers may be blended for various reasons. One reason is to achieve a material that has a combination of the properties of the constituents, e.g. a blend of two polymers, one of which is chemically resistant and the other tough. Another reason is to save costs by blending a high-performance polymer with a cheaper material. A very important use of blending is the combination of an elastomer with a rigid polymer in order to reduce the brittleness of the rigid polymer. It is in the latter use that the self-organising feature of some blends is seen to play a part. This feature arises when the components of the blend are incompatible, i.e. they do not mix, and they segregate into different regions within the material. Copolymerisation might at first sight be thought to be a route for compelling two polymers to mix. If, however, the polymer is a block copolymer it is still possible, and indeed usual, for the components to segregate, which can give rise to interesting and useful structures.

The materials of the third group to be discussed in this chapter, liquid-crystal polymers, possess the property of self-organisation into highly oriented local regions that can be further aligned very easily to give solid structures with high modulus and strength, which makes them important materials for many applications.

12.2   Polymer blends

12.2.1   Introduction

When any two materials are mixed together, or blended, the properties of the resulting mixture depend on the level at which intimate mixing takes place and on whether any chemical reactions between the components of the mixture take place. An example of a very coarse level of mixing is concrete, formed by the mixing of pebbles and mortar. The mortar itself is a somewhat finer mixture of sand and cement and the setting process involves chemical reactions between the cement and both the sand and the pebbles. The most intimate form of mixing is at the molecular level; two substances that have mixed at this level may each be said to have dissolved in the other. If the molecules have similar sizes, each type of molecule is then surrounded, on average, by molecules of the two types in proportion to the numbers of the two types in the mixture.

Some polymers can be mixed homogeneously with others at a molecular level over a wide range of compositions in such a way that there is only one phase present, but most cannot. Immiscible polymer compositions can, however, often be induced to coexist at a variety of coarser levels of mixing, similar in some instances to the structure of cement, but on a much finer scale. In the next section the factors that determine the mixing behaviour are considered. First it is useful to give some definitions.

A *miscible polymer blend* is one for which the miscibility and homogeneity extend down to the molecular level, so that there is no phase separation. An *immiscible blend* is one for which phase separation occurs, as described in the next section. An immiscible blend is called *compatible* if it is a useful blend wherein the inhomogeneity caused by the different phases is on a small enough scale not to be apparent in use. (Blends that are miscible in certain useful ranges of composition and temperature, but immiscible in others, are also sometimes called compatible blends.) Most blends are immiscible and can be made compatible only by a variety of *compatibilisation* techniques, which are described in section 12.2.4. Such compatibilised blends are sometimes called *polymer alloys*.

12.2.2   Conditions for polymer–polymer miscibility

The necessary condition for two substances A and B to mix is that the Gibbs free energy of the mixture must be less than the sum of the Gibbs free energies of the separate constituents, i.e. that ΔG_{mix}, the *increase* in free energy on mixing, should be < 0. The free energy is given by

$G = H - TS$, where H is the enthalpy and S is the entropy, so that the necessary condition for mixing at temperature T is

$$\Delta G_{mix} = \Delta H_{mix} - T \Delta S_{mix} < 0 \qquad (12.1)$$

where ΔH_{mix} is the enthalpy of mixing and ΔS_{mix} is the entropy of mixing. Unless there are special attractive forces, such as hydrogen-bonding, between molecules of the two different types A and B, there is usually a stronger attraction between molecules A and A and molecules B and B than there is between molecules A and B. This means that if A and B molecules mix randomly, $\Delta H_{mix} > 0$. For this (hypothetical) random mixing, ΔH_{mix} per mole is usually written in the form

$$\Delta H_{mix} = RT \chi_{AB} \upsilon_A \upsilon_B \qquad (12.2)$$

where υ_A and υ_B are the volume fractions of molecules A and B, respectively, in the mixture and χ_{AB} is the *Flory–Huggins interaction parameter*.

Consider first two small-molecule compounds, A and B, and assume that there are no special interaction forces, such as hydrogen-bonding, between A molecules, between B molecules and between A and B molecules. The change in enthalpy when the two compounds mix is then likely to be small. If there is no change of enthalpy, the mixing is said to be *athermal mixing* because it takes place without heat being emitted or absorbed. The free energy of mixing is then entirely due to the increase in entropy on mixing, which is always positive, so that mixing takes place for all concentrations. Even if the mixing is not athermal and ΔH_{mix} is positive, the increase in entropy is often sufficient to allow mixing at least for some concentration range, because there are enormously more ways in which the molecules can be arranged in the mixture than there are for the separated pure constituents. For polymers this is not so.

Consider the mixing of two polymers A and B and suppose for simplicity that each is mono-disperse and that the lengths, thicknesses and flexibilities of the two types of molecule are very closely the same. Imagine that equal amounts of the two polymers are mixed, say n molecules of each. If the polymers were truly indistinguishable, the entropy of the mixture would be equal to exactly twice the entropy of each of the component sets of n molecules before mixing. If, on the other hand, the molecules are different, however slightly, there will be an increase in entropy when they mix, because of the number of different arrangements that can be produced by interchanging molecules of the two different types. Imagine any particular state of the mixture, with each of the molecules in one specific conformation out of the many that it can take up. Whatever this state, it can be arrived at in $(2n)!/(n!)^2$ ways. The factor $(2n)!$ is the number of ways of arranging $2n$ distinguishable molecules on $2n$ sites, which must be divided by $(n!)^2$ to take account of the identity of the n molecules of type A and the n molecules of type B. The total number N of ways of arranging the $2n$ molecules in the mixture is thus given by

$$N = N_{\mathrm{o}}(2n)!/(n!)^2 \tag{12.3}$$

where N_{o} is the total number of conformational states available to the mixture when the difference between A and B molecules is neglected. If S_{o} is the total entropy of the unmixed polymers, then using Boltzmann's expression $S = k \ln N$ for the entropy of the mixture leads to

$$\Delta S_{\mathrm{mix}} = S - S_{\mathrm{o}} = k \ln N - S_{\mathrm{o}} = k \ln N_{\mathrm{o}} + k \ln\left(\frac{(2n)!}{(n!)^2}\right) - S_{\mathrm{o}}$$
$$= k \ln\left(\frac{(2n)!}{(n!)^2}\right) \tag{12.4}$$

because $S_{\mathrm{o}} = k \ln N_{\mathrm{o}}$. Using Stirling's approximation for large m, namely $\ln(m!) = m \ln m - m$, leads to

$$\Delta S_{\mathrm{mix}} = k[(2n) \ln(2n) - 2n - 2(n \ln n - n)] = k[2n \ln(2n) - 2n \ln n]$$
$$= 2nk \ln 2 \tag{12.5}$$

Suppose now that each polymer chain is made up of x monomer units joined together and consider what the corresponding value of ΔS_{mix} would be if the monomers were separate molecules, so that the monomers of A could mix totally freely with those of B when the substances were mixed. It is merely necessary to replace n in the above argument by xn, so that, for this mixture, $\Delta S_{\mathrm{mix}} = 2xnk \ln 2$, which is x times greater than for the polymer mixture. Because the value of x may easily be 1000 or much larger, the entropy of mixing of polymers is usually thousands of times smaller than it would be for the mixing of the corresponding monomers, so that even a small positive value of ΔH_{mix} is likely to prevent mixing from taking place.

The above argument somewhat overemphasises the immiscibility of polymers, because it considers only the mixing of polymers of high molar mass at equal concentrations of the two components. Considering the entropy of mixing more generally shows that mixing becomes more likely for low molar masses and small concentrations of the minority component and that all polymers are miscible at low enough concentrations of the minority component. It is therefore useful to consider how miscibility actually varies with composition in polymer blends before considering the modifications that need to be made to the theory to account for the fact that some pairs of polymers are miscible over ranges of composition rather larger than might be expected.

When ΔG_{mix} at a particular temperature is plotted against composition for a blend of two polymers, a curve of the type shown schematically in fig. 12.1(a) is obtained in the simplest case. The quantity ΔG_{mix} is usually taken to be the free energy per unit volume and represents the value that would be obtained if the two components could be uniformly molecularly mixed at the corresponding composition, which is usually specified in terms of the volume fraction v of one of the components. The points B represent the points where the straight line shown is tangential to the curve and are called the *binodal points*. The points S are the points of inflection, where the curvature changes sign, and are called the *spinodal points*.

In discussing the implications of this curve it is useful to consider the diagram shown in fig. 12.1(b). Suppose that a particular mixture corresponding to the composition v_o is made, so that ΔG_{mix} is represented by the point O. Assume that, by some mechanism, the mixture actually splits into two phases represented by the points P_1 and P_2, corresponding to compositions v_1 and v_2, respectively. Because the overall composition of the mixture cannot change, the actual volumes of the two phases depend on the values of v_o, v_1 and v_2. Provided that there are no changes of volume on mixing, it can be shown (see problem 12.1) that the overall value of ΔG_{mix} is now given by the point P on the chord P_1P_2 where it crosses a line parallel to the ΔG_{mix} axis drawn at composition v_o. For the particular compositions illustrated, this value of ΔG_{mix} would be greater than the value corresponding to a single phase of composition v_o.

It follows from the construction in fig. 12.1(b) that for any composition between one end of the curve of fig. 12.1(a) and the nearest point B, partition of the mixture into two phases of different compositions would lead to an increase in the total value of ΔG_{mix}. Compositions in the two corresponding regions are therefore stable against fluctuations of composition. In contrast, for any composition between the two points B where the common tangent

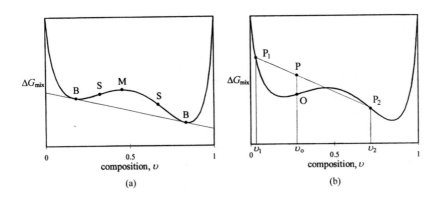

(a)

(b)

Fig. 12.1 (a) ΔG_{mix} plotted against composition for a partially miscible polymer. The points B are the binodal points, where a common tangent touches the two sections of the curve near the minima. The points S are the spinodal points, where the curve has points of inflection. M is the maximum between the spinodal points. (b) See the text.

touches the curve, the lowest free energy of mixing is obtained by the coexistence of two separate phases corresponding to the points B. This is actually an example of a more general thermodynamic rule for determining the equilibrium concentrations of phases in mixtures. When changes in volume take place on mixing, the argument holds if molar free energies and mole fractions, rather than volume fractions, are used. A further example of the application of this *common-tangent construction* is given in section 12.4.4 in connection with the theory of liquid-crystal polymers.

A closer look at the application of the construction of fig. 12.1(b) to the region between the two points B shows that any composition between the two points S, called the *spinodal region*, is unstable against small fluctuations in composition and the lowest free energy can be obtained by splitting into two phases with compositions corresponding to the points B. This is called *spinodal decomposition* (SD). The two regions between consecutive points B and S, called the *binodal regions*, are more interesting. The curvature in these regions is such that for *small* fluctuations in composition the free energy increases, but for larger fluctuations it can decrease if the mixture splits into two phases of compositions corresponding to the two points B. This kind of decomposition is called *binodal decomposition* (BD).

Unlike SD, BD is not spontaneous, because small variations of composition lead to an increase in the free energy and therefore tend to die out. If, however, a large enough fluctuation takes place, a concentration within the spinodal region may be achieved and then phase separation will take place spontaneously. The mechanism for phase separation in the binodal region is thus nucleation and growth. Droplets of the equilibrium phase corresponding to one of the points B form by nucleation and then grow by a process of diffusion of the polymer molecules from the surrounding matrix. Eventually coalescence of droplets takes place.

The phase separation in SD takes place in a completely different way. There is no nucleation stage because any slight change in the composition lowers the free energy. The process therefore takes place in three stages, diffusion, liquid flow and finally coalescence. In the early stages the different regions have compositions that are somewhere between those of the two points B and their boundaries are ill-defined, with the composition changing smoothly from one region to another.

It must be emphasised that the various stages of phase separation and the final structures obtained depend on many factors, some of which are considered in later sections. The previous paragraphs represent idealisations of what may happen during real polymer blend processing. One of the most important factors that determine the behaviour of a polymer blend is the way that temperature affects the ΔG_{mix} curve, which is now considered.

Example 12.1

A particular blend of polymers A and B has binodal points B_1 and B_2 at compositions containing volume fractions 0.20 and 0.85 of B, respectively. If a blend containing equal volumes of A and B is made, what fraction f of the total volume will be in phases corresponding to point B_1 in equilibrium?

Solution

The phase corresponding to B_1 contains a volume $0.20fV$ of B, where V is the total volume of the blend. Similarly the phase corresponding to B_2 contains a volume $0.85(1-f)V$ of B. The total volume of B is $V/2$, so that $0.20fV + 0.85(1-f)V = V/2$. Solving gives $f = 0.35/0.65 = 0.54$.

The curve obtained by plotting the points B on a temperature–composition plot, or *phase diagram*, is called the binodal curve or simply the *binodal*. A similar curve for the spinodal points is called the *spinodal*. There are two simple possible ways in which the curve of fig. 12.1(a) can change when the temperature T is changed. Either the points B get closer together as T decreases and finally coalesce or they get closer together as T increases and finally coalesce. In the first type of behaviour, illustrated in fig. 12.2(a), there is a *lower critical solution temperature* (LCST) below which there is a single phase for all compositions and above which there are two phases for a certain range of concentrations that increases in width with increasing T. In the second type of behaviour, illustrated in fig. 12.2(b), there is an *upper critical solution temperature* (UCST) above which there is a single phase for all compositions and below which there are two phases for a certain range of concentrations that increases in width with decreasing T. Many polymer systems have LCSTs, whereas most solutions of small molecules have UCSTs. More complicated types of

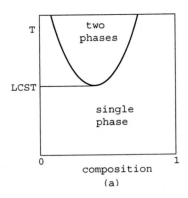
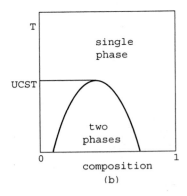

Fig. 12.2 Schematic phase diagrams for blends showing binodals corresponding to (a) lower and (b) upper critical solution temperatures.

behaviour are also observed for polymers, with more than one UCST or LCST or neither type of critical solution temperature.

As already explained, the simplest forms of the theory of mixing suggest that polymers of high molar mass should not be miscible over significant ranges of composition in the absence of specific attractive interaction forces between the molecules. The theories cannot explain the existence of a LCST either, because they suggest that the only temperature-dependent term in ΔG_{mix} is the entropic term $-T\Delta S_{mix}$ and that, if it plays any significant part, it must *increase* the likelihood of mixing as the temperature is raised. The fundamental assumption behind the simplest attempts to calculate ΔG_{mix} is that it can be neatly split into two contributions, the entropic contribution, which *does not* contain any part arising from the interaction between the different types of molecule, and a part that depends *only* on the interactions and is purely enthalpic. More advanced theories in which the interaction term contains both entropic and enthalpic parts have therefore been developed. In order to allow for the existence of a LCST, this term is also made temperature-dependent.

The physical origin of the contribution of the interaction term to the entropy of mixing is that the interaction causes one or both of the two constituent polymers to take up a different distribution of conformations from that taken up in the pure state. The temperature dependence is introduced most simply by allowing the mixture to have a larger volume than the sum of the volumes of its constituent molecules, i.e. by introducing the idea of *free volume* and allowing this free volume to change with temperature.

According to Koningsveld *et al.*, if the configurational entropy term is sufficiently small that it can be neglected, ΔG_{mix} per mole can be expressed in its simplest form as

$$\Delta G_{mix}/(RT) \approx \chi'_{AB}\upsilon_A\upsilon_B = (a_0 + a_1/T + a_2 T)(b_0 + b_1\upsilon_B + b_2\upsilon_B^2)\upsilon_A\upsilon_B$$

(12.6)

where T is the temperature, υ_A and υ_B are the volume fractions of components A and B and χ'_{AB} is the polymer–polymer interaction parameter, which now includes both entropic and enthalpic contributions. The parameters a_0, a_1, a_2, b_0, b_1 and b_2 depend on the nature of the polymers and must usually be determined by experiment.

12.2.3 Experimental detection of miscibility

A wide range of methods can be used to study polymer miscibility. One of the simplest relies on the fact that materials that do not absorb light also do not scatter it if they are homogeneous down to a scale below that of the wavelength but do scatter it if they are not. If a blend of two polymers that exhibit a LCST is made at a temperature at which the particular composition is miscible it will not scatter light. If it is then slowly heated until the binodal is reached phase separation will start to take place and the sample will become turbid, or cloudy. In a similar way, a solid sample consisting of a single phase, corresponding to a miscible blend, will be clear but one in which phase separation has taken place, corresponding to an immiscible blend, will appear cloudy.

A method that is frequently used because of its relative simplicity is the determination of the glass-transition temperature T_g of the blend, which can be determined by any of the methods discussed in chapter 7. DSC is often used.

Consider first two polymers that, at some temperature, are miscible at all compositions. Let a series of samples with various compositions be made at this temperature and then cooled quickly (quenched) below the glass-transition temperatures T_{g1} and T_{g2} of the two polymers. If DSC measurements are now made on these samples it will generally be found that, for each sample, there is a single T_g lying between T_{g1} and T_{g2} and, if the values of T_g are plotted against composition, a smooth curve something like that shown as 1 in fig. 12.3 will be produced.

The following semi-empirical equation has been found by Utracki and Jukes to describe the variation of T_g for some pairs of polymers that are miscible over the whole composition range:

$$w_1 \ln(T_g/T_{g1}) + w_2 \ln(T_g/T_{g2}) = 0 \qquad (12.7)$$

where w_1 and w_2 are the weight fractions of polymers 1 and 2 and T_{g1}, T_{g2} and T_g are the absolute glass-transition temperatures. There are, however, some pairs of polymer for which T_g lies above both T_{g1} and T_{g2}.

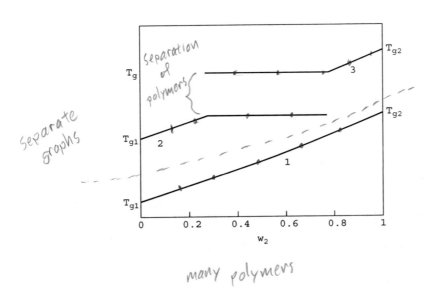

Fig. 12.3 The glass-transition temperatures T_g plotted against w_2, the weight fraction of polymer 2, for a blend of polymers 1 and 2. Curve 1 is for a pair of polymers miscible at all compositions; curves 2 and 3 are for a pair of polymers miscible only for high and low concentrations, respectively, of polymer 1 but immiscible for intermediate compositions.

Now consider a pair of polymers that exhibit a LCST and assume that a series of samples of various compositions is made at a temperature above the LCST, allowed to come to equilibrium and then quenched rapidly below T_{g1} and T_{g2}. For samples sufficiently rich in polymer 1 or polymer 2 the blend will consist of a single phase but for samples of intermediate compositions phase separation will have taken place before quenching and the phases formed will have been 'frozen in' on quenching. DSC measurements will reveal a single T_g for the single-phase samples and two T_gs for the two-phase samples. These two values of T_g will correspond to those of the limiting compositions for miscibility. A plot of T_g against composition will then ideally look like that shown in fig. 12.3 by curves 2 and 3. Note, however, that results for immiscible compositions will not generally be repeatable on the same sample unless it is heated to the original temperature and requenched. This is because, once a sample has been heated above the higher T_g in the DSC measurements, a new equilibrium phase structure will start to form, because the details of the phase separation depend on the temperature.

The experiment considered above is idealised. It was assumed that equilibrium could be obtained at the original temperature, leading to a well-defined phase structure with large regions of each separate phase. The glass transition involves perhaps 50–100 backbone carbon atoms, so that the size of the volume sampled by determining T_g is about 2–3 nm across. If, in some particular process, samples in which phase separation does not take place on a scale appreciably larger than this are made the values of the two T_gs obtained need not be those corresponding to the limiting compositions for miscibility and need not be independent of composition. The general conclusion remains, however, that for a miscible blend there is only one T_g, whereas for an immiscible blend there are two.

Example 12.2

A polymer blend that obeys equation (12.7) consists of equal fractions by weight of two polymers with glass-transition temperatures 50 and 190°C. Calculate the glass-transition temperature of the blend.

Solution

For the given conditions, equation (12.7) reduces to
$\ln(T_g/T_{g1}) + \ln(T_g/T_{g2}) = 0$, which implies that $\ln[T_g^2/(T_{g1}T_{g2})] = 0$, or $T_g = \sqrt{T_{g1}T_{g2}} = \sqrt{(50 + 273)(190 + 273)} = 387$ K, or 114 °C. Notice that, in this particular case, T_g is given by the geometric mean, not the arithmetic mean, of the transition temperatures of the two components.

Infrared spectroscopy is frequently used to study blends. As explained in section 2.6.4, the infrared spectrum of a mixture of substances can be used to determine the concentrations of the various components, provided that there is no change in the spectrum of either component on mixing. This will be the case for a polymer blend when the components are immiscible. Miscibility usually involves some strong interaction between the two polymers, so the spectrum of the blend is then not simply a weighted sum of the spectra of the individual polymers. One of the most important types of interaction in polymer blends is hydrogen-bonding, particularly when one of the polymers contains carbonyl groups, $>C=O$.

In polymers in which H-bonding does not take place, the carbonyl stretching mode usually occurs at about 1730 cm^{-1}, whereas it occurs at lower frequencies when H-bonding takes place. In a similar way, if —NH or —OH groups are involved in H-bonding, the corresponding bond-stretching frequencies are lowered. Figure 12.4 shows an example of the kind of effect that may be observed.

Some miscible polymer blends show evidence of interactions between the two polymers by more subtle changes in the spectra. Such changes are also often seen in the spectra of H-bonded blends and they are due to changes in the conformational states of the polymers due to the interactions between them.

Other methods that may be used to study miscibility in solid polymers include measurements of the velocity of sound. If the polymers are miscible

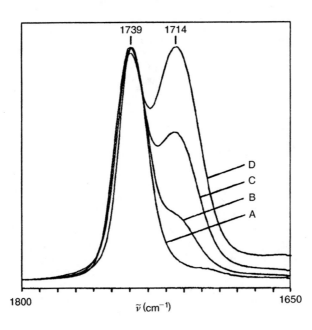

Fig. 12.4 The effect of hydrogen-bonding on the IR spectrum of cast blends of poly(vinyl phenol) $+(CH_2—CH(\bigcirc OH)+)_n$, where $\bigcirc$ represents the phenylene ring, and poly(vinyl acetate) $+(CH_2—CH(O.C=O.CH_3)+)_n$. The concentrations of poly(vinyl acetate) are A, 100; B, 80; C, 50; and D, 20 wt%. H-bonding takes place between the $>C=O$ and H—O—$\bigcirc$— groups. The figure shows the unbonded carbonyl peak at 1739 cm^{-1} and the H-bonded carbonyl peak at 1714 cm^{-1}. The effects in the O—H stretching region are more complicated because of association of the OH groups in pure poly(vinyl phenol). (Adapted by permission of Marcel Dekker Inc.)

the velocity is expected to be linearly related to the composition, whereas for immiscible blends there will be strong scattering at discontinuities between phases and consequent attenuation and deviation from linearity.

12.2.4   Compatibilisation and examples of polymer blends

If an attempt is made to blend two polymers that are immiscible over most of the composition range, a heterogeneous structure results, in which there are regions of one almost pure polymer and regions of the other almost pure polymer, generally giving a coarse-grained structure with little adhesion between the component regions, so that the blend is not useful. In order to make it useful the blend must be *compatibilised*. Compatibilisation may sometimes be achieved by suitable mechanical mixing of the polymers followed by a lowering of the temperature, so that the equilibrium structure cannot be attained. When this method cannot be used, compatibilisation is effected by incorporating macromolecules called *compatibilisers* into the blend.

Figure 12.5 shows some of the possible types of compatibiliser, which are all forms of *block* or *graft copolymers*. The basic principle is that the

Fig. 12.5 Possible types of compatibiliser: (a) diblock, (b) triblock, (c) multigraft and (d) single-graft copolymers at the interface of a heterogeneous polymer blend. A and B represent the two components of the blend; ○ and ● represent the monomers of the blocks C and D, respectively, of the compatibilising molecules. (Adapted with permission of Elsevier Science.)

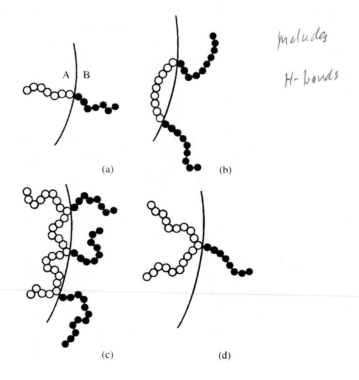

includes

H-bonds

blocks of type C are miscible with the polymer of type A and the blocks of type D are miscible with the polymer of type B, so that the adhesion between A and B regions is immensely improved. In the simplest case the C block is the same as the polymer A and the D block is the same as the polymer B, but this need not be so. The basic reason why immiscible blends tend to have coarse structures is that there is a *surface energy* associated with the interface between the different phases, where the immiscible polymers are constrained to coexist. Increasing the domain sizes reduces the overall surface area and hence the total surface energy. Compatibilisers effectively reduce the surface energy per unit area.

There are several methods whereby the compatibiliser molecules can be introduced into the blend. They are principally

(a) by the addition of pre-made molecules during the blending,

(b) by the addition of a pre-made linear polymer that dissolves in one component of the blend and reacts with groups on the second component to form a block or grafted polymer,

(c) by the addition of compounds of low molar mass during blending that react with both components of the blend to form a block, branched or grafted copolymer,

(d) by the introduction of catalysts to promote interchange reactions when the two components of the blend are polycondensates (see fig. 12.6) and

(e) by applying high shearing forces to the polymers during blending so that chain scission occurs, when the new reactive chain ends can cross-combine at the interface between the two component polymers to give block or graft copolymers.

Several other routes to compatibilisation include the following.

(f) The addition of ionomers. Ionomers are polymers that have a small number of pendant negatively charged ionic groups that are neutra-

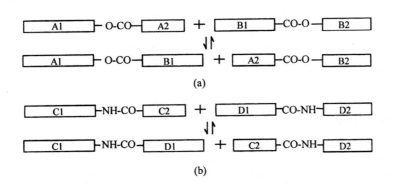

Fig. 12.6 Examples of interchange reactions: (a) transesterification; A1 and A2 represent different lengths of a polyester chain of type A and B1 and B2 represent different lengths of chain of a second type of polyester; and (b) transamidation; as for (a) except that C and D represent polyamides. The double-arrow symbol indicates a reversible reaction.

lised by the presence of positive counter-ions such as Na^+ or K^+. The ionic groups tend to form clusters and, if ionic groups are attached to a small number of the chains of both types of polymer in a blend, mixed clusters can be formed and act as physical cross-links.

(g) The chemical modification of one or both of the components of the blend by introducing groups into the chain that are attracted to groups present or introduced into the other type. By this means the enthalpy of mixing is reduced, which may result in an increase of the compatibility of the components or even in their miscibility.

(h) The addition of a third polymer that is at least partially miscible with both components of the blend. Depending on the concentrations, either a two-phase morphology or a single-phase ternary blend may result.

(i) The introduction of a cross-linking agent for one of the components of the blend, usually an elastomer forming the major component. Sufficiently high cross-linking prevents the elastomer forming the continuous phase that it otherwise would and it forms rather large irregular particles that are held together by a continuous phase of the minor component. The process is called *dynamic vulcanisation*.

Some of the methods (a)–(i) are used for the production of commercial blends, but others have so far been used mainly on an experimental scale. An example of the latter is (a), because the production of small quantities of the specific graft or block copolymer is very expensive.

One of the first polymer blends to be patented, in 1846, was the blend of natural rubber with gutta percha. Varying the composition allowed materials with properties between those of the soft and sticky natural rubber and the hard and brittle gutta percha to be produced. One of the first patents for a thermoplastic polymer blend was taken out in 1929, for poly(vinyl chloride) (PVC) and acrylonitrile rubber (NBR), where the NBR acted as a polymeric plasticiser for the PVC. There is now an enormous range of polymer blends and alloys and new grades are being introduced at the rate of hundreds a year. Utracki has divided the commercial types into a number of groups on the basis of the principal component whose properties are modified by blending, as shown in table 12.1.

12.2.5   Morphology

The morphology of a solid polymer blend depends upon many factors, which include both the precise composition of the blend and the processing conditions used in producing the final solidified material. In miscible blends that have a UCST, quenching from a high temperature at which

Table 12.1. *The principal polymer types on which commercial polymer blends are based*

Polystyrene and styrene copolymers	Polyamide
Poly(vinyl chloride)	Polyester
Poly(vinylidene halide)s	Polycarbonate
Poly(methyl acrylate *or* methacrylate)	Polyoxymethylene
Polyethylene	Poly(phenylene ether)
Polypropylene	Various for speciality engineering resins

the components are miscible will lead to phase separation, whereas in blends that have a LCST separated phases will begin to dissolve. Another important factor is whether one or both of the component polymers is semicrystalline. If at least one of the polymers is semicrystalline, phase segregation is likely to take place below the temperature at which this polymer crystallises. The morphology that results when phase separation takes place, as it does for most blends, depends on the modification of the equilibrium morphology by the processing conditions.

In general, when one of the components of a two-phase blend is present in a small overall concentration it tends to be present as approximately spherical droplets dispersed throughout the major phase, or matrix. The size of these regions is affected by the nature of the polymers, including the molar mass and molar-mass distribution, the processing conditions and the presence or absence of compatibilisers. As the concentration increases, more complex structures will be present. At a concentration called the *phase-inversion concentration* there is no longer a distinction between the matrix and dispersed phases and in equilibrium both phases would become continuous in order to lower the surface energy, giving a so-called *co-continuous morphology*. Processing conditions can markedly change the morphology of blends in a variety of ways and the morphologies obtained are always kinetically determined rather than being equilibrium structures.

All processing involves high hydrostatic or shear stresses, usually both. For miscible polymers these can lead to a raising of the LCST, so that a single-phase blend may form that decomposes spontaneously by spinodal decomposition when the stresses are removed. The final morphology, i.e. the fineness of the dispersion of the phases, can then be controlled by quenching at an appropriate stage. High shear rates can cause a decrease in the size of droplets of a minor phase and fibrils can be produced under appropriate conditions of steady-state shearing or elongation, when the dispersed droplets deform in the same way as does the matrix. Under certain conditions of flow, at temperatures near or above the melting point of semicrystalline dispersed phases, long fibrils of these phases can be formed within the matrix.

12.2.6   Properties and applications

Among the most important improvements in properties that have been claimed in patents for polymer blends are increases in impact strength, tensile strength, rigidity modulus and heat deflection temperature. Improvements in processability are also high on the list of reasons for blending, but are not considered further here.

An important category of blends is those designed to increase the impact strength of polystyrene (PS). High-impact polystyrene (HIPS) is produced in a number of different ways with a number of different compositions. One method leads to the production of HIPS containing 94 wt% of PS and 6 wt% of a styrene–butadiene graft copolymer (SBR) which itself contains 30 wt% styrene. The blend can be produced by dissolving 4 wt% polybutadiene in styrene and then polymerising the solution.

HIPS polymers consist of a matrix of polystyrene containing approximately spherical particles of rubbery polybutadiene ranging in size from about 0.1 to 10 μm depending on the precise composition and method of polymerisation. The rubber particles may themselves contain small regions of PS as shown in fig. 12.7. Another series of rubber-toughened blends is the acrylonitrile–butadiene–styrene (ABS) terpolymers, which are basically

Fig. 12.7 The morphology of high-impact polystyrene (HIPS): a transmission electron micrograph, showing dispersed elastomer particles (~ 22 μm in diameter) with polystyrene occlusions. (Courtesy of Dow Chemical Company.)

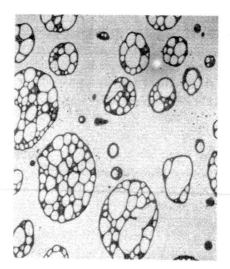

styrene-acrylonitrile (SAN) block copolymers toughened with either poly-butadiene or a polystyrene–polybutadiene copolymer. Again, there is a wide possible range of compositions and methods of production. The rubber-particle sizes cover a similar range to that for HIPS but the rubber content is usually rather higher. These particles may again contain occlusions of the matrix-type polymer.

The mechanism for the increase in impact strength, often called the *rubber toughening*, of these blends is the large increase in the number of sites for yielding or crazing caused by the rubber particles, together with the fact that the presence of the graft copolymer crossing the interface prevents any cracks there. Crazing or yielding takes place at the particles because they cause stress concentrations. The elastic modulus of the particles is so much lower than that of the matrix that the particles behave almost like cavities in the material and the stress concentration factor at the equator of a spherical cavity is approximately 2. The result is a somewhat lower break-ing stress but a greatly increased extension before breaking occurs, so that the energy required to cause breakage, which is represented by the area under the stress–extension curve, is greatly increased.

An example of the use of quite a different type of morphology is the production of bi-constituent fibre systems from blends based on polya-mides such as nylon-6 or nylon-6,6, using the idea mentioned in the pre-vious section that long fibrils of a minor constituent can be produced within a matrix of a major constituent under suitable conditions. Systems that have been found to produce fibres useful in car-tyre manu-facture include the combination of nylon-6 as the matrix with poly(ethy-lene terephthalate) as the minor component (PA–PET blend) and the combination of nylon-6,6 with poly(hexamethylene isophthalamide) (PA–PAI blend). In both cases the incorporation of the minor component raises the modulus and reduces the tendency to 'flat-spotting' of the tyres on standing. These two blends are incompatible, but in both cases there is the possibility of inducing interchange reactions of the type described in section 12.2.4, leading to improvements in properties.

A further very important type of morphology is the co-continuous mor-phology, which can lead to materials combining the useful properties of the components, for example, to combinations of high modulus and high impact strength. An important example is the blend of bisphenol-A poly-carbonate (PC) with ABS, which is itself already a blend, as described above. The blends can be either dispersed or co-continuous, the co-con-tinuity being observed for PC fractions in the region from about 35% to 65% by volume. The co-continuous blends have very good mechanical properties up to the glass-transition temperature of the PC, at about 150 °C, and good impact resistance is conferred by the ABS. The morphology actually consists of three phases, PC, SAN and the elastomeric component. Dipole–dipole interactions between the PC and the SAN act as the compatibiliser.

As noted earlier, there are now many hundreds, indeed thousands, of polymer blends in use or that have been described in the literature. The present section has merely given a few examples. It should also be remembered that even polymers often thought of as being single-component polymers are often blends, indeed all polymers are blends of molecules with different molar masses. More importantly, most polyethylenes (PEs) are blends. Those that are simply blends of different grades of PE to meet particular requirements for density or other properties, or are blends of PE with up to 15 wt% of other polymers to modify properties, are generally described simply as polyethylenes. Only those blends of PE with larger concentrations of a second polymer are usually described as blends.

12.3 Copolymers

12.3.1 Introduction and nomenclature

A *copolymer* is defined in section 1.3.1 as a polymer built from more than one type of chemical unit, and in section 4.1.1 a distinction between *random copolymers* of the type

$$-B-A-B-B-A-B-A-A-B-A-A-A-B-B-$$

and *block copolymers* of the type $-A_nB_m-$ is made, where A_n and B_m represent long sequences, or blocks, of A and B units, respectively. A molecule of such a block copolymer may consist at the simplest of just two blocks, one each of A-type and B-type units, in an *AB diblock polymer*, of three blocks, in an *ABA triblock polymer*, or of large numbers of alternating blocks of A and B. If the blocks of this last type consist of single A or B units the polymer is called an *alternating copolymer*. More complicated copolymers can have more than two different types of units or different types of blocks.

All these polymers have so far been assumed to be *linear*, but this is not the only possibility. Further important types of copolymers are the *graft copolymers* in which blocks of one type of polymer are attached as side-chains to another polymer. One use of the latter type of structure in compatibilising blends has already been indicated. More complicated structures are, for instance, the *cross-linked graft structures* in which chains of A units are cross-linked by chains of B units. Figure 12.8 shows some examples of possible types of copolymer.

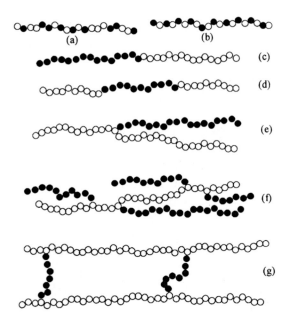

Fig. 12.8 Schematic representations of various types of copolymers: (a) random, (b) alternating, (c) diblock, (d) triblock, (e) single-graft, (f) multigraft and (g) cross-linked graft.

It is useful to introduce a simple nomenclature in order to indicate fairly concisely the type of copolymer involved in any material, as follows:

random copolymer of units A and B	poly(A-co-B)
alternating copolymer of units A and B	poly(A-alt-B)
graft copolymer of B grafted onto a backbone of A	poly(A-g-B)
block copolymer of A and B	poly(A-b-B)
diblock copolymer of A and B	AB diblock
triblock copolymer of A and B and A	ABA triblock
network of A cross-linked by grafted B	poly(A-cl-B)

There is a further category of polymers that are not copolymers in the sense that units of different types are chemically joined together, as in all the types considered so far, but consist of two *interpenetrating networks* (IPNs), one formed by each type of polymer. Thus the different types of polymer are constrained to coexist because the different types of chain would have to pass through each other for the polymers to segregate.

The tendency considered in section 12.2 for polymers of different types to segregate applies for the true copolymers, i.e. excluding IPNs, just as it does to blends. If one component is a very minor component or if the polymer is a random copolymer, segregation does not generally take place, but in block copolymers in which the minor components form more than a certain proportion of the whole, segregation generally takes place in some range of temperatures. The important difference here is that the components are chemically bonded together, so even a very strong tendency to segregate does not lead to any weakening of the structure. This allows copolymers consisting of highly incompatible polymers to produce interesting and useful structures. The different units within copolymers are compelled to mix at a fairly fine level; roughly speaking, in an AB copolymer units of type A cannot be further from units of type B than a few RMS lengths of the shorter block. For a blend of A and B there is no such restriction, because the A and B units are not chemically linked.

It is not possible within the scope of the present chapter to consider all forms of copolymer and further attention is therefore largely restricted to linear block copolymers.

12.3.2 Linear copolymers: segregation and melt morphology

The methods of testing for segregation in copolymers are similar to those for blends. Block copolymers in which there is sharp segregation of the different components have two glass-transition temperatures that are close to the values for the component homopolymers. Many random copolymers, however, have only a single glass-transition temperature T_g that varies smoothly with composition between the T_gs of the homopolymers. This shows that there is only one phase present, i.e. there is no segregation of the different types of unit into different regions of the material.

For these polymers the following equation for T_g (absolute) can be deduced from the free-volume theory of the glass transition (see section 7.5.3):

$$\frac{1}{T_g} = \frac{1}{w_1 + aw_2}\left(\frac{w_1}{T_1} + \frac{aw_2}{T_2}\right) \tag{12.8}$$

where w_1 and w_2 are the weight fractions of the two monomers whose homopolymers have glass-transition temperatures T_1 and T_2 (absolute), respectively, and a is, for a given pair of polymers, a constant that is usually not very far from unity.

In deriving equation (12.8) it is assumed that the fractional free volumes of the homopolymers and the copolymers at their respective glass-transition temperatures are the same and that both the specific volume and the occupied volume of the copolymer are obtained by multiplying the corresponding values for the homopolymers by their weight fractions and adding the results. Figure 12.9 shows an example of data for a type of copolymer to which this equation applies, with $a = 1.75$. The data for many pairs of vinyl polymers fit well to equation (12.8) with a equal to unity.

Example 12.3

A linear random copolymer of two types of monomer A and B contains 25% by weight of A units. Calculate its glass-transition temperature T_g given that the copolymer system obeys equation (12.8) with $a = 1.5$ (with monomer A corresponding to monomer 1) and that the glass-transition temperatures of homopolymers of A and B are $T_A = 90\,°C$ and $T_B = 0\,°C$ respectively.

Solution

The weight fraction of A is $w_A = 0.25$ and the weight fraction of B is $w_B = 0.75$. The absolute T_gs for the homopolymers are $T_A = 363$ K and $T_B = 273$ K. Thus

$$\frac{1}{T_g} = \left(\frac{1}{0.25 + (1.5 \times 0.75)}\right)\left(\frac{0.25}{363} + \frac{1.5 \times 0.75}{273}\right) = 3.50 \times 10^{-3}$$

so that $T_g = 286$ K, or $13\,°C$.

In discussing the morphology of block copolymers it is necessary to consider first what happens in equilibrium in the melt. As discussed for blends in section 12.2.2, the necessary condition for mixing of two substances A and B is that the Gibbs free energy of the mixture must be less than the sum of the Gibbs free energies of the separate constituents, i.e. that ΔG_{mix}, the increase in free energy on mixing, should be <0, leading to expression (12.1), i.e.

$$\Delta G_{mix} = \Delta H_{mix} - T\,\Delta S_{mix} < 0$$

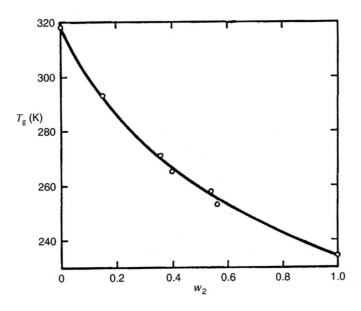

Fig. 12.9 Glass-transition temperatures T_g of copolymers of polychlorotrifluoroethylene with poly(vinylidene fluoride) plotted against the weight fraction w_2 of the latter. The curve is the theoretical curve equation (12.8) with $a = 1.75$. (Reproduced by permission from John Wiley & Sons Limited.)

where ΔH_{mix} is the enthalpy of mixing and ΔS_{mix} is the entropy of mixing. Similar considerations apply to the units in an AB copolymer and, in a similar way, ΔH_{mix} per mole can be written in the form of equation (12.2), i.e.

$$\Delta H_{mix} = RT\chi_{AB}\upsilon_A\upsilon_B$$

It is shown in section 12.2.2, that for a blend of two polymers in which the different types of polymer sequences are not chemically attached to each other, the entropy of mixing, ΔS_{mix}, is inversely proportional to the number of monomer units in the chains of the polymers. The same is true of block copolymers, so that $\Delta S_{mix} = C/N$, where N is the degree of polymerisation and C is a quantity that is constant for a particular overall composition of the copolymer, i.e. the ratio of A to B units. ΔS_{mix} is always positive, but becomes very small when N becomes very large, which means that long AB block copolymers will generally segregate because ΔH_{mix} is generally positive.

The tendency to segregation depends on the 'competition' between the enthalpic and entropic components of ΔG_{mix} and hence on the ratio $\Delta H_{mix}/(T\Delta S_{mix})$. It therefore depends on the value of the quantity $(R\upsilon_A\upsilon_B/C)\chi_{AB}N$. If this quantity is less than unity, there will be no segregation in equilibrium; if it is greater than unity there will be segregation. If segregation does take place, the value of the ratio will also govern the type of segregation. The quantity $R\upsilon_A\upsilon_B/C$ depends only on the volume fractions of the two components and therefore only on υ_A or υ_B (because $\upsilon_A + \upsilon_B = 1$), whereas $\chi_{AB}N$ is independent of the volume fractions but depends on the temperature, because χ_{AB} should really be replaced by the form of χ'_{AB} given in equation (12.6) to allow for an entropic contribution to ΔH_{mix}.

It is not difficult to guess the types of segregated structures to be expected for strongly segregating copolymers in two simple cases. When one of the component polymers is present at a concentration only slightly greater than that required to cause segregation, any small cluster formed by the minority-type units must of necessity be completely surrounded by units of the other type and the interactions between the two types of units will then be minimised if the cluster becomes spherical. On the other hand, if the two components are present in equal concentration the simplest way that they can be present in equivalent ways is for an alternating-layer, or lamellar, structure to be formed. These expectations are borne out by theory and experiment, both of which show that other types of structure can be present in composition ranges intermediate between those just considered.

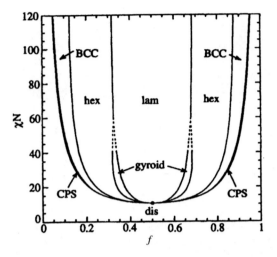

Fig. **12.10** The phase diagram for a conformationally symmetric AB diblock copolymer, calculated according to a theoretical model, where $f = v_A$ or v_B. Regions of stability of disordered, lamellar, gyroid, hexagonal, BCC and close-packed spherical (CPS) phases (very narrow regions) are indicated. (See fig. 12.11 for an illustration of the gyroid phase.) (Adapted with permission from the American Chemical Society.)

Figure 12.10 shows the phase diagram calculated according to a particular theoretical model for a diblock copolymer in which the statistical segment lengths (i.e. the lengths of the equivalent random links) of the A and B blocks are equal and fig. 12.11 shows the structure of one of the more complicated phases. The types of regions shown, as well as others, have been observed in real diblock copolymers and electron micrographs of some of them are shown in fig. 12.12. The precise phase diagram for a diblock copolymer depends on the chemical structure of the A and B blocks and, in particular, it will not be symmetric about $f = 0.5$ if the statistical segment lengths of the A and B blocks are unequal. Whatever type of segregated structure is present, the regions containing different types of chemical unit A and B are always separated by a transition region in which there is a continuous change of composition. For low values of

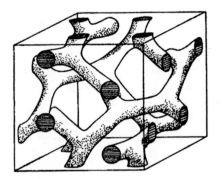

Fig. **12.11** The gyroid phase, in which both components of a diblock copolymer form continuous regions. (Reprinted with permission of Elsevier Science.)

Fig. 12.12 Some examples of diblock copolymer morphologies. Transmission electron micrographs of stained samples of (a) the lamellar phase of a polystyrene-polyisoprene (PS–PI) diblock with PS fraction 0.64; (b) the hexagonal phase of a PS-poly(2-vinylpyridene) diblock with $f_{PS} = 0.35$; and (c) and (d) the gyroid phase of a PS–PI diblock with $f_{PS} = 0.39$, showing projections with approximate threefold and fourfold symmetries, respectively. All samples were rapidly cooled from the melt and thus exhibit the melt-state morphology. (Reprinted with permission from the American Chemical Society.)

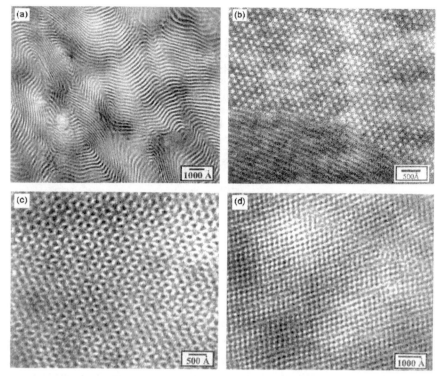

χN the segregation is weak and the transition region is wide, whereas for high values the segregation is strong and the transition region is narrow.

Figures 12.12(a) and (b) show clearly that there are different domains present in the melt in which the axes of the segregated structures are aligned in different directions. In fig. 12.12(b) this is particularly clear, with the rods of the hexagonal phase perpendicular to the plane of the figure in the top part and parallel to it in the lower left-hand part. It is possible by suitably shearing the melt to align the different domains in such a way as to obtain macroscopic regions with approximately common alignment throughout.

Turning now to the solid-state structure of linear block copolymers, it is in principle necessary to distinguish among polymers in which (a) both of the components A and B form glasses in the solid state, (b) one of the components is an elastomer and the other forms a glass or other rigid non-crystalline structure, (c) both components are elastomers and (d) one or both components are crystalline. Bearing in mind that one of the major reasons for making copolymers is to combine the properties of the two polymers in some way, it is not unreasonable to suppose that some of the most interesting polymers from a commercial point of view are to be found in category (b). Polymers in category (d) are interesting from the point of view of the additional complications of morphology that crystallinity brings about. Some of the properties of these two types of copolymer are now considered.

12.3.3   Copolymers combining elastomeric and rigid components

There are two important types of polymer in which an elastomeric and a glassy or non-crystalline rigid component are copolymerised. The simpler thermoplastic elastomers are usually triblock ABA copolymers with B as the elastomeric component and A as the glassy component, whereas the segmented polyurethane elastomers are multiblock copolymers in which alternate blocks are 'hard', i.e. relatively inflexible, and 'soft', i.e. relatively flexible.

Important examples of the ABA type are the styrene–butadiene–styrene (SBS) and styrene–isoprene–styrene (SIS) triblocks in which the outer blocks of the glassy styrene are much shorter than the inner elastomeric blocks. The usual phase arrangement of these materials is shown schematically in fig. 12.13. SBS copolymers have two glass-transition temperatures, as expected for a segregated structure, at -90 and $+90\,^{\circ}$C. The morphology depends on the precise composition and on the method of preparation; samples of SBS containing either cylinders in a hexagonal array or spheres in a body-centred cubic array have been obtained.

The materials are tough and elastic below the glass-transition temperature of the outer blocks, the glassy domains acting as both cross-links and fillers for the rubbery matrix, which would otherwise behave like an unvulcanised rubber. SBS compositions usually have tensile strengths in the

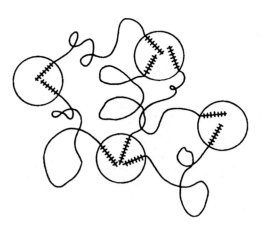

Fig. 12.13 A schematic diagram of the phase structure of an SBS-type copolymer: ——————, elastomeric component; and +++, glassy component. (Reprinted by permission of Kluwer Academic Publishers.)

region 300–350 MPa and they have been used as hot-melt adhesives and for shoe soles. Their chief use, however, is as components in blends.

There are two basic types of polyurethanes, the *polyether urethanes* and the *polyester urethanes*, depending on whether the soft segment is a polyether, such as $\text{-}(O(CH_2)_4\text{-})_n$, or a polyester, such as $\text{-}(O(CO)(CH_2)_4\text{-}(CO)O(CH_2)_4\text{-})_m$. Both of these structures are completely single-bonded along the backbone, which gives them their flexibility. In contrast, the urethane hard segments gain their stiffness from having aromatic or other rings along their backbones, although they are not completely rigid. A typical example is $\text{-}((CO)(NH)\textcircled{\odot}(CH_2)\textcircled{\odot}(NH)(CO)(CH_2)_4O\text{-})$, where $\textcircled{\odot}$ signifies a *para*-disubstituted benzene (phenylene) ring.

The NH groups of the urethane linkage can form hydrogen bonds with the oxygen atoms of the ether linkages or the carbonyl groups of the esters and also with the carbonyl groups of the urethane. When segregation takes place, as it usually does, the H-bonding between the urethane hard blocks often leads to a paracrystalline structure in which the chains are approximately parallel but not in crystal register, giving a structure illustrated schematically in fig. 12.14. This has the effect of stiffening the corresponding regions, which act as cross-links for the soft segments in a similar way to the polystyrene regions in SBS polymers. The domain size, which can be determined from small-angle X-ray scattering or electron microscopy, is much smaller than that of the SBS polymers, however, being of the order of 3-10 nm. The two-phase structure is also readily appreciated from modulus–temperature curves such as those of fig. 12.15, which clearly show two glass-transition regions.

12.3.4   Semicrystalline block copolymers

The morphology of the solids formed from block copolymers in which one component can crystallise often depends strongly on whether they are

Fig. 12.14 A schematic diagram of the domain structure of a segmented polyurethane. (Reprinted by permission of Kluwer Academic Publishers.)

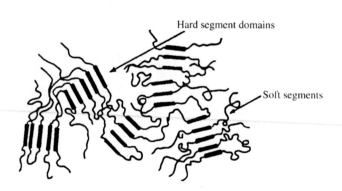

Hard segment domains

Soft segments

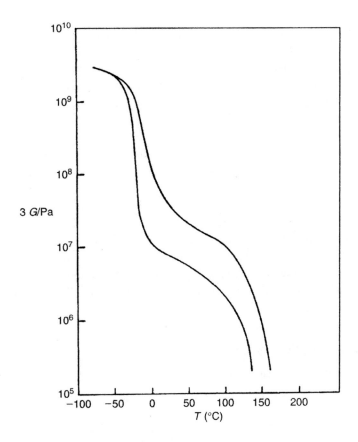

Fig. 12.15 Modulus–temperature curves for two polyurethanes with different compositions. Note the two glass-transition temperatures for each polymer, corresponding to those of the hard and soft segments. (Adapted by permission of John Wiley & Sons, Inc.)

crystallised from solution or from the melt and on whether they are crystallised from a segregated or unsegregated state. It can also depend on whether a segregated melt is strongly or weakly segregated and on the speed of cooling.

Crystallisation has been studied in a number of block copolymers that contain polyethylene (PE) as the crystallisable component, for instance high-molar-mass copolymers of PE with poly(ethylene propylene) (PEP), $-(CH_2CH(CH_3)CH_2CH_2-)_n$, which is rubbery. Whatever the composition within a wide range, irrespective of whether the polymers were diblock or triblock and, for the latter, whichever polymer formed the central block, the crystallinity was found to vary linearly with the PE content, showing that segregation had taken place. WAXS showed that the crystal structure was the normal orthorhombic form and light scattering and optical micrographs showed that spherulites were formed. Samples of PE–PEP–PE containing 27% PE have been prepared by casting from solution above the melting point of polyethylene to produce a hexagonal-packed microphase separation, whereas for all compositions between 12% and 25% PE, PE–PEP diblocks were found to crystallise from the disordered melt in a lamellar structure with a spherulitic superstructure.

An interesting question arises for those copolymers that crystallise with lamellar structures. Do the polyethylene chains fold so that the chain axes are perpendicular to the lamellar surface, as they do in the homopolymer, or parallel to it? The answer need not be the same in all cases, even for a single type of copolymer, but there is certainly evidence that the chains fold so that their axes are perpendicular to the lamellar planes in some cases. Such evidence has, for instance, been obtained for PE–PEP–PE triblocks by studying the development of WAXS and SAXS patterns as samples were oriented by drawing.

These results show that the crystalline block copolymers constitute an interesting class of materials, but their study is a relatively new area of polymer science and much work remains to be done before the various morphologies can be reproducibly obtained and fully understood.

12.4 Liquid-crystal polymers

12.4.1 Introduction

As indicated briefly in section 1.3.2, rod-like polymers, or polymers containing rod-like segments, can exhibit liquid-crystalline order, i.e. their chains line up approximately parallel to each other locally. Aligning these locally aligned regions, or *domains*, so that all chains in the sample are highly oriented with respect to a given direction produces materials of exceptional strength, rigidity and toughness. Such materials have many applications as fibres, films or moulded parts. These are the *main-chain liquid-crystal polymers (LCPs)*. Another class of polymers that can exhibit liquid-crystalline order is the comb-like polymers that contain liquid-crystal-forming side groups. These *side-chain LCPs* are often of importance for optical displays or for non-linear optical devices.

The term *liquid crystal* is in some ways unfortunate, since materials in this state are not crystalline and they may but need not be liquid. A preferable term that is often used is *mesophase*, which means 'middle phase' and attempts to indicate that the order in this state is between that of the liquid state and that of the crystalline state. *Mesogens* are then molecules that tend to form mesophases and the attribute of being liquid-crystalline-like is called being *mesomorphic*.

Before describing LCPs it is useful to describe briefly the various forms of liquid crystallinity exhibited by small molecules.

12.4.2 Types of mesophases for small molecules

There are two principal categories of mesophases, *thermotropic* and *lyotropic*. Thermotropic liquid crystals are formed within a particular range of temperature in a molten material, with no solvent present, whereas lyotropic liquid crystals are formed by some substances when they are dissolved in a solvent. Within each of these categories there are three distinct classes of mesophases, which were first identified by Friedel in 1922. The simplest of these to describe are the *nematic* and *smectic* classes, illustrated schematically in fig. 12.16. These phases are formed by long thin rigid molecules which tend to line up parallel to each other.

The word *nematic* means thread-like and in the nematic phase there is simply a tendency for the rods to become parallel to each other; unlike the molecules in a true crystalline phase, they are not exactly parallel and there is no long-range correlation between the positions of adjacent molecules. The average direction of the molecules is called the *director* and is indicated in fig. 12.16 by the arrow labelled n.

The word *smectic* describes the soapy feel of the smectic phase. The difference between the ordering in this phase and that in the nematic phase is that, in addition to the rods being nearly parallel, there is a very strong tendency for the ends of the molecules to lie in parallel planes, so that the structure becomes *layered*, as indicated in fig. 12.16. The layered structure is responsible for the soapy feel. As in the nematic phase, the rods are not exactly parallel and there is no long-range correlation between the lateral positions of the rods. There are several varieties of smectic structure. If the director is normal to the planes of the layers, as shown in fig. 12.16, the variety is *smectic A*; if the director is inclined at the same angle for all layers, it is *smectic C*.

The third class of liquid crystals is the *cholesteric* class, which is of rather less importance for synthetic polymers than the other two classes. In this

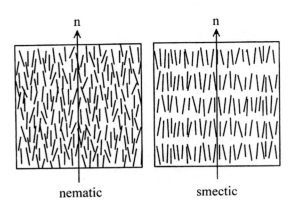

nematic   smectic

Fig. 12.16 Schematic illustrations of nematic and smectic liquid-crystalline order. The short lines represent long thin rigid molecules.

class the molecules are arranged locally like those in the nematic phase, but the director of the nematic is effectively rotated periodically around an axis normal to itself, so that its tip describes a helix. This new axis is the director for the cholesteric state. The cholesteric mesophase arises for *chiral* mesogens, i.e. for mesogens that do not have mirror symmetry, and was first observed for esters of cholesterol.

What has been described so far is the very local arrangement of the molecules in mesophases. It must not be imagined, however, that the director has the same orientation throughout a macroscopic sample. There are two ways in which the director can change orientation; it can reorient smoothly and continuously, i.e. by *distortion*, or it can reorient discontinuously at a *disclination* (or *singularity*). There are several types of both distortion and disclination; one type of each is illustrated in fig. 12.17. Boundaries between domains with different director orientations in nematic liquid crystals may be formed by *walls* in which the director rotates through the thickness of the wall in a similar way to the rotation of the director in a cholesteric liquid crystal, but the total rotation is limited to a finite angle. Such domain walls, which may be only tens of micrometres thick, are similar to the Bloch walls that separate domains in ferromagnetic materials, where the local magnetisation direction is the analogue of the director as the oriented feature.

Within any region of a liquid-crystalline material the degree of alignment of the molecules can be specified by the so-called *order parameter S*. This parameter is identical to the quantity $\langle P_2(\cos\theta)\rangle$ discussed in section 10.3. For a given domain or microscopic region within the material the angle θ is the angle between the axis of a typical molecule and the local director. For a macroscopic region of an oriented fibre LCP it could be the angle between a typical chain segment and the fibre axis.

Fig. 12.17 Schematic diagrams showing a distortion and a disclination. (a) A bend distortion. The thick line indicates how the director orientation changes with position. (b) A particular type of disclination. The lines indicate how the director orientation changes with position near the disclination and the dot indicates either a point singularity or a line singularity normal to the page. ((b) Reproduced from *Liquid Crystal Polymers* by A. M. Donald and A. H. Windle, © Cambridge University Press 1992.)

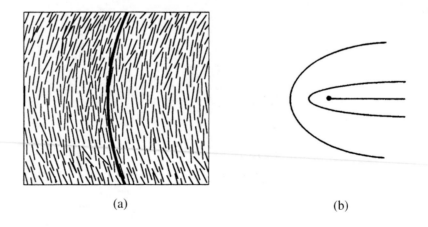

(a)   (b)

Example 12.4

In a particular domain of a liquid crystal the axes of the molecules are equally likely to lie anywhere within the range of angles $\theta = 0$ to θ_o from the director. Derive an expression for the order parameter S for the domain and evaluate S for $\theta_o = 20°$.

Solution

$S = \langle P_2(\cos\theta) \rangle = \frac{1}{2}(3\langle\cos^2\theta\rangle - 1)$. However,

$$\langle\cos^2\theta\rangle = \frac{\int_0^{\pi/2} 2\pi \cos^2\theta\, N(\theta)\sin\theta\, d\theta}{\int_0^{\pi/2} 2\pi N(\theta)\sin\theta\, d\theta}$$

where $N(\theta)\, d\omega$ is the fraction of molecular axes that lie within *any* infinitesimal solid angle $d\omega$ at θ to the director and $2\pi\sin\theta\, d\theta$ is the total solid angle within an infinitesimal range $d\theta$ at θ. $N(\theta)$ is constant for $0 < \theta < \theta_o$ and zero elsewhere, so that

$$\langle\cos^2\theta\rangle = \frac{\int_0^{\theta_o} \cos^2\theta\sin\theta\, d\theta}{\int_0^{\theta_o} \sin\theta\, d\theta} = \frac{1 - \cos^3\theta_o}{3(1 - \cos\theta_o)}$$

and

$$\langle P_2(\cos\theta)\rangle = \frac{1}{2}\left(\frac{1 - \cos^3\theta_o}{1 - \cos\theta_o} - 1\right) = \cos\theta_o(1 + \cos\theta_o)/2$$

which for $\theta_o = 20°$ is equal to 0.91.

12.4.3 Types of liquid-crystal polymers

There are two principal divisions of LCPs, as indicated in sections 1.3.2 and 12.4.1, *main-chain* and *side-chain* LCPs, as shown schematically in fig. 1.4. Figure 12.18 shows the chemical structures of some main-chain LCPs and illustrates the fact that both the rigid sections and the flexible spacers between them may vary considerably in length. Figure 12.18(a) shows a polymer that is very rigid, because the only bonds about which rotation can take place are parallel to, and in fact collinear with, the backbone of the molecule. This does not mean that there is no flexibility at all, because thermal energy allows some bending to occur. Figure 12.18(b) illustrates a polymer in which the chain consists of relatively rigid units containing two phenylene rings separated by short spacers, —$(CH_2)_2$—, with a moderate amount of flexibility, whereas fig. 12.18(c) illustrates a polymer in which quite long relatively rigid units are separated by fairly long flexible —$(CH_2)_{10}$— spacers.

In true thermotropic systems the liquid-crystalline phase exists over a range of temperatures between the melting point, or glass-transition tem-

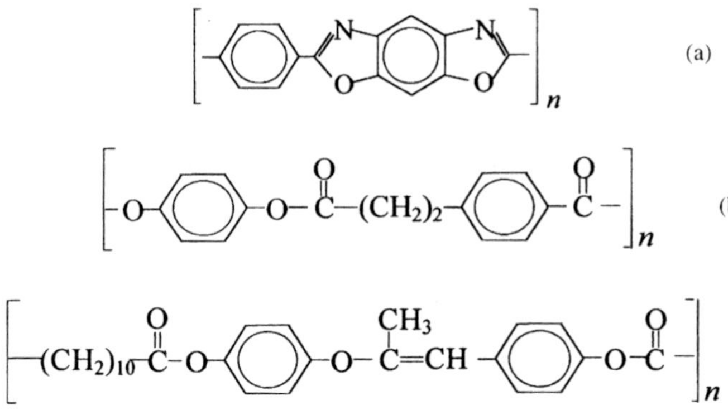

Fig. 12.18 Some examples
of main-chain liquid-
crystal polymers: (a)
poly(p-phenylene
benzobisoxazole) (PBO),
lyotropic; (b) poly(4-
carboxybenzene-propionic
acid), thermotropic; and
(c) a thermotropic polymer
with a long flexible
decamethylene spacer.

perature if the polymer is non-crystalline, and a higher temperature called the *isotropisation temperature* or the *liquid-crystal-to-isotropic transition temperature*, $T_{lc\rightarrow i}$. Above this temperature the liquid phase is isotropic. For some polymers that become liquid-crystalline above the melting or glass-transition temperatures the polymer degrades before $T_{lc\rightarrow i}$ is reached; they are nevertheless classed as thermotropic. When small mesogenic molecules are polymerised the melting and transition temperatures both generally rise and, unless spacers of the general types shown are incorporated, the liquid-crystal phase may be at too high a temperature to be reached before the polymer degrades. The melting temperature may, however, be lowered by the addition of a liquid of low molar mass and polymers such as that of fig. 12.18(a), which contain no spacers, are lyotropic rather than thermotropic LCPs.

Another way in which flexibility can be introduced so that mesogenic groups have sufficient freedom to associate to form a liquid-crystal phase is by attaching them as side groups to a fairly flexible polymer backbone, with a flexible *spacer* between each mesogenic group and the backbone. The backbones are usually poly-acrylates, -methacrylates, -siloxanes or -phosphazenes. The polymers are usually glassy on cooling and the latter two types of backbone usually give lower glass-transition temperatures. Figure 12.19 shows the general chemical structure of these types of polymer.

There is a further very important class of main-chain liquid-crystal polymers that differ from those discussed above by being random co-polyesters based on polymers such as poly(hydroxybenzoic acid) $+(O-\bigcirc\!\!\!\bigcirc-C=O)+_n$, where $-\bigcirc\!\!\!\bigcirc-$ represents a *para*-disubstituted benzene ring (*para*-phenylene unit). The random copolymers include units of various other forms containing phenylene or naphthylene rings and their introduction reduces the crystallinity and melting point considerably without necessarily lowering the value of $T_{lc\rightarrow i}$. Polymers of these types have

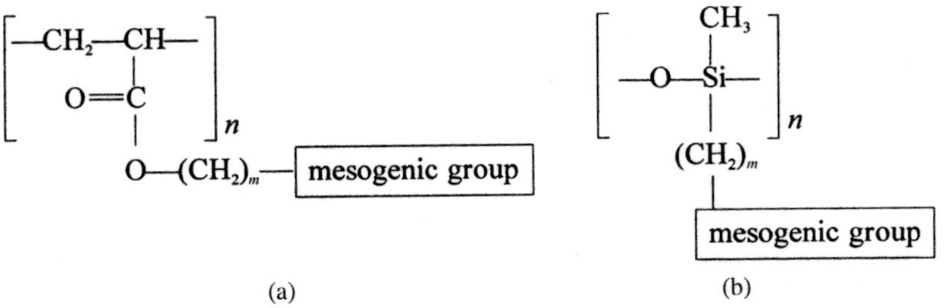

Fig. 12.19 Examples of side-chain LCPs: (a) a polyacrylate and (b) a polysilane.

been commercialised as the *Xydar* and *Vectra* series, the chemical structures of which are shown in fig. 12.20. The hydrobenzoic acid unit is often abbreviated to HBA and the hydroxynaphthoic acid unit to HNA so that the *Vectra* polymers are sometimes called HNA–HBA copolymers.

So far both melting and transition temperatures have been referred to without giving any actual values. Table 12.2 lists these important parameters for all the polymers mentioned in this section.

12.4.4 The theory of liquid-crystal alignment

The most important theoretical models for attempting to account for liquid-crystal alignment are the *lattice model*, introduced by Flory in 1956, and the *mean-field model* introduced about 1960 by Maier and Saupe. Of these, the former has proved to be more readily applicable to polymers. It is based on the rather obvious idea that, as the number of rigid impenetrable rods in a given volume is increased, they must eventually become aligned approximately parallel to each other, at least in local regions, and it attempts to predict the concentration at which this will tend to happen. In this model there is no specific interaction between the rods except for the short-range repulsive force that corresponds to their mutual impenetrability.

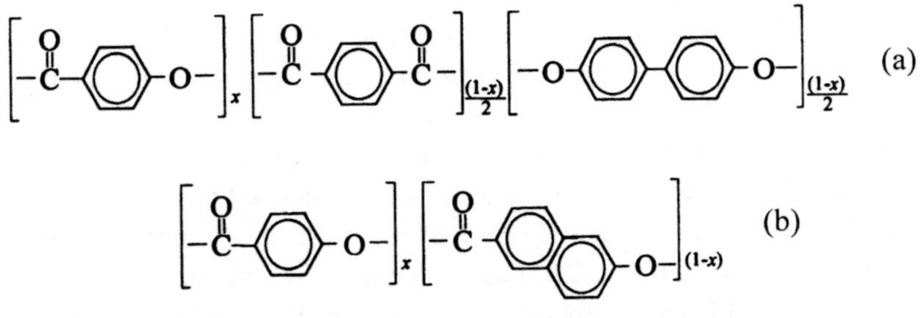

Fig. 12.20 Random co-polyesters: (a) Xydar series and (b) Vectra series. Unlike the usual chemical formulae, the subscripts merely give the fractions of the various types of unit, which are distributed randomly along the polymer chain.

Table 12.2. *Melting and transition temperatures for selected liquid-crystal polymers*

Polymer	Melting temperature T_m	Transition temperature $T_{lc \to i}$
Poly(p-phenylene benzobisoxazole)		Lyotropic
Poly(4-carboxybenzene-propionic acid)	425 °C	> 425 °C
Decamethylene-spaced polymer	199 °C	223 °C
Polyacrylate with $m = 8$ and mesogenic group —O—◎—◎—CN	About 50 °C	About 80 °C
Polysiloxane with $m = 3$ and mesogenic group —O—◎—(C=O)—O—◎—O—CH₃	$T_g \sim 290$ °C	About 330 °C
Xydar for $x = 0.33$	380 °C	
Vectra for $x = 0.6$	252 °C	

The Maier–Saupe theory, on the other hand, does postulate a specific interaction between the rods that tends to align them. This is represented by the assumption that each rod has an energy that depends only on its angle of misalignment from the average direction of alignment of all the rods, i.e. from the director, and on the degree of alignment of all the rods. It is a theory that has certain similarities with the mean-field theory of ferromagnetism. It will not be considered in any further detail because it has not found much application to the theory of LCPs, although the idea of an orientation-dependent energy has been used as an additional factor in refinements of the lattice model.

The lattice model concerns itself with finding the number of ways that a given number of rigid rods can be arranged within a given volume. In order to make the problem mathematically tractable the rods are idealised to have a square cross-section and the length of each rod is assumed to be an integral multiple x of its cross-section, so that x is the *aspect ratio* of the rod. The rod can thus be considered to consist of x units, each of which is a cube.

The volume in which the rods are placed is considered to be made up of cubic cells, or sites, of the same size as these units, arranged on a lattice with one edge parallel to the director of the liquid-crystal arrangement that arises when the rods are packed sufficiently densely. A rod at an angle ψ to the director is then represented in the simplest case by y sections each of x/y (assumed integral) units and each section is assumed to occupy x/y cells of the lattice that are adjacent to each other in the direction parallel to the director, as shown in fig. 12.21. The problem is to calculate how the number of ways that the rods can be placed on the lattice varies with the degree

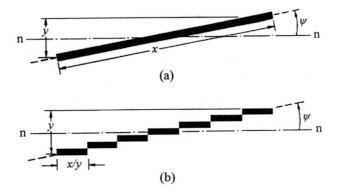

Fig. 12.21 (a) A rod oriented at angle ψ to the director nn of a domain and (b) the representation of the rod as y sequences of cubes each parallel to nn. (Re-drawn with permission from Taylor and Francis Ltd.)

of alignment of the rods, because this then leads to an expression for the free energy as a function of the degree of alignment. Minimisation of the free energy then gives the equilibrium state.

The problem can be divided into two parts by first assuming that the *orientation* of every rod is fixed within the lattice. This leads to a number N_{comb} of possible spatial arrangements of the rods on the lattice. For each of these there is a large number of ways N_{orient} in which the rods can be oriented. The total number of ways of arranging the rods is thus $N = N_{\text{comb}}N_{\text{orient}}$. The model assumes that the system consists of n_{p} rods and n_{s} solvent molecules and each of the solvent molecules is assumed to occupy a single cell or site on the lattice. It can be shown that to a very good approximation, for large numbers of rods and solvent molecules,

$$N_{\text{comb}} = \frac{(n_s + \bar{y}n_p)!}{n_s!n_p!(n_s + xn_p)^{n_p(\bar{y}-1)}} \tag{12.9}$$

where $\bar{y}$ is the average value of y.

The simplest assumption that can be made about the distribution of orientations of the rods is that all orientations within a cone of semi-angle ψ_{max} with its axis parallel to the director are equally likely. For ψ_{max} small it is easy to show (see problem 12.7) that N_{orient} is equal to $C(\bar{y}/x)^{2n_p}$, where C is a constant. The entropy S of the system is equal to $k \ln N$, so that

$$S = k \ln N = k(\ln N_{\text{orient}} + \ln N_{\text{comb}}) \tag{12.10}$$

and, if there is no energy of interaction between the rods, the Gibbs free energy G of the system is given by $G = -TS$. It is useful to consider the difference $G - G_{\text{o}}$, where G_{o} is the value of G for complete alignment of the rods parallel to the director. Thus

$$-\frac{G-G_o}{kT} = (\ln N_{\text{orient}} + \ln N_{\text{comb}}) - (\ln N_{\text{orient}} + \ln N_{\text{comb}})_o$$

$$= \ln(N'_{\text{orient}} N'_{\text{comb}}) \tag{12.11}$$

where the primed values indicate the corresponding quantities divided by their value for the fully aligned state. In this state all the values of y are equal to 1, so that $\bar{y} = 1$. Thus

$$\frac{G-G_o}{kT} = -\ln(N'_{\text{orient}} N'_{\text{comb}}) = -\ln\left(\frac{(n_s + \bar{y}n_p)!\bar{y}^{2n_p}}{(n_s + n_p)!(n_s + xn_p)^{n_p(\bar{y}-1)}}\right) \tag{12.12}$$

The logarithm can be evaluated by using Stirling's approximation, $\ln(m!) = m\ln m - m$, for the factorial of a large integer m. Equation (12.12) then leads directly to

$$\frac{G-G_o}{kT} = -(n_s + \bar{y}n_p)\ln(n_s + \bar{y}n_p) + n_p(\bar{y} - 1) + (n_s + n_p)\ln(n_s + n_p)$$

$$+ n_p(\bar{y} - 1)\ln(n_s + xn_p) - 2n_p \ln \bar{y} \tag{12.13}$$

It is convenient to express the free energy in terms of the volume fraction v of the system occupied by the rods, which is given by

$$v = \frac{xn_p}{n_s + xn_p} \tag{12.14}$$

It follows that $n_s/n_p = x(1 - v)/v$. Dividing both sides of equation (12.13) by n_p and substituting this expression for n_s/n_p leads, after some rearrangement, to

$$\frac{G-G_o}{n_p kT} = \frac{x}{v}(F_1 \ln F_1 - F_2 \ln F_2) + \bar{y} - 2\ln \bar{y} - 1 \tag{12.15}$$

where $F_1 = 1 - v + v/x$ and $F_2 = 1 - v + v\bar{y}/x$.

Figure 12.22 shows $(G - G_o)/(RT)$ plotted against $\bar{y}/x$ for a series of values of v for rods with an aspect ratio $x = 75$, where G now refers to the free energy per mole of rods, obtained by setting n_p equal to the Avogadro number in equation (12.15). At first sight it appears that the condition $\bar{y}/x = 1$ should correspond to all rods being aligned perpendicular to the director. By considering the representation of rods not lying in a single plane of cells, Flory and Ronca have argued, however, that this condition corresponds to random orientation. Accepting this rather strange result allows fig. 12.22 to be interpreted as follows: for values of v lower than a certain critical value v^* there is only one minimum in the free-energy curve, corresponding to random orientation, whereas for values of v higher than this there is a second minimum corresponding to a partially ordered phase. For $x = 75$, $v^* = 0.11$.

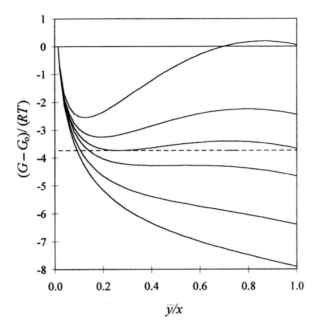

Fig. 12.22 A plot of
$(G - G_o)/(RT)$ against $\bar{y}/x$
for the following values of
υ reading from the
topmost curve: 0.22, 0.16,
0.13, 0.11, 0.06 and 0.02.
(Adapted from *Liquid
Crystal Polymers* by A. M.
Donald and A. H. Windle,
© Cambridge University
Press 1992.)

Figure 12.22 also shows that, for values of υ lower than about 0.13, the free energy for random orientation is lower than the free energy for the partially ordered phase. If the solution remained in a single phase, the phase would be disordered for concentrations lower than this value. It is, however, possible for the solution to have a lower total free energy by splitting into two phases of different compositions when the overall composition is close to the value $f_p = 0.13$.

In fig. 12.23 quantities proportional to the free energies G^o_{orient} and G^o_{rand} per site at the minima in the free-energy curves of fig. 12.22 are shown as functions of f_p for the ordered polymer and for the random polymer. The superscript on the Gibbs free energies signifies that the absolute values are shown; the free energy for the fully aligned state has not been subtracted, although it can be shown that subtracting it does not affect the following conclusions. It follows from the expressions for N_{orient} and N_{comb} given earlier that the appropriate equations for the free energies per site (proportional to the free energies per unit volume) of the oriented and random phases, respectively, are

$$x\frac{G^o_{orient}}{kT} = x(1 - \upsilon)\ln(1 - \upsilon) + \upsilon\ln\left(\frac{\upsilon}{x}\right) + \upsilon(\bar{y} - 1) - xF_2\ln F_2 - 2\upsilon\ln\left(\frac{\bar{y}}{x}\right)$$

$$(12.16a)$$

Fig. 12.23 Quantities proportional to the minimum free energies per unit volume G^o_{rand} (curve showing a minimum) and G^o_{orient} plotted against the volume fraction v for $x = 75$. See the text for discussion.

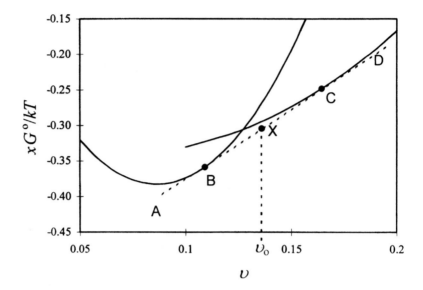

and

$$x\frac{G^o_{rand}}{kT} = x(1 - v)\ln(1 - v) + v\ln\left(\frac{v}{x}\right) + v(x - 1) \qquad (12.16b)$$

where equation (12.16b) is obtained by putting $\bar{y} = x$ in equation (12.16a). The curve in fig. 12.23 for the oriented phase is obtained from equation (12.16a) by finding the minimum value of the RHS with respect to varying $\bar{y}$ for a fixed value of x and each value of v. The line AD represents the common tangent to the two curves and it touches the curves for the random and oriented phases at B and C, respectively.

Minimisation of the total free energy ensures that any mixture with an overall concentration $v = v_o$ lying between the concentrations corresponding to the points B and C splits into two phases. One phase is oriented and has the concentration corresponding to the point C and the other is random and has the concentration corresponding to the point B. The amounts of each phase automatically adjust to give the correct overall composition v_o and the average free energy per site is given by the point X on the tangent line corresponding to this composition. Because the tangent line lies below both curves the average energy per site, and thus the total free energy for the system of two phases, is always less than it would be for a single phase, either ordered or random, corresponding to any composition in the range between B and C.

This is a further example of the common-tangent construction for equilibrium between two phases that has already been discussed in section

12.2.2 in connection with polymer–polymer miscibility. For this construction to work when the volume fraction is used as the abscissa it is necessary to plot the free energy per unit volume or a quantity proportional to it, as is done in fig. 12.23. Within the *biphasic range*, i.e. the range of concentrations between those corresponding to the points B and C in fig. 12.23, all that happens as the concentration is changed is that more of one of the phases is formed. Outside this range there is always a single phase corresponding to the curve with the lower free energy at that concentration.

Further details of the model, including the effect of allowing a small attractive force between the rods through the introduction of a non-zero Flory–Huggins parameter, the prediction of phase diagrams and the effect of introducing an orientation-dependent interaction of the Maier–Saupe form are given in reference (4) of section 12.5.

It is useful to introduce one further idea here. The condition for the minimum in the Gibbs free energy of the oriented phase is

$$\left(\frac{\delta}{\delta \bar{y}} G^{o}_{orient}\right)_{x} = 0 \qquad (12.17)$$

Setting $\upsilon = 1$ in equation (12.16a) gives the Gibbs free energy per site for a *thermotropic* liquid crystal. Application of equation (12.17) then gives the relationship between $\bar{y}$ and x at any minimum in G, whereupon the minimum possible value of x that satisfies this relationship can be found. The value obtained is $2e \cong 5.44$ (see problem 12.8). This suggests that, if a material is to exhibit thermotropic properties, the minimum aspect ratio of its constituent rods will be 5.44. This is slightly lower than the value 6.4 which is usually quoted and which is obtained from a more accurate version of the theory given by Flory and Ronca.

It remains to ask to what extent this theory for isolated rigid rods can be applied to polymer molecules. A fairly straightforward extension to the theory can be made by carrying out the analysis for sets of rods joined end to end by completely flexible joints. The result is that, for reasonably high aspect ratios, the joining of the rods in this way has only a negligible effect on the predictions of the theory. According to the ideas presented in section 3.3.4 any polymer chain can be represented statistically by an equivalent chain of rods, or links, with completely flexible joints, provided that the length and number of rods are appropriately chosen. It might therefore be expected that the theory would give results in reasonable agreement with experiment if the aspect ratio of the rods were taken to be the length of the equivalent link, or Kuhn length, divided by the chain diameter. As described in section 3.3.5, a persistence length can be defined, which for a long chain is equal to half the Kuhn length. It is in fact necessary to take the aspect ratio to be the *persistence ratio*, defined as the persistence length divided by the chain diameter. This leads to the idea of a *critical persistence ratio* for polymer molecules if they are to exhibit thermotropic properties.

12.4.5 The processing of liquid-crystal polymers

As indicated in section 12.4.1, one of the principal uses of LCPs is in the production of materials that have exceptional strength, rigidity and toughness. These properties arise through the high orientation of the molecules that can be achieved by orientation under flow during the processes of fibre spinning, film extrusion and moulding. LCPs can also be oriented by the application of electric or magnetic fields and this type of orientation is particularly important, though not only so, for their use as non-linear optical materials.

(i) *Fibre spinning*
Solution spinning is used, as described in section 10.2.3, for the production of oriented fibres from materials that decompose at high temperature and therefore cannot be melt-spun. An example is the aromatic polyamide LCP poly(*p*-phenylene terephthalamide) (PPTA, fig. 12.24), the polymer from which Kevlar and Twaron fibres are made. The polymer is dissolved in a solvent such as concentrated sulphuric acid, in which the molecules take up a lyotropic liquid-crystalline state, i.e. they line up parallel in domains. The directors of these domains become aligned parallel to the fibre axis as the solution passes through the spinneret. Spinning into air may take place, which permits stretching of the thread-line to increase the molecular orientation before the thread enters a coagulant bath where the solvent is removed. The alignment can be further improved by subsequent heat treatment under tension.

Melt-spinning of thermotropic polymers can be carried out using very similar equipment to that used for spinning conventional thermoplastics. High molecular orientation is obtained in the fibres at far lower spin–stretch ratios than is the case for conventional polymers. High-temperature heat treatment of the fibres spun from rigid-rod thermotropic polymers such as the co-polyesters described in section 12.4.3 generally leads to substantial changes in the mechanical properties. Some of these changes are due to physical reordering of the structure, but the effects observed are principally due to further polymerisation in the solid state.

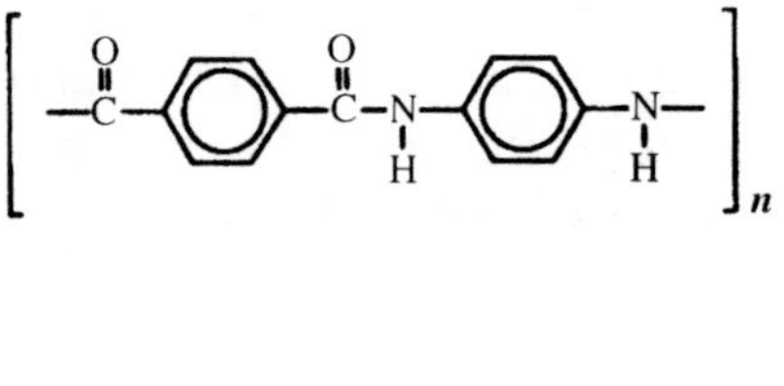

Fig. 12.24 The repeat unit of poly(*p*-phenylene terephthalamide), PPTA, from which Kevlar fibres are produced.

(ii) *Production of films and mouldings*

Extrusion of thermotropic polyesters through a slit die leads to the formation of a film, as for conventional thermotropic polymers, and the film is highly oriented in the extrusion direction. The films are usually rather weak in the transverse direction, so the production of films with similar mechanical properties parallel and perpendicular to the extrusion direction requires rather special techniques. Thicker sections, such as rods and tubes, can also be extruded.

Thermotropic LCPs have low melt viscosities because of the ready alignment of the molecules under flow and they are thus suitable for use in moulding, where they can fill complex moulds well. One problem, however, is that if the polymer is introduced into the mould through more than one 'gate' the separate flows of anisotropic polymer might not interpenetrate when they meet, which can give rise to weak 'weld lines'. This difficulty can be overcome satisfactorily by proper design of the moulds.

(iii) *Orientation in electric and magnetic fields*

Molecular polarisability is discussed in detail in chapter 9. Orientational polarisation involves the rotation, under the influence of an electric field, of molecules, or parts of molecules, that have permanent electric dipoles. This effect clearly leads to a possible mechanism for orienting molecules and, in particular, for reorienting the domains of liquid crystals or LCPs that have permanent dipoles. It is less obvious that deformational polarisation can lead to orientation. If, however, the polarisability of a molecule is anisotropic, as is usually the case, the polarisability and hence the induced electric dipole will differ for different orientations of the molecule in the field. The interaction energy with the field is therefore reduced if the molecule rotates so that its direction of maximum polarisability coincides with the field direction. Electric-field orientation is largely of importance in device applications.

Magnetic field orientation is analogous to the induced-polarisation effect for electric-field orientation. Here it is the anisotropy of the magnetic susceptibility of the molecule that leads to a lowest interaction energy for a particular orientation of the molecule in the field.

12.4.6   The physical structure of solids from liquid-crystal polymers

(i) *Nematic order*

Main-chain LCPs that exhibit nematic order tend towards pseudo-hexagonal packing in the liquid-crystalline state. Wide-angle X-ray-diffraction

Fig. 12.25 Schematic diagrams of X-ray diffraction patterns for (a) a nematic LCP and (b) a smectic LCP. (Reproduced from *Liquid Crystal Polymers* by A. M. Donald and A. H. Windle, © Cambridge University Press, 1992.)

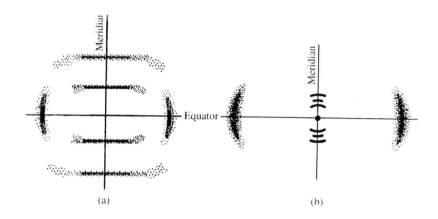

(a)   (b)

(WAXS) photographs of oriented samples therefore have fairly sharp diffraction spots along the equator (see fig. 12.25(a)). There is no crystal register of the chains along their length, so that there are no coherent interference effects along the other layer lines, which therefore do not show sharp spots and merely indicate the periodicity within the individual molecules. For perfect alignment the intensity distribution along all the layer lines depends only on the 'thickness' and structure of the chain perpendicular to its length: a 'thin' chain will give a wide distribution, a 'thick' chain will give a narrow distribution. For less perfect alignment the layer lines become curved and the equatorial spots are broadened, and both become rings for unoriented material.

When the polymer solidifies the chains may crystallise, so that corresponding units come into register both laterally and along their lengths, and the lateral order may change from pseudo-hexagonal. Crystallisation is never complete, so that WAXS photographs show features of both oriented crystalline and liquid-crystalline material. Crystallisation of a LCP does not lead to overall reorientation of the molecules in any region of the polymer, i.e. to changes in the orientation of the director, but it may increase the local value of the order parameter S.

It is difficult to imagine how the random co-polyesters can crystallise, but there is clear evidence from DSC and from WAXS that they can do so to a limited extent. WAXS photographs show fairly sharp, unequally spaced, diffraction maxima along the meridian, as well as equatorial spots indicative of pseudo-hexagonal packing of the chains. The meridional maxima tend to be concentrated towards the meridian rather than being widely spread perpendicular to it, which is believed to be due to the formation of small crystallites by the coming together of chain segments that happen to have matching sequences of the different components. The crystallinity is usually about 10%–20%.

(ii) *Smectic order*

Main-chain LCPs can exhibit smectic behaviour, particularly when they are of the type consisting of rigid segments separated by flexible spacers, and side-chain LCPs often do so. Figure 12.26 shows schematically how the mesogenic units may be arranged to give layers for both types of polymer and how the layers of mesogenic units on different chains in side-chain LCPs may be arranged with respect to each other to give the overall layer structure. Smectic alignment leads to low-angle X-ray scattering from the layers as well as wide-angle arcs along the equator, as shown in fig. 12.25(b).

(iii) *Macroscopic structures: morphology of fibres*

As a result of the presence of distortions, disclinations and domain boundaries a great variety of structures can be seen in LCPs in the liquid-crystalline state when they are viewed between crossed polarisers. A full description of these structures is given in the book by Donald and Windle cited in section 12.5. A primary interest in LCPs is, as has already been noted, that they can readily be induced to provide materials that are macroscopically oriented and have improved properties. Only structures present in the oriented solid forms are discussed here.

One of the most important features of extruded and moulded articles is the presence of a skin-core structure caused by high shear gradients in the flowing polymer close to the die or mould surface, which induce high orientation. In rods or other extrudates the skin-core effect can be removed by subsequent drawing to increase the overall orientation of the sample; this cannot, of course, be done for mouldings.

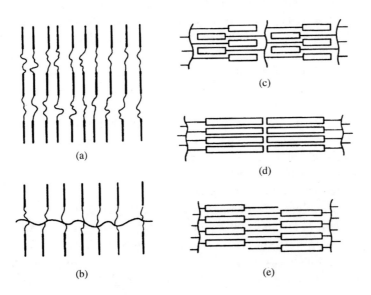

Fig. 12.26 Schematic diagrams showing various types of organisation in smetic LCPs: (a) alignment of main-chain mesogenic units where there is sufficient flexibility in the spacers; (b) alignment of side-chain mesogenic units where there is sufficient flexibility in the spacers and in the backbone; (c), (d) and (e) possible relationships between side-group mesogenic units on different chains. (e) Represents the overlap of flexible tails. ((a), (b) Reproduced from *Liquid Crystal Polymers* by A. M. Donald and A. H. Windle, © Cambridge University Press 1992; (c), (d) and (e) reproduced by permission from Springer Verlag.)

Fibres, particularly those spun from solution, may also exhibit skin effects, but these usually disappear during the process of drawing-down. The structure of the fibres is often rather complex, but is usually in the form of fibrils that are themselves composed of microfibrils. This is similar in general terms to the structure of highly oriented fibres from more conventional polymers. Detailed models have been given for fibres derived from various individual LCPs, but no general agreement has been reached even for a single polymer. One feature of some generality is a periodic variation of the director orientation with respect to the axis of the fibre, which leads to a banded appearance when the structure is viewed between crossed polarisers in the optical microscope. For Kevlar and other highly crystalline fibres this appearance is due to the fact that the axes of the molecules take up alternately two orientations each making the same angle with the fibre axis, giving the so-called *pleated-sheet* structure. In random co-polyesters the change of orientation takes place more continuously, giving a wavy structure.

12.4.7   The properties and applications of liquid-crystal polymers

For structural applications the most important properties of a polymer, apart from chemical resistance and long-term stability, are its tensile modulus and its tensile strength, or stress at breakage. As discussed in section 11.6.1, if the maximum values for these properties are to be achieved for a particular polymer, a minimum requirement is that all the molecules should be aligned parallel to the tensile axis. The maximum values then obtainable in theory are proportional to the modulus and strength of the individual molecules and to the number of them that cross unit cross-section normal to the tensile axis. What is often more important in applications than the absolute values of modulus or strength is the ratio of these quantities to the density or specific gravity of the material These ratios are called the *specific modulus* and *specific stress at breakage* and fig. 12.27 shows values of these quantities for some selected LCP and other fibres.

The fibre that best combines both high specific modulus and high specific stress at breakage is a particular type made from PBO, the formula for which is shown in fig. 12.18(a). It is clear that this molecule must be very stiff, because the only bonds about which rotation can occur are collinear with the molecular axis and there is also a relatively large number of bonds nearly parallel to the axis. Fibres of the Kevlar or Vectra type have significantly lower values both for stiffness and for strength, which is due at least in part to the fact that they have some backbone bonds that are not collinear with the molecular axis and they

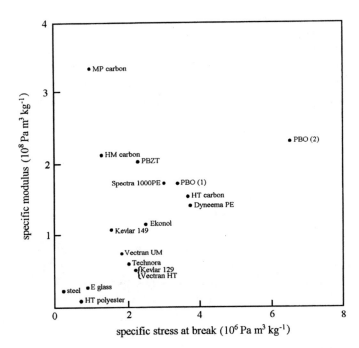

Fig. 12.27 Specific stress at breakage and specific modulus for some LCP and other selected fibres. (Adapted by permission of Kluwer Academic Publishers.)

have a lower density of bonds parallel or nearly parallel to the axis (see figs 12.24 and 12.20).

Polymers such as PBO in which there is no, or very little, inherent flexibility in the chain are called *rigid-chain* polymers and those such as the random co-polyesters that form the basis of Vectra fibres are called *semi-rigid*. The semi-rigid polymers have the advantage that they can usually be processed in the melt, as thermotropic materials, whereas the rigid polymers must be processed in solution as lyotropic materials. On the other hand, not only are the semi-rigid polymers less stiff than the rigid ones but also the stiffness falls off faster with increasing temperature. This is because at higher temperatures there are more departures of the chains from their fully extended form, due to rotations about the single bonds not collinear with the chain axis. Application of an external load then causes rotation around bonds to straighten the chains again, rather than causing increases in bond angles or lengths, which would require a larger force for a given degree of extension. An interesting associated effect is that these polymers have a negative axial coefficient of thermal expansion. This is due to the shortening effect of the bond rotations which more than compensates for the normal effects of increases in bond length and openings of bond angle with increasing temperature. The resulting negative expansion is usually much smaller than typical positive values for randomly oriented polymers.

So far it has tacitly been assumed that the molecular chains are highly aligned parallel to the fibre axis, but there must in reality be a distribution of orientations. If the degree of orientation is very high, $\langle \cos^4 \theta \rangle = \langle (1 - \sin^2 \theta)^2 \rangle \approx 1 - 2\langle \sin^2 \theta \rangle$, $\langle \sin^2 \theta \cos^2 \theta \rangle = \langle \sin^2 \theta (1 - \sin^2 \theta) \rangle \approx \langle \sin^2 \theta \rangle$ and $\langle \sin^4 \theta \rangle \approx 0$, where θ is the angle between the axis of a typical oriented unit and the fibre axis and the angle brackets $\langle \rangle$ denote the average value. Equation (11.15) thus reduces to

$$s_{3333} \approx s_{3333}^o + 2(-s_{3333}^o + s_{1133}^o + 2s_{2323}^o)\langle \sin^2 \theta \rangle \tag{12.18}$$

The units are, however, highly anisotropic, so that $s_{2323}^o \gg s_{1133}^o$, and equation (12.18) then reduces to $s_{3333}^o \approx s_{3333}^o + 4s_{2323}^o \langle \sin^2 \theta \rangle$. By again taking note of the high anisotropy of the fibre and making similar approximations about the orientation averages, it can be shown that $4s_{2323} = 1/G$, where G is the shear modulus governing the torsional rigidity both of the fibre and of the unit. Example 11.4 shows that $s_{3333} = 1/E_3 = 1/E$ and $s_{3333}^o = 1/E_{max}$, where E is the extensional modulus of the fibre and E_{max} is the modulus for fully aligned units. It follows that, to a good approximation, the expected relationship between the orientation of a LCP fibre and its modulus is given by

$$\frac{1}{E} = \frac{1}{E_{max}} + \frac{\langle \sin^2 \theta \rangle}{G} \tag{12.19}$$

This equation with fixed values of E_{max} and $\langle \sin^2 \theta \rangle$ has often been found to fit experimental data for the temperature dependences of E and G well, although slightly different models have sometimes been used. In this simple model the oriented units are assumed to be aggregates of fully aligned chains and it is the low value of G for these units and its temperature dependence that cause the modulus of the fibre to be lower than E_{max} and to be temperature-dependent.

Extrusions and mouldings usually have high orientation in the surface layers due to there being high shearing forces at the walls of the extrusion die or mould. The modulus of the surface layer of a rod is the most important factor determining its distortion under bending forces, so that the surface orientation improves the bending stiffness. In tension, however, what matters is the average modulus across the section. This can be improved for extrudates by designing the dies so that the material can be drawn down, i.e. stretched, subsequent to extrusion to produce the final cross-section required. This produces greater alignment of the polymer in the interior of the extrudate and thus improves its properties, but cannot, of course, be done for mouldings, for which the final shape is usually determined by the shape of the mould.

Example 12.5

The following values of the tensile modulus E and rigidity modulus G, expressed in gigapascals, were found for a random co-polyester fibre at two different temperatures: $E = 125$, $G = 1.1$ and $E = 62$, $G = 0.28$. Assuming that equation (12.19) applies, calculate the value of E_{max} and the order parameter S for the fibre.

Solution

Equation (12.19) can be written

$$\frac{1}{E} - \frac{1}{E_{max}} = \frac{\langle \sin^2 \theta \rangle}{G}$$

Inserting values E_1, G_1 and E_2, G_2 for the two temperatures into this equation and dividing one resulting equation by the other leads immediately to

$$\left(\frac{1}{E_2} - \frac{1}{E_{max}} \right) \bigg/ \left(\frac{1}{E_1} - \frac{1}{E_{max}} \right) = \frac{G_1}{G_2}$$

Substituting the data leads to $(0.0161 - x)/(0.008 - x) = 3.93$, where $x = 1/E_{max}$. Solving gives $x = 5.24 \times 10^{-3}$ or $E_{max} = 1/x = 191$ GPa. Substitution of this value and the data for the first temperature into the first equation gives $\langle \sin^2 \theta \rangle = 1.1 \times (0.008 - 0.005\,24) = 0.003\,04$. Now,

$$S = (3\langle \cos^2 \theta \rangle - 1)/2 = 1 - \tfrac{3}{2}\langle \sin^2 \theta \rangle$$

leading to $S = 0.995$.

LCP fibres are used for making heavy-duty ropes and belts and also as the reinforcing fibres of composite materials. An important application of Kevlar is in the fabrication of bullet-proof clothing. The easy mouldability, good mould filling and low thermal expansion of thermotropic LCPs, combined with their resistance to solvents and their stability at temperatures up to 120 °C or higher make them suitable for precision mouldings for use in a variety of components such as electrical connectors.

A further major use for LCPs is in the area of *opto-electronics*. Their usefulness here arises partly from their ability to incorporate the properties of more conventional liquid crystals, which are fluid at ambient temperatures, within a material that does not need to be confined in a cell. When a small-molecule liquid-crystalline material is cooled to form a solid it usually crystallises in such a way that any structure within it, i.e. orientation of the molecules, is lost. LCPs can, however, form glasses in which orientation induced above the glass transition can be retained. This leads to the possibility, among others, of information-storage devices. For instance, the heating effect of a laser beam, with or without the simultaneous application of an electric field, can be made to change, within a very small region, a state of orientation that has previously been produced.

Another potentially important area of applications in opto-electronics is in the production of *non-linear optical* (NLO) *devices*. In section 9.2.1 the molecular electrical polarisability α is defined, with the assumption that the polarisation of a molecule is proportional to, or linear in, the applied electric field E. For sufficiently high electric fields, such as those that can be produced in high-power laser beams, the polarisation is no longer linear in E, but depends also on higher powers of E. This leads to various possible effects, some of which are now described.

If the derivation of the relationship between the refractive index and the polarisability in section 9.2.1 is followed through, it becomes clear that the refractive index is no longer independent of the electric field strength when the approximation of linearity no longer holds. Important possibilities arise because the molecular dipoles induced by the sinusoidally oscillating electric field of a light wave do not vary in strength sinusoidally under non-linear conditions. The non-sinusoidal oscillation can be analysed as a Fourier series of sinusoidal oscillations at the frequency of the oscillating field and at higher harmonic frequencies, which means that there is the potential for some of the energy in the incident light beam to be converted into light at these higher frequencies. Furthermore, if two light beams pass through a linear material they do not affect each other because the response of the material is the sum of the responses for the separate beams; for a non-linear material this is not so, and the beams interact with each other. All these effects can be used to produce opto-electronic devices.

So far no consideration has been given to the precise way that the effects produced by individual molecules contribute to the properties of the bulk medium. Just as the linear polarisability α is really a second-rank tensor, so that the polarisation produced in a molecule depends on the orientation of the molecules with respect to the electric field, the non-linear coefficients are also tensors of higher rank, so that the magnitudes of the non-linear effects depend also on the orientation of the molecule with respect to the electric field. The ease of alignment of the molecules in LCPs thus gives them advantages over other non-linear materials, provided that the individual molecules exhibit the required non-linear effects. Fortunately, the requirements for high NLO parameters are the same as those that lead to the straight rigid molecules or side groups that are conducive to liquid-crystal-linity. Both main-chain and side-chain LCPs can exhibit appropriate properties, but most attention has been paid to side-chain polymers.

Further discussion of the requirements for NLO activity, how they can be met using LCPs and the types of device that can be produced is beyond the scope of the present book but can be found in the book referred to as (4) in section 12.5.

12.5 Further reading

(1) *Polymeric Multicomponent Materials: An Introduction*, by L. H. Sperling, John Wiley & Sons Inc., New York, 1997. This book has about 80% of its content relating to the fields of polymer blends, blocks, grafts and interpenetrating networks and about 20% relating to polymer composites of various types.

(2) *Multicomponent Polymer Systems,* edited by I. S. Miles and S. Rostami, Longman Scientific and Technical, London, 1992. Three of the ten chapters of this book are devoted to blends, block copolymers and rubber-toughened polymers.

(3) *The Physics of Block Copolymers*, by I. W. Hamley, Oxford Science Publications, Oxford, 1998. Most of this book is devoted to melt and solution properties, but there is one chapter on solid-state structure.

(4) *Liquid Crystalline Polymers* by A. M. Donald and A. H. Windle, Cambridge University Press, 1992. This book covers the field of LCPs at a slightly more advanced level than the present chapter and, naturally, in much more detail.

(5) *Polymers: Chemistry and Physics of Modern Materials*, by J. M. G. Cowie. See section 1.6.2.

12.6 Problems

12.1. A particular sample of a blend of polymers A and B separates into two phases of volumes V_1 and V_2, with volume fractions v_1 and v_2, respectively, of A. If the values of ΔG_{mix} per unit volume for the two phases are ΔG_1 and ΔG_2, respectively, prove that the construction of fig. 12.1(b) correctly gives the value of ΔG_{mix} per unit volume of the whole sample, provided that there is no change in volume on mixing.

12.2. In the regions near the binodal points for a particular polymer blend ΔG_{mix} (in arbitrary units) is given approximately by
$$\Delta G_{mix} = 2085(f - 0.200)^2 - 316 \quad \text{and} \quad \Delta G_{mix} = 5720(f - 0.842)^2 - 412,$$
where f is the volume fraction of one of the polymers in the blend. Calculate approximately the values of f that correspond to the binodal points by noting that the gradient of the common tangent to the two regions of the curve of ΔG_{mix} against composition will be approximately equal to that of the line joining the two minima of ΔG_{mix}.

12.3. A blend of polymers A and B has a binodal that can be represented in the temperature range 100–200 °C by the expression $T = a(v - v_0)^2 + b$, where T is the temperature and v the volume fraction of A in the blend, with v_0, a and b equal to 0.45, 1250 °C and 100 °C, respectively. A sample of volume 10 cm^3 consisting of equal volumes of A and B is allowed to come to equilibrium at 150 °C. (a) Assuming that there is no change of volume on mixing the polymers, calculate the volume fractions of A in the two phases of the phase-separated blend and the total volume of each phase. (b) Assuming that the phase segregation leads to the formation of macroscopic regions of each phase, that the morphology is unchanged by quenching below the glass transition, and that the polymers obey an equation equivalent to equation (12.7) with volume fractions instead of weight fractions, calculate the two observable glass-transition temperatures, given that the glass-transition temperatures for A and B are 90 and 20 °C, respectively.

12.4. A linear random copolymer of two types of monomer A and B containing 20% by weight of A units has a glass-transition temperature $T_{20} = 15$ °C. The glass-transition temperatures of homopolymers of A and B are $T_A = 100$ °C and $T_B = 5$ °C, respectively. Assuming that the copolymer obeys equation (12.8), calculate the glass-transition temperature T_{80} for a random copolymer of A and B containing 80% of A units.

12.5. A particular graft copolymer of polystyrene (PS) and polyisoprene (PI) containing 17% by volume of PS segregates to form a body-centred-cubic array of PS spheres. Assuming that the segregation is virtually complete and that the length a of the cube edge of the array is 33.2 nm, calculate the radius of the spheres.

12.6. By assuming that the length of the C—C bonds in a phenylene ring —◎— is the average of the lengths for singly and doubly bonded carbon atoms, calculate the lengths of the O—◎—O and C—◎—C units (the distances between the centres of the oxygen and outer carbon atoms, respectively). Hence estimate the length of the fully extended repeat unit shown in fig. 12.18(b). Simplify by assuming that all bond angles in the molecule are equal to 120°.

12.7. Assuming that all orientations of n rods each of length x within a cone of semi-angle ψ_{max} are equally likely, show that, for small ψ_{max}, the number of ways that the rods can be arranged is equal to $C(\bar{y}/x)^{2n}$, where C is a constant and $\bar{y}$ is the average value of the projections of the length of the rods perpendicular to the axis of the cone.

12.8. Starting from equation (12.16a), show that according to the version of the theory of liquid-crystal alignment presented in section 12.4.4 the minimum aspect ratio for the rods of a material that exhibits thermotropic liquid crystallinity is 5.44.

12.9. The table gives corresponding values of the tensile modulus E and rigidity modulus G of a liquid-crystal copolymer at various temperatures. Show that these data are consistent with equation (12.19) and deduce values for E_{max} and the order parameter S.

E (GPa)	104	52	41	26	21
G (GPa)	2.56	1.33	0.81	0.56	0.38

Appendix:
Cartesian tensors

Definitions

Consider a homogeneous, *isotropic* electrically conducting medium. If an electric field is applied to such a material the relationship between the current density J and the applied electric field E can be written

$$J = \sigma E \tag{A.1}$$

The current density is parallel to the applied field and the constant of proportionality is the conductivity, σ.

For a homogeneous *anisotropic* medium, each component of J depends linearly on each component of E, so that

$$J_1 = \sigma_{11}E_1 + \sigma_{12}E_2 + \sigma_{13}E_3$$
$$J_2 = \sigma_{21}E_1 + \sigma_{22}E_2 + \sigma_{23}E_3$$
$$J_3 = \sigma_{31}E_1 + \sigma_{32}E_2 + \sigma_{33}E_3 \tag{A.2}$$

If the *Einstein summation convention* is used, whereby any expression is summed over every repeated subscript, this becomes

$$J_i = \sigma_{ij}E_j \qquad \text{for } i \text{ and } j = 1, 2, \text{ or } 3 \tag{A.3}$$

The quantities σ_{ij} are the *components of a second-rank tensor* (two subscripts) $[\sigma_{ij}]$. The components of a second-rank tensor relate the components of two *vectors*. Vectors may be considered to be first-rank tensors (one subscript) and scalars to be tensors of rank zero.

For any two vectors V and v related by the tensor $[T_{ij}]$

$$V_i = T_{ij}v_j \tag{A.4}$$

Rotation of axes

A vector has a 'value' (its magnitude and direction) independent of the reference axes. A tensor also has a meaning independent of the choice of axes, but its components depend on the axes chosen, just as the components of a vector do.

Let a_{ij} be the direction cosines of a 'new' axis Ox_i' with respect to an 'old' axis Ox_j. Then the nine direction cosines a_{ij} form a matrix that relates co-ordinates with respect to the new axes to co-ordinates with respect to the old axes:

$$\begin{array}{cc} & \text{old} \\ & \begin{array}{ccc} x_1 & x_2 & x_3 \end{array} \\ \text{new}\ \begin{array}{c} x_1' \\ x_2' \\ x_3' \end{array} & \begin{pmatrix} a_{11} & a_{12} & a_{13} \\ a_{21} & a_{22} & a_{23} \\ a_{31} & a_{32} & a_{33} \end{pmatrix} \end{array}$$

All nine values of a_{ij} are not independent: the following relationships must hold:

$$\sum_j a_{ij}^2 = \sum_i a_{ij}^2 = 1 \text{ (three direction cosines squared for any axis sum to unity) (A.5)}$$

$$\sum_j a_{ij}a_{kj} = \sum_i a_{ij}a_{ik} = 0 \text{ (any two axes of either set are mutually perpendicular)}$$
$$\text{(A.6)}$$

It can be shown that tensor components transform as follows (using the Einstein convention):

rank 0 (scalar)	T	$= T$
first rank (vector)	T_i'	$= a_{ij}T_j$
second rank	T_{ij}'	$= a_{ik}a_{jl}T_{kl}$
general rank	$T_{abcd\ldots}'$	$= a_{ap}a_{bq}a_{cr}a_{ds}\ldots T_{pqrs\ldots}$

where T_i' etc. refer to the components with respect to the new axes and T_i etc. to the components with respect to the old axes.

Symmetric tensors

All the second-rank tensors needed in this book are symmetric in the sense that:

$$T_{ij} = T_{ji} \qquad \text{for all } i,j \tag{A.8}$$

For a symmetric tensor there is a set of *principal axes* for which $T_{ij} = 0$ unless $i = j$, i.e.

$$\begin{bmatrix} T_{11} & T_{12} & T_{13} \\ T_{21} & T_{22} & T_{23} \\ T_{31} & T_{32} & T_{33} \end{bmatrix} \quad \text{becomes} \quad \begin{bmatrix} T_{11}^{\text{o}} & 0 & 0 \\ 0 & T_{22}^{\text{o}} & 0 \\ 0 & 0 & T_{33}^{\text{o}} \end{bmatrix}$$

when it is expressed with respect to the principal axes: T_{11}^{o}, T_{22}^{o} and T_{33}^{o} are its *principal components*, sometimes written T_1, T_2 and T_3.

Property and field tensors

The principal axes of second-rank tensors are often fixed by symmetry if the tensors represent properties of materials. For an orthorhombic crystal, for example, they must be parallel to the crystallographic axes.

The second-rank tensors used in this book are *property tensors*, such as polarisability, and *field tensors*, such as stress and strain. The latter are also symmetric tensors, but their principal axes are not necessarily parallel to the crystallographic axes.

Tensors for the small-strain elastic properties of materials

Consider a homogeneously stressed material and imagine a small surface area δS somewhere within it (see fig. A.1). Let the direction of the vector δS represent the normal to the area and let the length of δS represent the magnitude of the area. The material on the side of δS towards which the vector δS points exerts a force on the material on the opposite side of δS. Let this force be δF. Because the stress is homogeneous, δF must be proportional to δS, but it must also depend on the direction of the normal to δS. Thus the two vectors δF and δS must be related by a second-rank tensor $[\sigma_{ij}]$, called the *stress tensor*, so that $\delta F_i = \sigma_{ij}\,\delta S_j$. (The symbol σ is conventionally used both for electrical conductivity and for stress, but no confusion should arise in this book because the two properties are never both discussed in the same chapter.)

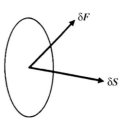

Fig. A.1

The applied stress causes a state of strain in the material and it is assumed here that the strain is small. Imagine a line element in the material represented by the vector l before the stress is applied. After the application of the stress the line element will in general have changed its length by an amount proportional to the original length and will have rotated. If the change in l is δl, it follows that δl must be related to l by a second-rank tensor $[e_{ij}]$, so that $\delta l_i = e_{ij}l_j$. This tensor does not, however provide a suitable description of the state of strain of the material, because its components would not be zero for a simple rotation of the material without any applied stress. It is obvious that, for a very small pure rotation, the three quantities e_{ii} are zero, and it is not difficult to show that the three quantities of the type $e_{ij} + e_{ji}$ for $i \neq j$ are zero as well. The tensor $[e_{ij}]$ can thus in general be considered as the sum of two tensors, an antisymmetric tensor $[r_{ij}]$ that describes any rotational component, with $r_{ij} = 0$ for $i = j$ and $r_{ij} = \frac{1}{2}(e_{ij} - e_{ji})$ for $i \neq j$, and a symmetric tensor $[\varepsilon_{ij}]$, the *strain tensor*, with components $\varepsilon_{ii} = e_{ii}$ and $\varepsilon_{ij} = \frac{1}{2}(e_{ij} + e_{ji})$ for $i \neq j$.

Now consider a cube with faces of unit area normal to the co-ordinate axes $OX_1X_2X_3$ embedded in a homogeneously stressed material. It follows immediately that σ_{ii} represents the force applied in the direction OX_i to the face of the cube normal to OX_i, i.e. it represents the *normal stress* parallel to OX_i. In a similar way ε_{ii} ($= e_{ii}$) represents the change in the length of an element of length parallel to OX_i divided by the original length of the element, i.e. it represents the tensile strain in the direction OX_i. The quantity σ_{ij} for $i \neq j$ represents the force applied in the direction OX_i on the face normal to OX_j, i.e. it represents a *shear stress*. Similarly, for ε_{ij} for $i \neq j$ represents a *shear strain*, i.e it represents half the change in the angle between the projections of lines originally parallel to OX_i and OX_j onto the OX_iX_j plane caused by the application of the stress. If there are no forces exerted on the cube other than directly by the material immediately adjacent to it, it follows from the conditions for static equilibrium that $\sigma_{ij} = \sigma_{ji}$, i.e. that the tensor $[\sigma_{ij}]$ is symmetric.

For small strains, it is assumed that the material behaves linearly, i.e. that every component of strain is linearly related to every component of stress. Using the Einstein notation for summation, this relationship can be written

$$\sigma_{ij} = c_{ijkl}\varepsilon_{kl} \qquad \text{or} \qquad \varepsilon_{ij} = s_{ijkl}\sigma_{kl} \tag{A.9a,b}$$

Table A.1. *Relationships among the non-zero components of* $[c_{ijkl}]$ *for various symmetries*

Orthorhombic	Uniaxial	Isotropic
c_{1111}	$c_{1111} = c_{2222}$	$c_{1111} = c_{2222} = c_{3333}$
c_{2222}		
c_{3333}	c_{3333}	
$c_{1122} = c_{2211}$	$c_{1122} = c_{2211}$	$c_{1122} = c_{2211} = c_{1133}$
$c_{1133} = c_{3311}$	$c_{1133} = c_{3311}$	$= c_{3311} = c_{2233} = c_{3322}$
$c_{2233} = c_{3322}$	$= c_{2233} = c_{3322}$	
$c_{(12)(12)}$	$c_{(12)(12)} = \frac{1}{2}(c_{1111} - c_{1122})$	$c_{(12)(12)} = c_{(23)(23)} = c_{(13)(13)}$
$c_{(23)(23)}$	$c_{(23)(23)} = c_{(13)(13)}$	$= \frac{1}{2}(c_{1111} - c_{1122})$
$c_{(13)(13)}$		

The quantities c_{ijkl} are the components of a fourth-rank tensor called the *stiffness tensor* and the quantities s_{ijkl} are the components of a fourth-rank tensor called the *compliance tensor* (note the confusing relationship between the names and the symbols). Specification of all values of either c_{ijkl} or s_{ijkl} allows all values of the other set to be calculated.

For an orthorhombic material, only nine of the 81 components c_{ijkl} are independent and, if the axes $OX_1X_2X_3$ are chosen to coincide with the symmetry axes of the material, these are c_{1111}, c_{2222}, c_{3333}, $c_{1122} = c_{2211}$, $c_{1133} = c_{3311}$, $c_{2233} = c_{3322}$, $c_{(12)(12)}$, $c_{(23)(23)}$ and $c_{(13)(13)}$ and all the remaining components are zero. Here the notation $c_{(ij)(ij)}$ means that all the four components obtained by interchanging subscripts within a bracket are equal. These relationships are shown in the left-hand column of table A.1. The nine independent non-zero components of s_{ijkl} correspond to the same sets of subscripts.

For a material with uniaxial symmetry there are the further relationships shown in the second column of table A.1, so that there are only four independent elastic constants for this symmetry. For an isotropic material the further relationships shown in the third column apply and there remain only two independent elastic constants for an isotropic material, as stated in section 6.2.1.

Further reading

(1) *Physical Properties of Crystals: Their Representation by Tensors and Matrices*, by J. F. Nye, Revised Edition, Clarendon Press, Oxford, 1985. This book gives an excellent introduction to the use of tensors in physical science.

Solutions to problems

Where not specified otherwise, all notation in the solutions is the same as that used in the corresponding chapter.

Chapter 3

3.1. Again it is obvious that $M_n = 300\,000$ g mol^{-1}. M_w is given by

$$\frac{\left[\frac{1}{2}(2)^2 + (2.5)^2 + 2 \times (3)^2 + (3.5)^2 + \frac{1}{2}(4)^2\right] \times 10^{10}}{\left[(\frac{1}{2} \times 2) + 2.5 + (2 \times 3) + 3.5 + (\frac{1}{2} \times 4)\right] \times 10^5} = 3.100 \times 10^5 \text{ g mol}^{-1}$$

The polydispersity index is thus 1.033 and hence greater than before.

3.2. If $n(m)\,dm$ is the fraction of molecules, $\int_0^\infty n(m)\,dm = 1$, hence $\int_0^\infty ae^{-bm}\,dm = 1$.

Thus $[-(a/b)e^{-bm}]_0^\infty = 1$ and hence $a = b$.

$$M_n = \int_0^\infty mae^{-bm}\,dm \Big/ \int_0^\infty ae^{-bm}\,dm$$

$$= \int_0^\infty mae^{-bm}\,dm = \left[-\frac{ma}{b}e^{-bm}\right]_0^\infty + \frac{a}{b}\int_0^\infty e^{-bm}\,dm = \frac{a}{b^2} = \frac{1}{b}$$

$$M_w = \int_0^\infty m^2 ae^{-bm}\,dm \Big/ \int_0^\infty mae^{-bm}\,dm$$

$$= \left(\left[-\frac{m^2}{b}e^{-bm}\right]_0^\infty + \frac{2}{b}\int_0^\infty me^{-bm}\,dm\right)\Big/ \int_0^\infty me^{-bm}\,dm = \frac{2}{b}$$

Hence the polydispersity index is 2. The molar mass that contributes most is the one corresponding to the maximum in the curve of weight fraction against M and therefore to the maximum in $w = ame^{-bm}$. However, $dw/dm = a(1 - mb)e^{-bm}$, so that the maximum is when $m = 1/b$.

3.3. Drawing in those face diagonals of the cube that join atoms produces the tetrahedron; hence the required equivalence. Join the central atom to the centre of one of these face diagonals. It can then be seen by inspection that the sine of half the tetrahedral angle is given by half the face diagonal divided by half the body diagonal and is therefore equal to $\sqrt{2}/\sqrt{3} = 0.8165$.

Thus $\theta_t = 2\sin^{-1} 0.8165 = 109.47°$.

3.4.

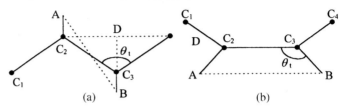

(a)   (b)

(a) Shows the *trans* conformation viewed normal to the plane of the backbone. A and B represent the projections onto the plane of two hydrogen atoms, H_1 and H_3, respectively, on the same side of the plane, say the near side. The required length is then AB, where $C_2A = C_3B = l_{CH}\cos(\theta_t/2)$, $C_2D = l_{CC}\sin(\theta_t/2)$ and $C_3D = l_{CC}\cos(\theta_t/2)$. θ_t is the tetrahedral angle and l_{CH} and l_{CC} are the lengths of the C—H and C—C bonds. Thus $AB = \sqrt{(C_2D)^2 + (C_3D + 2C_2A)^2} = 0.249$ nm. If the *gauche* conformation is imagined to arise from rotating the right-hand half of the molecule around C_2C_3 so that the other hydrogen atom H_4 attached to C_3 now lies on the near side of the plane, both hydrogen atoms H_3 and H_4 are equidistant AB from H_1 and thus the smallest separation is still $AB = 0.249$ nm. (b) Shows the molecule in the *eclipsed* conformation projected onto a plane containing the central C_2—C_3 bond and two hydrogen atoms (C_1 and C_4 do not lie in this plane). The separation AB between nearest H atoms is $l_{CC} - 2l_{CH}\cos\theta_t = 0.227$ nm.

3.5. The molar mass per CH_2 group is $14\,\mathrm{g\,mol^{-1}}$. Hence the number n of random links is $2.5 \times 10^5/(14 \times 18.5) = 9.65 \times 10^2$. The length l_r of a random link is $18.5b\sin\theta$, where θ is half the tetrahedral angle and $b = 0.154$ nm so that $l_r = 2.33$ nm and $r_{rms} = l_r\sqrt{n} = 72.4$ nm.

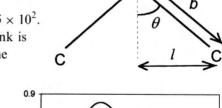

3.6. To a first approximation the fraction of chains having lengths *outside* the specified range is given by the sum of the two triangles indicated by the

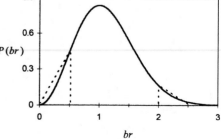

dashed lines, i.e. by $\frac{1}{4}(P(\frac{1}{2}) + P(2))$, so that the required estimate is given by

$$1 - \frac{1}{4}(P(\frac{1}{2}) + P(2)) = 1 - (1/\pi^{1/2})(\frac{1}{4}e^{-1/4} + 4e^{-4}) = 0.85.$$

A more exact answer, 0.874, is easily obtained by using a spreadsheet to perform an approximate integral or by looking up error functions.

3.7. Let n and l be the number and length of the equivalent random links and let the fully extended length of both chains be L. Then

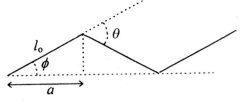

$$L = nl = n_0 a = n_0 l_0 \cos \phi \tag{1}$$

Equations (3.5) and (3.10) then show that

$$r_{rms}^2 = nl^2 = n_0 l_0^2(1 + \cos \theta)/(1 - \cos \theta) \tag{2}$$

Dividing equation (1) squared by equation (2) gives $n = n_0(1 - \cos \theta) \cos^2 \phi/(1 + \cos \theta)$. However, $\phi = \theta/2$, so that $n/n_0 = (1 - \cos \theta) \cos^2(\theta/2)/(1 + \cos \theta)$.

When the angle between bonds is the tetrahedral angle θ_t, $\theta = \pi - \theta_t$, $\cos^2(\theta/2) = \sin^2(\theta_t/2) = \frac{2}{3}$ and $\cos \theta = 2\cos^2(\theta/2) - 1 = \frac{1}{3}$, so that $n/n_0 = \frac{1}{3}$. Equation (1) shows that $l = (n_0/n)l_0 \cos \phi = \sqrt{6} l_0$.

3.8. The plot shown is a straight line through the origin. This shows the data to be consistent with equation (3.13) because n is directly proportional to M.

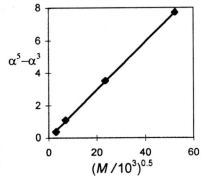

3.9. See the figure to solution 3.5. $l = b \sin \theta$. Setting $b = 0.154$ nm, $\theta = 109.5°/2$ gives $l = 0.126$ nm, corresponding to there being about 100 C atoms in the lamellar thickness.

3.10. The molar mass of the CH_2 group is $14\,\mathrm{g\,mol^{-1}}$.
Hence $160\,000/14 = 11\,430$ C atoms in a typical chain. Hence a typical chain could cross ~ 110–120 times.

Chapter 4

4.1. *Syndiotactic.*

Looking down on the chain
parallel to the plane of the zig zag.   Looking along the chain.

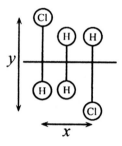
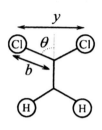

$x = c$ in example 4.3 $y = 2b \sin \theta$, $b = 0.177$ nm, $\theta \approx 109.5°/2$
    $= 0.251$ nm   giving $y = 0.289$ nm.

The separation of the Cl atoms is $(x^2 + y^2)^{1/2} = 0.383$ nm. For *iso-tactic*, the separation of the Cl atoms is simply $x = 0.251$ nm and the planar zig zag is thus not possible for the isotactic chain.

4.2. The possible tetrads are rrr, rmr, (rrm + mrr), (rmm + mmr), mmm and mrm. Proceeding as in example 4.1 and using a similar notation: $P_{rrr} = P_r^3$, $P_{rmr} = P_r^2(1 - P_r)$, $P_{rrm} = 2P_r^2(1 - P_r)$, $P_{rmm} = 2P_r(1 - P_r)^2$, $P_{mmm} = (1 - P_r)^3$ and $P_{mrm} = P_r(1 - P_r)^2$. (These probabilities again add up to unity.) The ratios of the intensities are thus $P_{rrr} : P_{rmr} : P_{rrm} : P_{rmm} : P_{mmm} : P_{mrm} = 0.166 : 0.136 : 0.272 : 0.223 : 0.091 : 0.111 = 0.61 : 0.50 : 1.00 : 0.82 : 0.33 : 0.41$.

4.3. A simple 3_1 helix has alternating *trans* and *gauche* bonds with the *gauche* bonds parallel to the helix axis. The *trans* bonds therefore make the angle $\pi - \theta_t$ with the helix axis, where θ_t is the tetrahedral angle. The length d of the repeat unit is thus $d = 3l[1 + \cos(\pi - \theta_t)]$, where l is the backbone bond length. Taking $l = 0.154$ nm leads to $d = 0.616$ nm. The layer-line angle ϕ_l is given by $d \sin \phi_l = l\lambda$, so that $\phi_2 = \sin^{-1}(2\lambda/d) = 39.0°$.

4.4. In the diagram the circle, of radius R, represents the cross-section of the camera, O is the position of the sample, OC is the direction of the incident X-ray beam and OP is the direction of the equatorial point for Bragg scattering with Bragg angle θ. The length CP′ is then equal to the meridional 'radius' of

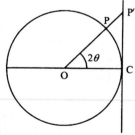

the pattern and the arc length CP is equal to the equatorial 'radius'. The ratio is thus $2R\theta/R \tan(2\theta) = 2\theta/\tan(2\theta)$, which, for $2\theta = 30°$, is 0.907.

4.5. For the lth layer line $c \sin \phi = l\lambda$ and for the (00l) powder ring $2d_{00l} \sin \theta = \lambda$. It follows that $c \sin \phi = 2ld_{00l} \sin \theta$. Now, $d_{00l} = c/l$, so that $\sin \phi = 2 \sin \theta = \sin(2\theta)/\cos \theta$. However, $\cos \theta < 1$ unless $\theta = 0$, which implies that $\phi > 2\theta$ unless $\phi = \theta = 0$.

4.6. The 'blobs' P and P$'$ represent scattering points related by the glide. ($h0l$) reflections are those from planes parallel to the b-axis, with h wavelengths path difference for planes separated by a along the a-axis. Because of the glide, there are planes half way between these planes containing exactly the same number of equivalent scattering points, so that, if h is odd, they will scatter an odd number of half-wavelengths out of phase with the first set. The total intensity of scatter will then be zero. A similar argument applies for the ($0kl$) planes.

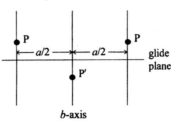

4.7. The dots represent the chain axes of PE. For any orthorhombic, $\tan \zeta = b/a$ and $d_{110} = a \sin \zeta$, so that $d_{110} = a \sin[\tan^{-1}(b/a)]$. Similarly, $\tan \xi = a/b$ and $d_{110} = b \sin \xi$, so that $d_{110} = b/\sin[\tan^{-1}(a/b)]$. However, $\sin \zeta$ is also equal to $b/\sqrt{a^2 + b^2}$, so that $d_{110} = ab/\sqrt{a^2 + b^2}$. For PE, $d_{110} = 0.742 \times 0.495/\sqrt{0.742^2 + 0.495^2} = 0.412$ nm. However, $d_{200} = a/2 = 0.371$ nm. Thus d_{110} and d_{200} differ by only 10% and the packing is pseudo-hexagonal.

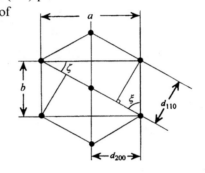

4.8. $d_{110} = 0.412$ nm, see the solution to problem 4.7, $d_{200} = a/2 = 0.371$ nm and $d_{020} = b/2 = 0.248$ nm. $2d_{hkl} \sin \theta = \lambda$.

Thus $2\theta_{hkl} = 2 \sin^{-1}[\lambda/(2d_{hkl})]$ and $2\theta_{110} = 27.2°$; $2\theta_{200} = 30.3°$ and $2\theta_{020} = 46.0°$.

$c \sin \phi = \lambda$. With $c = 0.255$ nm, $\phi = \sin^{-1}(0.194/0.255) = 49.5°$.

4.9. Bragg angles θ for (110) and (200) are $21.6°/2$ and $24°/2$. $2d_{hkl} \sin \theta_{hkl} = \lambda$ and hence $d_{110} = 0.411$ nm; $d_{200} = 0.370$ nm and $a = 2d_{200} = 0.740$ nm. However, see solution to problem 4.7, $d_{110} = a \sin[\tan^{-1}(b/a)]$, giving $b = a \tan[\sin^{-1}(d_{110}/a)] = 0.494$ nm. The layer line gives $c = \lambda/\sin 37.2° = 0.255$ nm. The spectral splitting suggests that there are two chains per unit cell; thus the molar mass per c-axis repeat $= N_A \times 10^3 \times abc \times \text{density}/2 = 28$ g mol^{-1}. A reasonable guess would be polyethylene, with molar mass per chemical repeat unit $14 \, \mathrm{g \, mol^{-1}}$ and two chemical repeats per c-axis repeat unit. The cell dimensions are also in reasonable agreement with those in section 4.4.1.

4.10. Bragg angles θ for (110), (200) and (020) are half the given angle and, similarly to the solution of problem 4.9, lead to $d_{110} = 0.405 \, \mathrm{nm}$; $d_{200} = 0.238 \, \mathrm{nm}$; $d_{020} = 0.383 \, \mathrm{nm}$. Hence $a = 2d_{200} = 0.476 \, \mathrm{nm}$ and $b = 2d_{020} = 0.766 \, \mathrm{nm}$. However (see the solution to problem 4.7), d_{110} should be given by $d_{110} = b \sin[\tan^{-1}(a/b)] = 0.404 \, \mathrm{nm}$. The indexing is thus consistent. The value of c is given by $\lambda / \sin 25.6° = 0.356 \, \mathrm{nm}$ and the mass of the unit cell is $4(12 + 2 + 16) = 120 \, \mathrm{amu}$.

The density is thus $120 \times 1.66 \times 10^{-27}/(abc) = 1.53 \times 10^{3} \, \mathrm{kg \, m^{-3}}$.

4.11. For orthorhombic, $d_{hkl} = 1/\sqrt{h^2/a^2 + k^2/b^2 + l^2/c^2}$ (see section 2.5.1). Thus $d_{221} = 1/\sqrt{4/1.03^2 + 4/0.54^2 + 1/0.50^2} = 0.2157 \, \mathrm{nm}$ and the corresponding diffraction angle $2\theta = 2\sin^{-1}[\lambda/(2d_{221})] = 41.83°$. The distance of the spot from the centre of the film is thus $8.0 \tan 41.83° = 7.16 \, \mathrm{cm}$. The height of the first layer line at the meridian is $8.0 \tan[\sin^{-1}(\lambda/c)] = 2.59 \, \mathrm{cm}$. The separation of the spots is thus approximately $2\sqrt{7.16^2 - 2.59^2} = 13.4 \, \mathrm{cm}$. Taking the curvature into account leads to a separation of $12.7 \, \mathrm{cm}$.

Chapter 5

5.1. By definition, $\chi_v = V_c/(V_a + V_c)$ and $\chi_m = M_c/(M_a + M_c)$. Hence

$$\frac{\chi_m}{\chi_v} = \frac{M_c}{V_c} \frac{V_a + V_c}{M_a + M_c} = \frac{\rho_c}{\rho_s}$$

which is equivalent to equation (5.1).

5.2 $M_c = M_s - M_a$ leads to $\rho_c V_c = \rho_s V_s - \rho_a V_a$, where M_s and V_s are the mass and volume of the sample, respectively. Subtracting $\rho_a V_c$ from both sides leads to

$$\rho_c V_c - \rho_a V_c = \rho_s V_s - \rho_a V_a - \rho_a V_c = \rho_s V_s - \rho_a(V_a + V_c)$$
$$= \rho_s V_s - \rho_a V_s$$

Thus $V_c(\rho_c - \rho_a) = V_s(\rho_s - \rho_a)$. However, $\chi_v = V_c/V_s$ and equation (5.2) follows

5.3. Equations (5.1) and (5.2) lead immediately to

$$\chi_m = \frac{\rho_c}{\rho_s}\left(\frac{\rho_s - \rho_a}{\rho_c - \rho_a}\right) = \left(\frac{\rho_c}{\rho_c - \rho_a}\right)\left(\frac{\rho_s - \rho_a}{\rho_s}\right) = A\left(1 - \frac{\rho_a}{\rho_s}\right) \qquad \text{(a)}$$

with $A = \left(\dfrac{\rho_c}{\rho_c - \rho_a}\right)$, which is independent of the sample.

Multiplying equation (a) by ρ_s gives

$$\chi_m \rho_s = A(\rho_s - \rho_a) \tag{b}$$

Substituting into this equation for the two samples of known χ_m and dividing one resulting equation by the other gives

$$\frac{0.1 \times 1346}{0.5 \times 1392} = \frac{1346 - \rho_a}{1392 - \rho_a}$$

and solving gives $\rho_a = 1335 \, \mathrm{kg\,m^{-3}}$. Substitution into equation (b) for the second sample gives $1/A = 0.5 \times 1392/(1392 - 1335) = 12.21$. However, $1/A = 1 - \rho_a/\rho_c$. Solving gives

$$\rho_c = \frac{\rho_a}{1 - 1/A} = \frac{1335}{1 - 0.0819} = 1454 \, \mathrm{kg\,m^{-3}}$$

χ_m for the third sample is given by

$$\chi_m = A\left(1 - \frac{\rho_a}{\rho_s}\right) = 12.21(1 - 1335/1357) = 0.198$$

or 19.8%.

5.4. The condition for the Gibbs free energy on melting can be written $\Delta g = \Delta h - T_m^o \Delta s = 0$, where the notation is as in the text, section 5.6.1. Thus $\Delta s = \Delta h/T_m^o$ and, if Δh and Δs are independent of temperature, then at any other temperature $\Delta g = \Delta h(1 - T/T_m^o)$. The *effective* value of the Gibbs free energy per unit mass for a lamellar crystal of density ρ_c, thickness l and surface energy σ per unit area is, however, higher by $2\sigma/(l\rho_c)$ (two surfaces on a mass $l\rho_c$ per unit area). Thus, at the melting temperature T_m,

$\Delta h(1 - T_m/T_m^o) + 2\sigma/(l\rho_c) = 0$, which is equivalent to equation (5.7).

5.5. This problem is most easily solved by using a spreadsheet to plot $R = M/M_{max}$ against $\sqrt{t}$. This should first be done for the experimental data and then, on the same graph, for an assumed value of b, the mean thickness of the amorphous layers. A sensible value to assume is 5–10 nm. This value can then be refined manu-

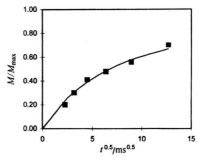

ally, by adjusting the value until a visually, good fit to the data is obtained. Alternatively a least-squares fit can be done, again by manual adjustment or by using a fitting routine such as the Solver tool in Microsoft Excel. The best fit is found for $b = 8.1$ nm and the graph shows the fit.

5.6. The angle θ_{max} corresponding to the maxima is given by $\tan^{-1}(21 \times 10^{-3}/25 \times 10^{-2}) = 4.8°$. Thus, according to equation (5.6), $R = 0.326\lambda/\sin 2.4° = 4.9\,\mu m$.

5.7. The Arrhénius equation (5.24) can be written $\nu = 1/\tau = \nu_o \exp[-\Delta E/(RT)]$. Thus $\ln \nu = \ln \nu_o - \Delta E/(RT)$. The plot is shown, with the best-fitting equation.

Thus $\Delta E/R = 2453$ K, giving

$\Delta E = 2453R = 2453 \times 8.31$ J mol^{-1}
$= 2.04$ kJ mol^{-1}.

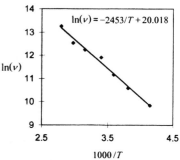

5.8. The c-axis repeat distance in polyethylene is 0.255 nm and corresponds to two CH_2 groups. The lamellar thickness thus corresponds to $30/0.255 = 118$ CH_2 groups. If the chain makes n random jumps of unit length the RMS distance moved is $n^{1/2}$, according to the standard theory for a random walk in one dimension. Thus, if the distance to move is 118 units, the required number of jumps is $118^2 = 13\,924$. The time required is thus $13\,924t$, where t is the mean time between jumps. For $t = 1/(2 \times 10^3)$s the time is thus $13\,924/(2 \times 10^3)$s $= 7.0$ s. According to the Arrhénius equation (5.24), t is proportional to $\exp[\Delta E/(RT)]$, so that, at 300 K, the time is increased by the factor $\exp[(\Delta E/R)(1/300 - 1/360)] = 1118$ and becomes 130 min.

5.9. The graph is as shown and the maximum is approximately at $\omega\tau \cong 0.616$, which means that the minimum in T_1 occurs for this value of $\omega\tau$. Differentiation and setting to zero leads immediately to

$$\frac{1 - \omega^2\tau^2}{(1 + \omega^2\tau^2)^2} + \frac{4(1 - 4\omega^2\tau^2)}{(1 + 4\omega^2\tau^2)^2} = 0.$$

Substitution of x for $(\omega\tau)^2$ and rationalising leads to $5 - x - 20x^2 - 32x^3 = 0$, which may be solved numerically to give $x = 0.3792$, which leads to $\omega\tau \cong 0.616$.

5.10. Graphs showing the original correlation function, the sum of the exponentials and the difference between these two multiplied by a suitable factor should be produced and should look as shown: (a) is the original curve and (b) is the difference curve times 10. (The sum of the exponentials has been omitted because it is so close to the original curve on the scale shown that it is difficult to see the difference.)

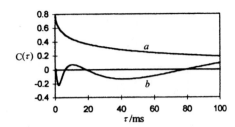

Chapter 6

6.1. Figure (a) shows the strain very much exaggerated. Figure (b) shows the unit cube as if it were undistorted, because

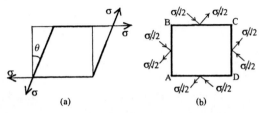

the strains are actually vanishingly small. Consider the strains in the directions of the diagonals AC and BD. To first order in θ

$$AC = \sqrt{(1+\theta)^2 + 1^2} = \sqrt{2(1+\theta)} = \sqrt{2} \times (1 + \theta/2); \quad \text{strain} = \theta/2$$

$$BD = \sqrt{(1-\theta)^2 + 1^2} = \sqrt{2(1-\theta)} = \sqrt{2} \times (1 - \theta/2); \quad \text{strain} = -\theta/2$$

The shear strain is thus equivalent to a tensile and a compressive strain of $\theta/2$ each at $45°$ to the original axes. Figure (b) shows that the shear stress is equivalent to a tensile and a compressive force parallel to these directions equal to $2 \times \sigma/\sqrt{2} = \sqrt{2}\sigma$. However, the area over which these act is the diagonal plane area $= \sqrt{2}$, so that the stresses are σ. Thus, using the condition of linearity,

$$\frac{\sigma}{2G} = \frac{\theta}{2} = \frac{\sigma}{E} + \frac{-\nu(-\sigma)}{E} \qquad \text{or} \qquad G = \frac{E}{2(1+\nu)}$$

6.2. $e_1 = \sigma_1/E - \nu\sigma_2/E - \nu\sigma_3/E = \sigma_1/E - \nu(\sigma_2 + \sigma_3)/E$

However, $\sigma_1 + \sigma_2 + \sigma_3 = -3p$, so that

$$e_1 = \sigma_1/E + \nu\sigma_1/E + 3\nu p/E = [(1+\nu)\sigma_1 + 3\nu p]/E$$

Similar results follow for e_2 and e_3.

6.3. From first principles, each stress produces its own strain and strains due to all stresses are additive. Thus
$e_1 = \sigma_1/E - \nu\sigma_2/E = 2.4 \times 10^{-3}$. Similarly, $e_2 = 1.1 \times 10^{-3}$ and
$e_3 = -(\nu\sigma_1/E + \nu\sigma_2/E) = -1.5 \times 10^{-3}$. Alternatively,
one could use equation (6.7), with $p = -(\sigma_1 + \sigma_2 + \sigma_3)/3$.

6.4. $e_1 = \sigma_1/E - \nu(\sigma_2/E + \sigma_3/E) = [\sigma_1 - \nu(\sigma_2 + \sigma_3)]/E$

but $\nu = \frac{1}{2}$ for an incompressible solid, hence

$e_1 = [\sigma_1 - (\sigma_2 + \sigma_3)/2]/E = [8 - (5 + 7)/2]/(4 \times 10^3) = 5 \times 10^{-4}$

Incompressibility means that equal changes in the three stresses have no effect on strain. Thus one requires $\sigma_1 = 3\,MPa$ and $\sigma_2 = 2\,MPa$.

6.5. Under uniform strain the modulus is the same as the effective modulus of a rod of unit total cross-section consisting of two cross-sections equal to the volume fractions of the crystalline and amorphous materials placed in parallel. To produce a unit strain in such a rod requires a stress σ equal to $\chi_v E_c + (1 - \chi_v)E_a$, which is therefore equal to the modulus. Substitution gives $E = 3.1 \times 10^9\,N\,m^{-2}$.

6.6. $U = C(\lambda_1^2 + \lambda_2^2 + \lambda_3^2 - 3)$, where $C = G/2 = E/6$. If the stretch direction is Ox_3, $\lambda_3 = \lambda$ and $\lambda_1\lambda_2\lambda_3 = 1$. Thus $\lambda_2 = \lambda_1 = 1/\sqrt{\lambda}$, where $\lambda = 2$. Let the original diameter be d and the original area be A.

  (i) The stretched diameter $= d\lambda_1 = 2/\sqrt{2} = \sqrt{2} = 1.41\,mm$.
  (ii) The nominal stress $= F/A = (E/3)(\lambda - 1/\lambda^2)$, (equation (6.21a)),
$= 1.2 \times 1.75/3 = 0.7\,N\,mm^{-2}$.
  (iii) True stress $=$ nominal stress$/(\lambda_1\lambda_2) = 0.7/(\lambda_1\lambda_2) = 0.7\lambda$
$= 1.4\,N\,mm^{-2}$.
  (iv) $F =$ nominal stress $\times$ original area $= 0.7\pi = 2.2\,N$.

6.7. On the analogy with surface tension, the excess pressure p inside is given by $p = 2F/r$, where F is the tension per unit length and r is the radius. If t is the original thickness, $F = G(1 - 1/\lambda^6)t$ (equation (6.30)), where $G = E/3$ and λ is the ratio of the diameters. Pressures inside are thus atmospheric ($10^5\,Pa$) plus 0.317, 0.472, 0.487, 0.318 and $0.160 \times 10^5\,Pa$. Note that p has a maximum value at low expansion. (It is unlikely that a real rubber would be neo-Hookeian at the larger diameters.)

6.8. 2×10^{25} cross-links per m^3 implies $N = 4 \times 10^{25}$ chains per m^3, assuming that four chains meet at every cross-link point and that both ends of every chain are linked, i.e. there are no dangling chains. A 50% increase means $\lambda = 1.5$. With $T = 300\,K$, equation (6.49) gives the true stress σ as

$\sigma = 4 \times 10^{25} \times 1.38 \times 10^{-23} \times 300(1.5^2 - 1/1.5) = 2.62 \times 10^5\,Pa$

This acts over the area $\pi r^2/\lambda = 10^{-6}\pi/1.5\,m^2$ and thus requires a force $2.62 \times 10^5 \times 10^{-6}\pi/1.5 = 0.55\,N$.

6.9. True stress $\sigma = NkT(\lambda^2 - 1/\lambda)$. Force $F = \sigma A/\lambda$, with A the original cross-section. Thus $F = NkTA(\lambda - 1/\lambda^2)$. For $\lambda = 2$,
$F_2 = NkTA(2 - \frac{1}{4}) = 7NkTA/4$; and for $\lambda = 3$,
$F_3 = NkTA(3 - \frac{1}{9}) = 26NkTA/9$.

Thus $F_3 - F_2 = NkTA(26/9 - \frac{7}{4}) = 1.139NkTA = 1$ N. Substitution gives $N = 2.12 \times 10^{26}$ m^{-3}. If the new temperature is T_N, then $F_3 = 26NkTA/9 = 7NkT_NA/4$. Thus $T_N = (26/9) \times \frac{4}{7} = 495.2$ K and the rise in temperature is 195.2 K.

6.10. Carrying out the corresponding steps to those leading from equation (6.26) to equation (6.30) leads to $F = \frac{3}{2}(\mu/\alpha)(1/\sqrt{\lambda} - 1/\lambda^5)t$, where F is the force acting on unit length of strained material. For area doubling, $\lambda = \sqrt{2}$. Substitution then leads to $F = 1.5 \times 4.2 \times 10^5 \times 0.664 \times 10^{-3}$ N $= 418$ N. The strained edge length is $0.1 \times \sqrt{2}$ m, so that $f = 0.1 \times \sqrt{2} \times 418$ N $= 59.2$ N.

Chapter 7

7.1. $\tau = \eta/E$, with τ the relaxation time, η the viscosity of the dashpot $= 10^{12}$ Pa s and E the modulus of the spring $= 10^{10}$ Pa. Thus $\tau = 100$ s.

Equation (7.16) leads to $\sigma = \sigma_o \exp(-t/\tau) = eE \exp(-t/100)$. With $e = 10^{-2}$, $t = 50$ s,
$\sigma = 10^{-2} \times 10^{10} \exp(-50/100) = 10^8 \exp(-0.5) = 0.61 \times 10^8$ Pa.

7.2. Equation (7.16) shows that immediately before t_2 the stress is $\sigma = \sigma_1 \exp[-(t_2 - t_1)/\tau]$, where $\sigma_1 = e_1 E$. If the strain is now increased by e_2 the spring extends by a further strain e_2 because the dashpot cannot instantaneously change its displacement. The stress therefore increases by $\sigma_2 = e_2 E$ and immediately afterwards has the value $\sigma' = \sigma + \sigma_2 = \sigma_1 \exp[-(t_2 - t_1)/\tau] + \sigma_2$. Because the absolute position of the dashpot plunger has no significance, equation (7.16) can be applied again with σ' replacing σ_1 and $t - t_2$ replacing t, leading to

$$\sigma = \{\sigma_1 \exp[-(t_2 - t_1)/\tau] + \sigma_2\} \exp[-(t - t_2)/\tau]$$
$$= \sigma_1 \exp[-(t - t_1)/\tau] + \sigma_2 \exp[-(t - t_2)/\tau]$$

which is the expression obtained if the BSP is assumed valid from the start.

7.3. $e(t) = \sum_i J(t - t_i)\sigma_i$, where σ_i is the step in stress at time t_i and the sum is over all σ_i applied before the time t at which e is measured.

(a) $e(1500) = J(1500 - 0)\sigma_0 + J(1500 - 1000)\sigma_{1000}$
$= 1.2 \times 10^{-9}(1 \times 1500^{0.1} + 0.5 \times 500^{0.1}) \times 10^6$
$= 3.61 \times 10^{-3}$

(N.B. the step in stress at 1000 s is 0.5 MPa, *not* 1.5 MPa.)

(b) $e(2500) = J(2500 - 0)\sigma_0 + J(2500 - 1000)\sigma_{1000}$
$+ J(2500 - 2000)\sigma_{2000}$
$= 1.2 \times 10^{-9}(2500^{0.1} + 0.5 \times 1500^{0.1} - 1.5 \times 500^{0.1}) \times 10^6$
$= 0.52 \times 10^{-3}$

7.4. $E = 5\exp(-t^{1/3})$ GPa, Hence

(a) $\sigma/e = 5\exp(-t^{1/3})$ GPa   (σ is stress, e is strain).

Thus the immediate stress σ_0 (at $t = 0$) $= 5e$ GPa. However,

$$\sigma_0 = 10^3/[\pi(4 \times 10^{-3})^2] = 1.989 \times 10^7 \text{ Pa} \rightarrow e = 3.98 \times 10^{-3}$$

(b) $\sigma = \sigma_0 \exp(-t^{1/3}) \rightarrow F(t) = F_0\exp(-t^{1/3})$, where F is the force. $F(24) = 10^3 \exp(-2.8845) = 55.9$ N.

7.5. The equilibrium stress is eG and the rate of change of stress at time t is $d\sigma/dt$. Thus $d\sigma/dt = -C(\sigma - eG)$, where C is a positive constant because the stress will decrease with time if $\sigma > eG$. Setting $C = 1/\tau$ gives the required equation.

Slight rearrangement then leads to $d\sigma/dt + \sigma/\tau = eG/\tau$, which is equivalent to equation (7.15) with $\tau = \eta/E$ only if $G = 0$.

7.6. For a unit cube the increase in strain δe is equal to the change in separation of the forces σ, so that the work done is $\delta\mathcal{E} = \sigma\,\delta e$. The work done on the unit cube per cycle is thus $\Delta\mathcal{E} = \int \sigma\,de$, where the integral is over a complete cycle. Writing $de = (de/dt)\,dt$ gives

$$\Delta\mathcal{E} = \int_0^{2\pi/\omega} \sigma(de/dt)\,dt$$

Substitution of $\sigma = \sigma_o \cos(\omega t + \delta)$, $e = e_o \cos(\omega t)$ leads to

$$\Delta\mathcal{E} = -\sigma_o e_o \int_0^{2\pi/\omega} \cos(\omega t + \delta)\sin(\omega t)\,dt = \pi\sigma_o e_o \sin\delta$$

Equation (7.21) shows that $\sin\delta = G_2 e_o/\sigma_o$ and the required result follows.

7.7. $de/dt = -e_1\omega\sin(\omega t) + e_2\omega\cos(\omega t)$;

$$\frac{\sigma/E - e}{\tau} = \frac{\sigma_o \cos(\omega t)/E - e_1 \cos(\omega t) - e_2 \sin(\omega t)}{\tau}$$

Equating the sin and cos components of the RHSs gives

$$e_2 = \omega\tau e_1 \tag{1}$$

and

$$e_2\omega\tau = \sigma_o/E - e_1 \tag{2}$$

Substituting e_2 from equation (1) into equation (2) gives

$$\omega^2\tau^2 e_1 = \sigma_o/E - e_1 \quad \text{or} \quad J_1 = \frac{e_1}{\sigma_o} = \frac{1/E}{1 + \omega^2\tau^2}$$

Equation (1) then leads to

$$J_2 = \frac{e_2}{\sigma_0} = \frac{\omega\tau/E}{1+\omega^2\tau^2}$$

The graph shows the results of an accurate calculation, the table gives the calculation for suggested values.

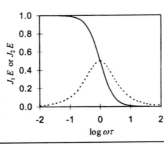

$\omega\tau$	0.01	0.10	0.316	1	3.16	10	100
$\log(\omega\tau)$	−2	−1	−0.5	0	0.5	1	2
$J_1 E$	1	0.99	0.91	0.5	0.09	0.01	10^{-4}
$J_2 E$	0.01	0.10	0.29	0.5	0.29	0.10	0.01

7.8. This problem is probably best solved by using a spreadsheet. The values of $\log J_2$ should first be plotted against $\log f$ for each temperature, on the same graph, as in the plot on the left below. The shift factors can then be guessed and the new $\log f$ values deduced for each temperature.

T (°C)	$\log a_T$	
90	1.65	9.32
80	0.95	8.53
70	0.0	7.57
60	−1.05	6.37
55	−1.7	5.65
50	−3.0	4.83

A new plot of $\log J_2$ for each temperature using the appropriate new $\log f$ values gives an approximate master curve. The shift factors can then be adjusted to get a better fit of all the data to the master curve. The shifts in the second column of the table give the master curve shown below in the plot on the right

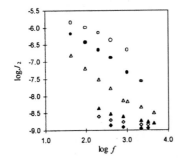
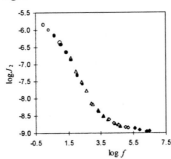

The shift factors calculated from the WLF equation are shown in the right-hand column of the table. These are measured from the curve that would in theory be obtained at T_g and should differ by a constant amount from those found with 70 °C as the reference T. Subtracting the values in the second column from those in the first leads to the approximate value 7.6 in all cases, confirming the applicability of the WLF equation.

7.9. In the WLF equation a_T is defined as $\tau(T_g)/\tau(T)$. Making the substitutions listed below equation (7.33) into equation (7.32) leads immediately to

$$\frac{\tau(T_g)\exp(-C_1/\log_{10} e)}{\tau(T)} = \exp\left(\frac{-C_1 C_2/\log_{10} e}{T - T_g + C_2}\right)$$

so that

$$a_T = \exp\left(\frac{-C_1 C_2/\log_{10} e}{T - T_g + C_2}\right) \frac{1}{\exp(-C_1/\log_{10} e)}$$

$$= \exp\left(\frac{[-C_1 C_2 + C_1(T - T_g + C_2)]/\log_{10} e}{T - T_g + C_2}\right)$$

which leads to

$$(\log_{10} e)(\ln a_T) = \log_{10} a_T = \frac{C_1(T - T_g)}{C_2 + (T - T_g)}$$

7.10. (a) The plot is as shown at the left below.

(b) $\log_{10} a_{100} = 17.4 \times (100 - 50)/[52 + (100 - 50)]$
$= 8.53 \log_{10} J_1(T_g, \omega) = 5 + 4/[\exp(L + 8.53 - 6) + 1]$

(c) $\log_{10} J_1(T, \omega) = 5 + 4/[\exp(L + 8.53 - L_T - 6) + 1]$ where
$L_T = \log_{10} a_T = 17.4 \times (T - 50)/[52 + (T - 50)]$ and
$L = \log_{10} 1 = 0$ for $\omega = 1$ s^{-1}. The plot is shown at the right below.

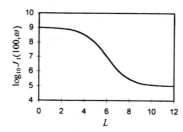

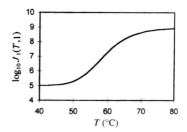

Chapter 8

8.1. Expanding the LHS of equation (8.6), viz.
$(\sigma_1 - p')^2 + (\sigma_2 - p')^2 + (\sigma_3 - p')^2$ gives
$C = \sigma_1^2 - 2\sigma_1 p' + p'^2 + \sigma_2^2 - 2\sigma_2 p' + p'^2 + \sigma_3^2 - 2\sigma_3 p' + p'^2$
Substituting $p' = \frac{1}{3}(\sigma_1 + \sigma_2 + \sigma_3)$ gives

$$C = \sigma_1^2 - 2\sigma_1(\sigma_1 + \sigma_2 + \sigma_3)/3 + \sigma_2^2 - 2\sigma_2(\sigma_1 + \sigma_2 + \sigma_3)/3$$

$$+ \sigma_3^2 - 2\sigma_3(\sigma_1 + \sigma_2 + \sigma_3)/3 + (\sigma_1 + \sigma_2 + \sigma_3)^2/3$$

$$= 2(\sigma_1^2 + \sigma_2^2 + \sigma_3^2)/3 - 2(\sigma_1\sigma_2 + \sigma_2\sigma_3 + \sigma_3\sigma_1)/3$$

$$= \frac{1}{3}[(\sigma_1^2 + \sigma_2^2 - 2\sigma_1\sigma_2) + (\sigma_2^2 + \sigma_3^2 - 2\sigma_2\sigma_3) + (\sigma_3^2 + \sigma_1^2 - 2\sigma_3\sigma_1)]$$

$$= \frac{1}{3}[(\sigma_1 - \sigma_2)^2 + (\sigma_2 - \sigma_3)^2 + (\sigma_3 - \sigma_1)^2]$$

The required result follows immediately from the definition of τ_{oct}, equation (8.7).

8.2. For the first condition, equation (8.7) leads to
$9\tau_{\text{oct}}^2 = (2\sigma)^2 + \sigma^2 + \sigma^2 = 6\sigma^2$ and for the second condition it leads to
$9\tau_{\text{oct}}^2 = (0)^2 + \sigma_2^2 + \sigma_1^2 = 2\sigma_1^2$. Hence $\sigma_1^2 = 3\sigma^2$ so that $\sigma_1 = \sqrt{3}\sigma$.

8.3. Let the rod have cross-sectional area A. The tensile force acting on the rod is then σA. Resolving this parallel to the plane with the normal making angle θ to the axis gives a shearing force $\sigma A \sin\theta$. This acts over the cross-section $A/\cos\theta$ of the rod parallel to the plane, so that the shear stress $= \sigma A \sin\theta/(A/\cos\theta) = \sigma \sin\theta\cos\theta$. The normal force over this area is $\sigma A \cos\theta$ so that the normal stress is $\sigma_n = \sigma\cos^2\theta$. Thus $\tau_c = \tau + \mu\sigma_n = \sigma(\sin\theta\cos\theta + \mu\cos^2\theta)$. Differentiation shows that this has a minimum value when $\cos(2\theta) - \mu\sin(2\theta) = 0$, i.e. when $\mu\tan(2\theta) = 1$. Writing $\mu = \tan\phi$ leads to $\tan\phi\tan(2\theta) = 1$, which has the solution $\theta = \pi/4 - \phi/2$. (Note that, if a *compressive* stress of magnitude σ is applied, the shear stress along any plane is reversed and so is the direction of yielding. The angle of yielding is not, however, the same because the normal stress is now compressive. This is equivalent to replacing the term $\mu\cos^2\theta$ by $-\mu\cos^2\theta$, which leads to $\theta = \pi/4 + \phi/2$.)

8.4. Let σ (positive) be the magnitude of the compressive stress, so that $\sigma_1 = -\sigma$. The stresses acting on the polymer are then as shown. Equation (8.9) then leads to
$$(\sigma + \sigma_2)\sin\theta\cos\theta = \tau_c + \mu(\sigma\sin^2\theta - \sigma_2\cos^2\theta)$$
Reordering gives

$$\sigma = \frac{\tau_c - \sigma_2(\mu\cos^2\theta + \sin\theta\cos\theta)}{\sin\theta\cos\theta - \mu\sin^2\theta} = \frac{\tau_c - \sigma_2(\mu - \mu\sin^2\theta + \sin\theta\cos\theta)}{\sin\theta\cos\theta - \mu\sin^2\theta}$$

$$= \frac{\tau_c}{\sin\theta\cos\theta - \mu\sin^2\theta} - \left(1 + \frac{\mu}{\sin\theta\cos\theta - \mu\sin^2\theta}\right)\sigma_2$$

$$= \frac{\tau_c - \sigma_2\mu}{\sin\theta\cos\theta - \mu\sin^2\theta} - \sigma_2 \qquad (1)$$

The minimum value of σ that will cause yield for a given value of σ_2 is thus the one corresponding to the maximum value of
$\sin\theta\cos\theta - \mu\sin^2\theta$.
Differentiating $\sin\theta\cos\theta - \mu\sin^2\theta$ and setting equal to zero gives
$\mu\tan(2\theta) = 1$. Substituting 0.25 for μ gives $\theta = 38.0°$ and substitution of this value gives

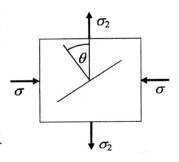

$\sin\theta\cos\theta - \mu\sin^2\theta = 0.39$. At yield $\sigma_1 = -\sigma = a + b\sigma_2$ and comparison with the third expression for σ given in equation (1) shows that $a = -5 \times 10^7/0.39 = -1.28 \times 10^8$ Pa and $b = 1 + 0.25/0.39 = 1.64$.

8.5. Equation (b) of example 8.2 shows that the angle θ of yield in compression is given by $\theta = \pi/4 + \phi/2$, where $\mu = \tan\phi$. Thus $\mu = \tan(2\theta - \pi/2)$. Substituting $\theta = 47°$ gives $\mu = \tan 4° = 0.070$. For tensile stress, $\theta = \pi/4 - \phi/2 = 45° - 2° = 43°$.

8.6. Expanding equation (8.17) leads to

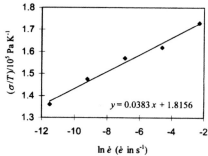

$$\frac{\sigma}{T} = \frac{R}{\upsilon}\ln\dot{e} + \frac{R}{\upsilon}\left(\frac{\Delta H}{RT} - \ln C\right)$$

so that a plot of σ/T against $\ln\dot{e}$ should yield a straight line with gradient R/υ. The plot is shown and the gradient is 3.83×10^3 Pa K^{-1}, so that the activation volume per site is

$$R/(3.83 \times 10^3 N_A) = k/(3.83 \times 10^3) = 3.6 \times 10^{-27} \text{ m}^3 = 3.6 \text{ nm}^3$$

Alternatively, plot σ against $\ln\dot{e}$ and divide the gradient by T.

8.7. $E = $ stress/strain $= $ (force/area)/(extension/length). Substituting gives

$$E = \left[200/\{\pi \times (2.5 \times 10^{-3})^2\}\right]/(0.25 \times 10^{-3}/0.1) \text{ Pa} = 4.07 \times 10^9 \text{ Pa}$$

Replacing γ by $G/2$ in equation (8.19) leads to $G = \pi l\sigma_B^2/(2E) = \pi l(F_B/A)^2/(2E)$, where F_B is the breaking force on a sample of cross-sectional area A. Substituting gives $G = \pi \times 1.0 \times 10^{-3}[1600/(15 \times 10^{-3} \times 1.5 \times 10^{-3})]^2/(2 \times 4.07 \times 10^9)$ $= 1.95 \times 10^3$ J m^{-2}.

8.8. Equation (8.28) can be rewritten $l^3 = EBb^3x/(64F)$ and equation (8.29) can be written $l^4 = 3x^2b^3E/(128G)$. Raising the first of these two equations to the fourth power and the second to the third power, equating the new RHSs and rearranging leads immediately to $G^3 = 216F^4x^2/(EB^4b^3)$. Substituting the given data then gives $G = 1.38 \times 10^2$ J m^{-2}.

8.9. Equation (8.21) shows that the critical stress intensity factor, K_c is given by $K_c = \sigma_B\sqrt{\pi l/2}$, where l is the length of the crack, and equation (8.30) shows that

$$\sigma_{cr}^2 = \frac{\pi}{8R}K_c^2 = \frac{\pi^2 l\sigma_B^2}{16R} \quad \text{or} \quad \sigma_{cr} = \frac{\pi\sigma_B}{4}\sqrt{\frac{l}{R}}$$

Substituting the given data leads to $\overset{\circ}{\sigma}_{cr} = 272$ MPa.

Chapter 9

9.1. In the figure the circle represents a cross-section of the hole and OO is a diameter parallel to the polarisation P. ABC represents a section of the surface of the sphere between the two cones formed by rotating the two lines separated by the angle $d\theta$ about OO. Each small element dA of this area has a surface charge $P \, dA \cos\theta$ and produces a field

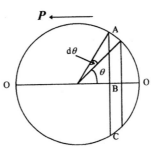

$(P \, dA \cos^2\theta)/(4\pi\varepsilon_0 r^2)$ parallel to OO at the centre of the sphere. The total field produced at the centre by all the area ABC is thus $(2\pi r^2 \sin\theta \, d\theta)(P\cos^2\theta)/(4\pi\varepsilon_0 r^2) = [P/(2\varepsilon_0)]\cos^2\theta \sin\theta \, d\theta.$

Integrating from $\theta = 0$ to $\theta = \pi$ then gives the required result.

9.2. For a polymer the Lorentz–Lorenz equation (9.9) may be rewritten

$$\frac{n^2 - 1}{n^2 + 2} = \left(\frac{N_A \, \alpha_d}{3\varepsilon_0 \, M}\right)\rho$$

where the quantity in parentheses is the molar refraction per repeat unit divided by the molar mass of the repeat unit. The repeat unit of POM has relative molar mass $12 + 2 \times 1 + 16 = 30$ and molar mass 3.00×10^{-2} kg. There are two C—H bonds and two C—O bonds per repeat unit, giving a refraction per repeat unit of $(2 \times 1.676 + 2 \times 1.54) \times 10^{-6} = 6.432 \times 10^{-6}$ m^3. Thus

$$(n^2 - 1) = 6.432 \times 10^{-6} \times 1.25 \times 10^3 (n^2 + 2)/(3.00 \times 10^{-2})$$

$$= 0.268(n^2 + 2)$$

which leads to $n = 1.449$.

9.3. Equation (9.28b) gives

$$\varepsilon'' = \frac{a\omega\tau}{1 + \omega^2\tau^2}$$

with a constant and equal to $\varepsilon_s - \varepsilon_\infty$. Thus the maximum value of ε'' occurs when

$$\frac{d\varepsilon''}{d\omega} = a\frac{(1 + \omega^2\tau^2)\tau - \omega\tau \times 2\omega\tau^2}{(1 + \omega^2\tau^2)^2} = 0$$

i.e. when $\omega\tau = 1$, so that the maximum value is $a/2$. The half-value points are therefore given by

$$\frac{\omega\tau}{1 + \omega^2\tau^2} = \frac{1}{4}$$

or $(\omega\tau)^2 - 4\omega\tau + 1 = 0$. Solving gives $\omega\tau = 2 \pm \sqrt{3} = 0.268$ or 3.732. The ratio of these is 13.93 and $\log_{10} 13.93 = 1.14$.

9.4. Summing three values of ε' and ε'' using the suggested values of the quantities and plotting ε'' against ε' leads to the solid curve in the figure.

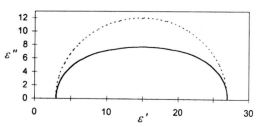

The dotted curve shows the semi-circle expected for a single Debye-type relaxation. To plot this do not forget that the effective values of ε_s and ε_∞ are the sums of those for the individual relaxations.

9.5. The length of chain is Nd and the allowed values of wavelength λ for electron waves are $\lambda = Nd/n$, where periodic boundary conditions are assumed and n is an integer. However, $p = h/\lambda$ and $E = p^2/(2m)$, where p is the momentum and E is the energy of the electron. Thus the allowed values of E are $E = n^2h^2/(2N^2md^2)$. The first $N/4$ energy levels are populated by N electrons in pairs with opposite spin for waves travelling in each direction. Thus the difference in energy between the bottom unfilled level and top filled level is $[(N/4+1)^2 - (N/4)^2]h^2/(2mN^2d^2) = (N+2)h^2/(4mN^2d^2)$.

For $N \gg 1$ this becomes $h^2/(4mNd^2)$. (Note that a treatment in terms of standing electron waves in a box of length L gives an answer half of this. Each standing-wave energy level can then be occupied by only two electrons, however, so that, if a few electrons are excited just above the Fermi level, the average energy required per electron is the same for the two methods of calculation.)

9.6. In equation (9.7) $N_0\alpha$ is the polarisability per unit volume, so that, for each direction Ox_i, the equation can immediately be written $(\varepsilon_i - 1)/(\varepsilon_i + 2) = x_i$, with x_i defined as given. Substituting n_i^2 for ε_i then gives the Lorentz–Lorenz equation in the form $(n_i^2 - 1)/(n_i^2 + 2) = x_i$, or $(n_i^2 - 1) = x_i(n_i^2 + 2)$. Solving for n_i immediately gives the required result.

9.7. The angle between the C—C bonds and the chain axis is $(180° - 112°)/2 = 34°$. Choosing axes as in example 9.2, $\theta_{CC} = 34°$ and $\phi_{CC} = 0$ in equations (9.40). The C—H bonds are perpendicular to the OZ axis and make angles $\pm 109.5°/2$ with the OX axis, so that $\theta_{CH} = 90°$ and $\phi_{CH} = \pm 54.75°$. Substitution in equations (9.40) and using the bond polarisabilities from table 9.5 then gives for a single C—C bond

$$\alpha_{33} = (2.8\sin^2 34° + 10.8\cos^2 34°) \times 10^{-41} \text{ F m}^{-2}$$
$$= 8.29 \times 10^{-41} \text{ F m}^{-2}$$

$$\alpha_{11} = (2.8\cos^2 34° + 10.8\sin^2 34°) \times 10^{-41} \text{ F m}^{-2}$$
$$= 5.30 \times 10^{-41} \text{ F m}^{-2}$$
$$\alpha_{22} = 2.8 \times 10^{-41} \text{ F m}^{-2}$$

and, for each pair of H—C bonds in a CH_2 group,

$$\alpha_{33} = 2 \times 6.7 \times 10^{-41} \text{ F m}^{-2} = 13.4 \times 10^{-41} \text{ F m}^{-2}$$
$$\alpha_{11} = 2(6.7 \times \sin^2 54.75° + 9.1 \times \cos^2 54.75°) \times 10^{-41} \text{ F m}^{-2}$$
$$= 15.00 \times 10^{-41} \text{ F m}^{-2}$$
$$\alpha_{22} = 2(6.7 \times \cos^2 54.75° + 9.1 \times \sin^2 54.75°) \times 10^{-41} \text{ F m}^{-2}$$
$$= 16.60 \times 10^{-41} \text{ F m}^{-2}$$

A translational repeat unit contains two C—C bonds and two CH_2 groups and therefore the polarisability components are obtained by adding the two sets above and doubling the results, to give $\alpha_{33} = 4.33 \times 10^{-40}$ F m^{-2}, $\alpha_{11} = 4.06 \times 10^{-40}$ F m^{-2} and $\alpha_{22} = 3.88 \times 10^{-40}$ F m^{-2}.

The unit cell contains two repeat units with their backbone planes at right angles to each other, so that the polarisabilities parallel and perpendicular to the chain axis become $2\alpha_{33} = 8.66 \times 10^{-40}$ F m^{-2} and $\alpha_{11} + \alpha_{22} = 7.94 \times 10^{-40}$ F m^{-2}, respectively. The volume of the unit cell of polyethylene is (see section 4.4.1) 0.0937 nm. the value of x_i as defined in problem 9.6 then becomes 0.348 for the chain-axis direction and 0.319 for the perpendicular direction, giving $n = 1.613$ and 1.551, respectively.

Chapter 10

10.1. $\langle\cos^2\theta\rangle = 0.2 \times \frac{1}{3} + 0.5 \times \cos^2 30° + 0.3 \times 1 = 1/15 + 3/8 + 0.3$
$$= 0.742$$
$$\langle\cos^4\theta\rangle = 0.2 \times \frac{1}{5} + 0.5 \times \cos^4 30° + 0.3 \times 1 = 1/25 + 9/32 + 0.3$$
$$= 0.621$$

Hence $\langle P_2(\cos\theta)\rangle = (3 \times 0.742 - 1)/2 = 0.61$ and
$$\langle P_4(\cos\theta)\rangle = (3 - 30 \times 0.742 + 35 \times 0.621)/8 = 0.31.$$

10.2. Equation (10.2) becomes

$$\langle P_2(\cos\theta)\rangle = \frac{\int_0^{\pi/2}(3\cos^2\theta - 1)(1 + \cos\theta)\sin\theta d\theta}{2\int_0^{\pi/2}(1 + \cos\theta)\sin\theta d\theta}$$
$$= \frac{\int_0^{\pi/2}(3\cos^3\theta + 3\cos^2\theta - \cos\theta - 1)\sin\theta d\theta}{2\int_0^{\pi/2}(1 + \cos\theta)\sin\theta d\theta}$$

$$= \left[\frac{3}{4}\cos^4\theta + \cos^3\theta - \frac{1}{2}\cos^2\theta - \cos\theta\right]_{\pi/2}^0 \Big/ \left(2\left[\cos\theta + \frac{1}{2}\cos^2\theta\right]_{\pi/2}^0\right)$$

$$= 1/12$$

10.3. For low birefringence $\langle P_2(\cos\theta)\rangle = \Delta n/\Delta n_{max} = 0.6$. If all chains make the same angle θ with the fibre axis, this is the value of $P_2(\cos\theta)$, so that

$$\cos^2\theta = [2P_2(\cos\theta) + 1]/3 = 0.733$$

leading to $\theta = 31°$.

10.4. Equation (10.12) leads to

$$\langle\alpha_{33}\rangle = 1.9 \times 10^{-39} + 0.65 \times (2.6 - 1.9) \times 10^{-39} = 2.36 \times 10^{-39} \text{ F m}^2$$

and equation (10.14) leads to

$$\langle\alpha_{11}\rangle = 1.9 \times 10^{-39} + 0.5 \times (1 - 0.65) \times (2.6 - 1.9) \times 10^{-39}$$
$$= 2.02 \times 10^{-39} \text{ F m}^2.$$

Equation (10.5) leads to

$$n_i = \sqrt{\frac{2\langle\alpha_{ii}\rangle + 3\varepsilon_0/N_0}{3\varepsilon_0/N_0 - \langle\alpha_{ii}\rangle}}$$

However, $3\varepsilon_0/N_0 = 3 \times 8.85 \times 10^{-12}/(4.2 \times 10^{27}) = 6.32 \times 10^{-39}$. Substitution gives

$$n_1 = \sqrt{(2 \times 2.0225 + 6.32)/(6.32 - 2.0225)} = 1.55$$

and similarly $n_3 = 1.67$, so that the birefringence is $1.67 - 1.55 = 0.12$.

10.5. For full alignment the absorbances $A_\parallel$ and $A_\perp$ for radiation polarised parallel and perpendicular to the draw direction, respectively, are proportional to $\langle\cos^2\theta_\mu\rangle = \cos^2\theta_M$ and $\langle\sin^2\theta_\mu\rangle/2 = \sin^2(\theta_M/2)$.

Thus the dichroic ratio $D = A_\parallel/A_\perp = 2(\cos 25°/\sin 25°)^2 = 9.2$.

10.6. $A_\parallel = \log(I_0/I_{\parallel}) = \log(100/50) = 0.3010 = C\langle\cos^2\theta_\mu\rangle$

Similarly

$$A_\perp = \log(100/75) = 0.1249 = \tfrac{1}{2}C\langle\sin^2\theta_\mu\rangle$$

Hence $C = 0.3010 + 2 \times 0.1249 = 0.5508$

$$\rightarrow \langle\cos^2\theta_\mu\rangle = 0.3010/0.5508 = 0.5465$$

$$\rightarrow \langle P_2(\cos\theta_\mu)\rangle = (3\langle\cos^2\theta_\mu\rangle - 1)/2 = 0.3198$$

However,

$$P_2(\cos\theta_M) = (3\cos^2\theta_M - 1)/2 = (3\cos^2 20° - 1)/2 = 0.8245$$

Thus

$$\langle P_2(\cos\theta)\rangle = \langle P_2(\cos\theta_\mu)\rangle/P_2(\cos\theta_M) = 0.3879$$

and

$$\langle \cos^2 \theta \rangle = [2\langle P_2(\cos \theta)\rangle + 1]/3 = 0.59$$

Simplifying assumptions: a highly oriented polymer has the simplest form of uniaxial distribution and is fully oriented, there is no correction for reflection at the surface of the sample, no internal field corrections and no scattering of radiation.

10.7.   For the first film

$$A_\parallel = \log(I_o/I_\parallel) = \log(100/20) = 0.6990 = C\langle \cos^2 \theta_\mu \rangle$$

Similarly

$$A_\perp = \log(100/95) = 0.0223 = \tfrac{1}{2}C\langle \sin^2 \theta_\mu \rangle$$

Hence

$$C = 0.6990 + 2 \times 0.0223 = 0.7436$$

$$\rightarrow \langle \cos^2 \theta_\mu \rangle = 0.6990/0.7436 = 0.9400$$
$$\rightarrow \langle P_2(\cos \theta_\mu)\rangle = (3\langle \cos^2 \theta_\mu \rangle - 1)/2 = 0.910$$

For the second film a similar calculation gives

$$\langle P_2(\cos \theta_\mu)\rangle = (3\langle \cos^2 \theta_\mu \rangle - 1)/2 = 0.5944$$

However, the first film is fully aligned, so that $\langle P_2(\cos \theta_\mu)\rangle = P_2(\cos \theta_M)$. Thus, for the second film $\langle P_2(\cos \theta)\rangle = \langle P_2(\cos \theta_\mu)\rangle/[P_2(\cos \theta_M)] = 0.65$ or, equivalently, $\langle \cos^2 \theta \rangle = [2\langle P_2(\cos \theta)\rangle + 1]/3 = 0.77$.

10.8.   Birefringence gives $\langle P_2(\cos \theta)\rangle = 0.85$ (assuming that the birefringence is small)

$$P_2(\cos \theta_M) = (3 \cos^2 15° - 1)/2 = 0.8995$$

Thus

$$\langle P_2(\cos \theta_\mu)\rangle = \langle P_2(\cos \theta)\rangle P_2(\cos \theta_M) = 0.7646$$

and

$$\langle \cos^2 \theta_\mu \rangle [2\langle P_2(\cos \theta_\mu)\rangle + 1]/3 = 0.8431$$

$A_\parallel = \log(I_o/I_\parallel) = \log(100/25) = 0.6021 = C\langle \cos^2 \theta_\mu \rangle$. Hence $C = 0.7142$. However,

$$A_\perp = \tfrac{1}{2}C\langle \sin^2 \theta_\mu \rangle = \tfrac{1}{2}C(1 - \langle \cos^2 \theta_\mu \rangle) = 0.0560$$

Thus $I_\perp/I_o = 1/10^{A_\perp} = 0.88$ and the transmission is thus 88%.

10.9. (a) For a highly uniaxially oriented sample the polarisabilities per unit parallel and perpendicular to the draw direction are equal, respectively, to the polarisability of an individual orienting unit parallel to the axis that orients and to the mean polarisability of the unit perpendicular to this axis. The mean polarisability of a unit is thus $[(2.3 + 2 \times 1.7) \times 10^{-39}]/3$ F m^2 = 1.9×10^{-39} F m^2.

(b) $C\langle\cos^2\theta_\mu\rangle = A_\parallel = \log(I_o/I_\parallel) = \log 2 = 0.3010$. If $I_\perp = I_o$, then $\langle\sin^2\theta_\mu\rangle = 0$ and hence $\langle\cos^2\theta_\mu\rangle = 1$. Thus $C = 0.3010$. For random orientation $\langle\cos^2\theta_\mu\rangle = \frac{1}{3}$ so that $A_\parallel = A_\perp = \log(I_o/I) = \frac{1}{3}C$. Thus $I_o/I = 10^{C/3} = 10^{0.1003} = 1.26$ and $I/I_o = 0.79$, so that the transmission is 79%.

10.10. Assume that the birefringences are additive, i.e. that equation (10.31) holds and that the form birefringence is zero. Then

$$\Delta n = f_1 \Delta n_{1,\text{max}} \langle P_2(\cos\theta)\rangle_1 + f_2 \Delta n_{2,\text{max}} \langle P_2(\cos\theta)\rangle_2$$

Taking the subscript 1 to refer to the crystalline material gives

$$0.109 = 0.35 \times 0.132 \times 0.91 + (1 - 0.35)\Delta n_{2,\text{max}} \times 0.85$$

which leads to $\Delta n_{2,\text{max}} = 0.121$. With the assumption given, the expected value is $133.5 \times 0.132/145.5 = 0.121$, whence consistency.

10.11. The plot is as shown, with □ for set A and ○ for set B. The orientation distribution of set A corresponds to uniaxial distribution of chain axes towards the OX_3 axis with no specific preference for chains to lie in the plane of the film. The orientation distribution of set B corresponds to uniaxial orientation away from the OX_2 axis, which is normal to the plane of the film, and hence to a uniform distribution of the projections of the chain axes onto the plane of the film. Both sets span a range from moderately to fairly highly oriented orientations of their respective types.

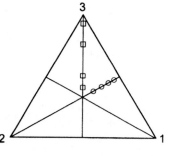

Chapter 11

11.1. It follows from the figure, which assumes constant volume and shows the unstretched sample on the left and the stretched sample on the right, that $\tan\theta_o = a/b$ and $\tan\theta = (a/b)/\lambda^{3/2}$, so that $\tan\theta = \lambda^{-3/2}\tan\theta_o$. It follows by differentiation that

$$\sec^2\theta \, d\theta = \lambda^{-3/2} \sec^2\theta_o \, d\theta_o \tag{1}$$

The solid angle $2\pi \sin \theta_o \, d\theta_o$ transforms on stretching into the solid angle $2\pi \sin \theta \, d\theta$ and each contains the fraction $[1/(4\pi)]2\pi \sin \theta_o \, d\theta_o$ of chains. The fraction of chains per unit solid angle in the drawn material is thus

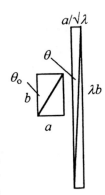

$$N(\theta) = \frac{1}{4\pi}\frac{\sin \theta_o \, d\theta_o}{\sin \theta \, d\theta} = \frac{\lambda^{3/2} \sin \theta_o \sec^2 \theta}{4\pi \sin \theta \sec^2 \theta_o}$$

using equation (1). Insert the expressions $\sin \theta = 1/\sqrt{1+\cot^2 \theta}$ and $\sec^2 \theta = 1 + \tan^2 \theta$, substitute for $\tan \theta_o$ in terms of $\tan \theta$ and rearrange using the second of these expressions. The required result then follows if $\sec \theta = 1/\cos \theta$ is substituted and some further slight rearrangement is done.

11.2.
$$\frac{d}{d\theta}\left(\frac{-\cos \theta}{\left[\lambda^3 - (\lambda^3-1)\cos^2 \theta\right]^{1/2}}\right) = \frac{\sin \theta}{\left[\lambda^3 - (\lambda^3 - 1)\cos^2 \theta\right]^{1/2}}$$

$$+ \frac{(\lambda^3 - 1)\sin \theta \cos^2 \theta}{\left[\lambda^3 - (\lambda^3 - 1)\cos^2 \theta\right]^{3/2}}$$

$$= \frac{\lambda^3 \sin \theta}{\left[\lambda^3 - (\lambda^3 - 1)\cos^2 \theta\right]^{3/2}} = 4\pi N(\theta) \sin \theta$$

where $N(\theta)$ is the pseudo-affine distribution function. Thus the fraction of units making angles less than θ_o with the draw direction is (see example 11.3).

$$2\int_0^{\theta_o} 2\pi N(\theta) \sin \theta \, d\theta = \left[\frac{\cos \theta}{\left[\lambda^3 - (\lambda^3 - 1)\cos^2 \theta\right]^{1/2}}\right]_{\theta_o}^{0}$$

$$= 1 - \frac{\cos \theta_o}{\left[\lambda^3 - (\lambda^3 - 1)\cos^2 \theta_o\right]^{1/2}}$$

The 2 outside the integral allows for units pointing 'up' and 'down' at a given angle θ. Substitution of $\lambda = 5$ leads to fractions 0.55 and 0.76 for $\theta_o = 10°$ and $20°$, respectively.

11.3. $\Delta n/\Delta n_{max} = [1/(5n)](\lambda^2 - 1/\lambda)$ so that $\Delta n_{1.5} = \Delta n_3 (1.5^2 - 1/1.5)/(3^2 - \frac{1}{3})$ and thus $\Delta n_{1.5} = 1.4 \times 10^{-3}$, where $\Delta n_{1.5}$ and Δn_3 are the birefringences for $\lambda = 1.5$ and 3, respectively. The number of random links, n, per chain is given by

$$n = \Delta n_{\text{max}}\,(\lambda^2 - 1/\lambda)/(5\,\Delta n) = 0.045 \times (3^2 - \tfrac{1}{3})/(5 \times 7.65 \times 10^{-3}) = 10.2.$$

11.4. Examination of the data shows immediately that Δn increases faster with λ the higher the value of λ, which accords with the affine rather than the pseudo-affine model. Plotting Δn against $\lambda^2 - 1/\lambda$ gives a straight line through the origin for λ less than about 3.2, with slope 1.54×10^{-3}. Thus the number n of links per chain $= \Delta n_{\text{max}}/(5 \times \text{slope}) = 10.1$.

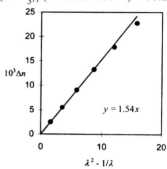

This is consistent with the application of the pseudo-affine model up to $\sqrt{n}$. For a sample with draw ratio 3,

$$\langle P_2(\cos\theta)\rangle = [1/(5n)](\lambda^2 - 1/\lambda) = 0.171.$$

11.5. $$\langle \sin^4\theta \rangle = \langle (1 - \cos^2\theta)^2 \rangle = 1 - 2\langle \cos^2\theta \rangle + \langle \cos^4\theta \rangle$$

and

$$\langle \sin^2\theta \cos^2\theta \rangle = \langle (1 - \cos^2\theta)\cos^2\theta \rangle = \langle \cos^2\theta \rangle - \langle \cos^4\theta \rangle$$

so that

$$s^{\text{o}}_{1111}\langle \sin^4\theta \rangle + s^{\text{o}}_{3333}\langle \cos^4\theta \rangle + 2(s^{\text{o}}_{1133} + 2s^{\text{o}}_{2323})\langle \sin^2\theta \cos^2\theta \rangle$$
$$= s^{\text{o}}_{1111} + 2(s^{\text{o}}_{1133} + 2s^{\text{o}}_{2323} - s^{\text{o}}_{1111})\langle \cos^2\theta \rangle$$
$$+ (s^{\text{o}}_{1111} + s^{\text{o}}_{3333} - 2s^{\text{o}}_{1133} - 4s^{\text{o}}_{2323})\langle \cos^4\theta \rangle$$

For a fully uniaxially oriented sample, $\langle \cos^2\theta \rangle = \langle \cos^4\theta \rangle = 1$, so that this expression reduces to s^{o}_{3333}, as it should. For a randomly oriented sample $\langle \cos^2\theta \rangle = \tfrac{1}{3}$ and $\langle \cos^4\theta \rangle = \tfrac{1}{5}$, so that

$$s_{3333} = (1 - \tfrac{2}{3} + \tfrac{1}{5})s^{\text{o}}_{1111} + \tfrac{1}{5}s^{\text{o}}_{3333} + 2(\tfrac{1}{3} - \tfrac{1}{5})(s^{\text{o}}_{1133} + 2s^{\text{o}}_{2323})$$
$$= (8/15)s^{\text{o}}_{1111} + \tfrac{1}{5}s^{\text{o}}_{3333} + (4/15)(s^{\text{o}}_{1133} + 2s^{\text{o}}_{2323})$$
$$= 11.1 \times 10^{-10}\ \text{Pa}^{-1}$$

11.6. For the OX_3 direction all σ_{ij} are zero except $\sigma_{33} = \sigma$. Thus $\varepsilon = \varepsilon_{33} = s_{3333}\sigma_{33} = s_{3333}\sigma$. Hence $E_3 = \sigma/\varepsilon = 1/s_{3333} = 9.1$ GPa. The direction cosine matrix for rotation of axes by $45°$ around OX_1 is

$$\begin{pmatrix} 1 & 0 & 0 \\ 0 & \sqrt{2} & \mp\sqrt{2} \\ 0 & \pm\sqrt{2} & \sqrt{2} \end{pmatrix}$$

where the signs depends on the direction of rotation. In general (see the appendix)

$$s'_{3333} = \sum_{ijkl} a_{3i}a_{3j}a_{3k}a_{3l}s_{ijkl} = \sum_{ij} a_{3i}^2 a_{3j}^2 (s_{iijj} + s_{ijij}(i \neq j) + s_{ijji}(i \neq j))$$

where the second form arises from the symmetry of the tensor for any orthorhombic or uniaxial material. From the form of the direction cosine matrix it follows that

$$s'_{3333} = \frac{1}{4} \sum_{i,j=2,3} (s_{iijj} + s_{ijij}(i \neq j) + s_{ijji}(i \neq j))$$

$$= \tfrac{1}{4}(s_{2222} + s_{2233} + s_{3322} + s_{3333} + s_{2323} + s_{3232} + s_{2332} + s_{3223})$$

$$= \tfrac{1}{4}(s_{2222} + 2s_{3322} + s_{3333} + 4s_{2332}) = 16.27 \times 10^{-10} \text{ Pa}^{-1}$$

Thus $E_{45} = 1/s'_{3333} = 0.61_5$ GPa.

11.7. (a) Considering the amorphous and crystalline material in parallel with each other, equation (11.17) leads to

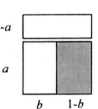

$$E_\parallel = (1-b)E_a + bE_c$$

Considering this section in series with the remainder, replacing E_a by $E_\parallel$ in equation (11.16) leads to

$$\frac{1}{E} = \frac{1-a}{E_c} + \frac{a}{E_\parallel} = \frac{1-a}{E_c} + \frac{a}{(1-b)E_a + bE_c} \tag{a}$$

(b) Considering the amorphous and crystalline material in series with each other, equation (11.16) leads to

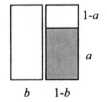

$$\frac{1}{E_s} = \frac{a}{E_a} + \frac{1-a}{E_c}$$

Considering this section in series with the remainder, equation (11.17) leads, with replacement of E_a by E_s, to

$$E = (1-b)E_s + bE_c = bE_c + \frac{E_a E_c(1-b)}{aE_c + (1-a)E_a} \tag{b}$$

For the simple series model with $a = 0.25$, equation (11.16) becomes $1/E = 0.25/E_a + 0.75/E_c$ and substitution of the values gives $E = 3.9 \times 10^9$ Pa. For the simple parallel model with $a = 0.25$, equation (11.17) becomes $E = 0.25E_a + 0.75E_c$ and substitution gives $E = 7.5 \times 10^{10}$ Pa. For the series–parallel model with $a = b = 0.25$, equation (a) above becomes $1/E = 0.5/E_c + 1/(E_a + E_c)$ and substitution gives $E = 6.7 \times 10^{10}$ Pa. For the parallel–series model with $a = b = 0.25$, equation (b) above becomes $E = 0.5E_a + E_a E_c/(E_a + E_c)$ and substitution gives $E = 5.1 \times 10^{10}$ Pa.

Chapter 12

12.1. The total volume of the sample is $V_1 + V_2$. Assuming that there are no changes of volume on mixing, the overall volume fraction v of A is

$$v = (V_1 v_1 + V_2 v_2)/(V_1 + V_2)$$

and the overall average value of ΔG_{mix} is

$$\Delta G_{mix} = (V_1 \Delta G_1 + V_2 \Delta G_2)/(V_1 + V_2)$$

From these two equation it follows that

$$\frac{\Delta G_{mix} - \Delta G_1}{v - v_1} = \frac{\Delta G_2 - \Delta G_1}{v_2 - v_1}$$

which shows that the point $(v, \Delta G_{mix})$ lies on the straight line joining the points $(v_1, \Delta G_1)$ and $(v_2, \Delta G_2)$.

12.2. The minima in ΔG_{mix} are -316 at $f = 0.200$ and -412 at $f = 0.842$, so that the gradient of the line joining the minima is $(-412 + 316)/(0.842 - 0.200) = 149.5$. Differentiating the expressions for ΔG_{mix} shows that the gradients near the binodal points are given by $4170(f - 0.200)$ and $11\,440(f - 0.842)$. Equating each of these in turn to 149.5 leads to $f = 0.164$ and $f = 0.829$ as the approximate compositions at the binodal points. Calculation of the gradient of the line joining these two points shows that it is -146.9, very close to the original gradient assumed. Substitution of this value produces hardly any change in the values of f.

12.3. (a) At $150\,°C$, $150 = 1250(v - 0.45)^2 + 100$ and solving gives $v = 0.25$ or 0.65 for the volume fractions of A in the two phases. If V is the volume in cm^3 of the phase with $v = 0.25$, the total volume of A in cm^3 is $0.25V + 0.65(1 - V)$. Setting this equal to half the total volume, i.e. to 5 cm^3, and solving for V gives $V = 3.75$ cm^3. The volume of the other phase is thus 6.25 cm^3.
(b) The modified version of equation (12.7) for miscible blends of A and B can be written $v\ln(T_g/T_A) + (1 - v)(T_g/T_B) = 0$, where v is the volume fraction of A in the blend and T_A and T_B are the absolute

glass-transition temperatures of A and B, respectively. The observed glass-transition temperatures, T_g are obtained by substituting these values and the values $v = 0.25$ and 0.65 into the equation, which, remembering that $x \ln y = \ln y^x$ and $\ln 1 = 0$, can be rewritten $T_g = T_A^v T_B^{1-v}$. Substitution of the two values of v gives $T_g = 309.4$ K $(36.1\,°C)$ or 337.1 K $(63.8\,°C)$.

12.4. Converting the temperatures to absolute gives $T_A = 373$ K, $T_B = 278$ K and $T_{20} = 288$ K. Solving equation (12.8) for a gives

$$a = \frac{w_A}{w_B}\left(\frac{1}{T_A} - \frac{1}{T_g}\right)\bigg/\left(\frac{1}{T_g} - \frac{1}{T_B}\right) = 1.584$$

Substitution as in example 12.3 then leads to $T_{80} = 340$ K $= 67\,°C$.

12.5. If segregation is complete, the ratio of the total volume of spheres per cube to the volume of the cube must equal the fractional volume of PS, i.e. 0.17. A body-centred cubic array has two spheres per unit cube and the ratio of the volume of the spheres to that of the cube is thus $(8\pi/3)(r/a)^3 = 0.17$, so that $r = \sqrt[3]{3 \times 0.17/(8\pi)}\,a = 9.1$ nm.

12.6. Table 3.1 shows that the average bond length $l_{\circledcirc}$ for singly and doubly bonded carbon atoms is 0.144 nm

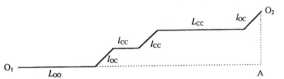

and the length of the C—O bond l_{CO} is 0.143 nm. The length of the O—⊚—C unit is thus $2l_{CO} + (1 + 2\cos 60°)l_{\circledcirc} = 0.574$ nm. Similarly, the length of the C—⊚—C unit is 0.596 nm. Denoting these lengths by L_{OO} and L_{CC}, respectively, and other bond lengths by suitable notation leads to the fully extended skeleton looking as shown, with fully extended length O_1O_2. The length O_1A is equal to $L_{OO} + L_{CC} + l_{CC} + (2l_{OC} + l_{CC})\cos 60° = 1.544$ nm and the length O_2A is equal to $(2l_{OC} + l_{CC})\sin 60° = 0.381$ nm. Thus $O_1O_2 = \sqrt{1.544^2 + 0.381^2} = 1.59$ nm.

12.7. The solid angle of the cone is $2\pi(1 - \cos\psi_{max}) = 4\pi\sin^2(\psi_{max}/2) = 4\pi[y_{max}/(2x)]^2$ for small ψ_{max}, where y_{max} is the value of y corresponding to ψ_{max}. The number of possible sites that rods can occupy is thus $A(y_{max}/x)^2$, where A is a constant. However, each rod can occupy any site at random, so that each rod can be oriented in $A(y_{max}/x)^2$ ways and the total number of ways N_{orient} that the n rods can be oriented is $[A(y_{max}/x)^2]^n = A^n(y_{max}/x)^{2n}$. For small values of ψ_{max}, $\bar{y} \propto y_{max}$, so that $N_{orient} = C(\bar{y}/x)^{2n}$, where C is a constant.

12.8. For a thermotropic, $\upsilon = 1$, and the RHS of equation (12.16a) becomes $\ln(1/x) + \bar{y} - 1 - \bar{y}\ln(\bar{y}/x) - 2\ln(\bar{y}/x)$. Differentiating this with respect to $\bar{y}$ for a fixed value of x and setting the result equal to zero gives the condition for a minimum in $G^{\mathrm{o}}_{\mathrm{orient}}$ for a particular value of x, i.e. it gives the relationship between $\bar{y}$ and x at the minimum. Differentiating term by term and setting to zero gives $1 - \ln(\bar{y}/x) - 1 - 2/\bar{y} = 0$, or $x = \bar{y}e^{2/\bar{y}}$. The minimum value of x is thus given by $dx/d\bar{y} = e^{2/\bar{y}} - 2e^{2/\bar{y}}/\bar{y} = 0$, i.e. $\bar{y} = 2$ and $x = 2e \cong 5.44$.

12.9. Plot $1/E$ against $1/G$. The intercept on the $1/E$ axis gives $1/E_{\max}$ and the gradient gives $\langle\sin^2\theta\rangle$. The plot is as shown. The intercept is 4.80×10^{-3} GPa^{-1}, leading to $E_{\max} = 1/(4.80 \times 10^{-3}) = 208$ GPa. The gradient is $16.97 \times 10^{-3} = \langle\sin^2\theta\rangle$. Thus

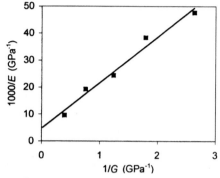

$$S = (3\langle\cos^2\theta\rangle - 1)/2 = 1 - \tfrac{3}{2}\langle\sin^2\theta\rangle = 0.97.$$

Index

In the alphabetisation brackets, hyphens and bond symbols are ignored. Items starting with numbers, superscripts or Greek characters are indexed by the roman characters of the item, except for σ and π , which are listed under 'sigma' and 'pi'. All abbreviations used in the index are also found as entries. Some entries concern material in tables, figures or figure captions. Material in problems is not indexed, with a few exceptions, but problems are frequently referred to in the text.